Community Psychology

Series Editors

The Community Psychology book series is envisaged as a space to review the established assumptions and knowledge economy underlying community psychology, and encourage writings that recognize the plurality of people and the many geographical, psychological and sociological locations that they occupy. The book series will enable contributors to stimulate thought that questions that which is constructed as critical knowledges, community psychology, and the meanings of liberation and community. Contributions to the book series draw attention to the applications of community psychology in the Global South and the Global North as they relate to such issues as violence, socio-economic inequality, racism, gender, migration, dispossession, climate change, and disease outbreaks. In do so, it centers community psychology as focused on the well-being of collectives, and dealing with such focal issues as deploying psychology to support social justice, the relevance and appropriateness of its internal logic, and methods that deal with the range of psychological, social, cultural, economic, political, environmental, epistemic, and local and global influences that bear on the quality of life of individuals, communities and society. The book series concentrates thus on the following three key areas of focus: 1) decoloniality, power and epistemic justice, 2) knowledge production, contestation and community psychology, and 3) community psychology in context. The series is of vital and immediate relevance to researchers, practitioners, faculty and students from the intervention sciences, including anthropology, sociology, public health, development studies, social work and urban studies.

More information about this series at https://link.springer.com/bookseries/15965.

Shose Kessi • Shahnaaz Suffla
Mohamed Seedat
Editors

Decolonial Enactments in Community Psychology

Springer

Editors
Shose Kessi
Department of Psychology
University of Cape Town
Cape Town, South Africa

Mohamed Seedat
Institute for Social and Health Sciences
University of South Africa
Johannesburg, South Africa

South African Medical Research Council-
University of South Africa Masculinity and
Health Research Unit
Cape Town, South Africa

Shahnaaz Suffla
Institute for Social and Health Sciences
University of South Africa
Johannesburg, South Africa

South African Medical Research Council-
University of South Africa Masculinity and
Health Research Unit
Cape Town, South Africa

ISSN 2523-7241 ISSN 2523-725X (electronic)
Community Psychology
ISBN 978-3-030-75200-2 ISBN 978-3-030-75201-9 (eBook)
https://doi.org/10.1007/978-3-030-75201-9

This Springer imprint is published by the registered company Springer Nature Switzerland AG
The registered company address is: Gewerbestrasse 11, 6330 Cham, Switzerland

Foreword

The Community Psychology Book Series: A Dialogical Decolonising Space

The Springer Community Psychology Book Series holds an ambitious vision that approaches community psychology as a site of knowledge and knowledge-making. The Series is imagined as a dialogical space for critical and situated knowledges and liberatory praxes. Through interrogations of decolonising psychologies' obligations in the era of the modern knowledge economy, and the embrace of alternative, disruptive, and new imaginings, the Book Series intends to elaborate on collective critical-liberatory projects in community psychology and cognate areas of scholarship. The Series is however alert to the limitations imposed on creative thought and practices by hierarchical and homogenising influences in and outside of the academy.

The Series encourages contributions that focus on community psychology as knowledge and the contestation for representation and authority. We invite contributors to examine the politics and archaeology of knowledge. Likewise, the Series draws together contributions that review how particular concepts and theories have gained ascendency in community psychology, and that offer insights into the benefits and limitations of creative methodologies applied in research, intervention, and analysis.

The Series encourages work that centrally engages with epistemicide—the deracination of other-than western modes of knowing and knowledge systems—and its manifestations in scholarship on community, community-making, community resistance, and the formations of social arrangements that seek to overturn racism, racialisation, heteropatriarchy, classism, and other forms of oppressive social relations. The Series is aligned with, and situated within the larger body of praxes of the South, borne out of struggles for self-determination, epistemic independence and epistemic agency, and visions and imaginations of radical humanism. The Series thus seeks to animate conversations about what it means to create and live in human formations that challenge 'race' and racism, gendered and patriarchal arrangements, inequitable economic and material arrangements, and problematic notions of sex and sexuality, as well as a range of other exclusionary -isms.

This book, *Decolonial Enactments in Community Psychology*, the fourth in the Series and led by Shose Kessi, highlights the centrality of social justice, subaltern knowledges, epistemic freedom, and decolonial aesthetics in decolonising community psychology towards radical social change. Contributions to the book interrogate and revise our understandings of community engagement and represent pluriversal enactments of community psychology in the Global South and the Global North as they relate to such issues as violence, race, gender and identity, dispossession and displacement, health and healing, memory, human rights and democracy, and solidarity-making. The contributors explore the diverse and encroaching historical, social, and political environments and processes that impact on community-making and community practices. This edited text is a germinal intervention that reconsiders how we think about community, community praxes, and methodologies of participation. *Decolonial Enactments in Community Psychology* expands the boundaries of what we may comprehend as knowledges of decolonial community practice.

Institute for Social and Health Sciences,
University of South Africa, Johannesburg, South Africa

South African Medical Research Council-University
of South Africa Masculinity and Health Research Unit
Cape Town, South Africa

Mohamed Seedat
Shahnaaz Suffla

Acknowledgements

As editors, we wish to acknowledge all the individuals and communities from across the globe whose stories have appeared in the pages of this volume. We thank them for sharing their knowledge and are inspired by their courage and struggles that continue to pave the way towards decolonial enactments in all of our lives. The idea for this volume came from the Sixth International Conference on Community Psychology (ICCP), the first to take place on African soil, in Durban, South Africa, in 2016. Many of the authors in this volume were present at that conference. The conference was co-organised by the Institute for Social and Health Sciences (ISHS) at the University of South Africa (UNISA) and the Psychological Society of South Africa (PsySSA). We wish also to acknowledge the reviewers who dedicated their time and offered critical insights which greatly enhanced the scholarly impact that these contributions will have. Finally, we thank Josephine Cornell for her magnificent project management and technical skills.

Contents

List of Contributors

Devin G. Atallah Psychology Department, University of Massachusetts, Boston, MA, USA

Richard Barber Free Theatre Company, Melbourne, VIC, Australia

Floretta Boonzaier Department of Psychology, University of Cape Town, Cape Town, South Africa

Rosa Cordillera A. Castillo Institute for Asian and African Studies, Humboldt Universität zu Berlin, Berlin, Germany

Diego Cifuentes Victoria University, Melbourne, VIC, Australia

Yvonne Clark SAHMRI Women and Kids Theme, South Australian Health and Medical Research Institute, Adelaide, SA, Australia

Shari Daya University of Cape Town, Cape Town, South Africa

Malual Deng South Sudan First Child Education, Melbourne, VIC, Australia

Urmitapa Dutta Department of Psychology, University of Massachusetts Lowell, Lowell, MA, USA

Isla Emery-Whittington SHORE and Whariki Research Centre, Massey University- Te Kunenga ki Pūrehuroa, University of New Zealand, Wellington, New Zealand

Jesica Siham Fernández Santa Clara University, Santa Clara, CA, USA

Sarah Malotane Henkeman Social Law Project, University of the Western Cape, Cape Town, South Africa

Tanja Hirvonen Flinders University, Adelaide, SA, Australia

Pilar Kasat Curtin University, Perth, WA, Australia

Shose Kessi Department of Psychology, University of Cape Town, Cape Town, South Africa

Siew Fang Law Victoria University, Melbourne, VIC, Australia

Institute for Social and Health Sciences, University of South Africa, Johannesburg, South Africa

University of Western Sydney, Penrith, NSW, Australia

Nick Malherbe Institute for Social and Health Sciences, University of South Africa, Johannesburg, South Africa

South African Medical Research Council-University of South Africa's Masculinity and Health Research Unit, Cape Town, South Africa

Simone M. Peters Department of Psychology, University of Cape Town, Cape Town, South Africa

Amy F. Quayle Victoria University, Melbourne, VIC, Australia

Elelwani L. Ramugondo Health Sciences Faculty, University of Cape Town, Cape Town, South Africa

Kopano Ratele Psychology Department, Stellenbosch University, Stellenbosch, South Africa

Mohamed Seedat Institute for Social and Health Sciences, University of South Africa, Johannesburg, South Africa

South African Medical Research Council-University of South Africa Masculinity and Health Research Unit, Cape Town, South Africa

Christopher C. Sonn Victoria University, Melbourne, VIC, Australia

University of the Witwatersrand, Johannesburg, South Africa

Shahnaaz Suffla Institute for Social and Health Sciences, University of South Africa, Johannesburg, South Africa

South African Medical Research Council-University of South Africa Masculinity and Health Research Unit, Cape Town, South Africa

Ruth Urson University of Cape Town, Cape Town, South Africa

Part I
Conceptions of Engagement for Community Psychology

Towards A Decolonial Community Psychology: Derivatives, Disruptions and Disobediences

Shose Kessi, Shahnaaz Suffla, and Mohamed Seedat

This edited collection was inspired by the presentations given at the sixth International Conference on Community Psychology (ICCP) held in Durban, South Africa in 2016. The conference was co-hosted by two of the editors of this collection, Shahnaaz Suffla and Mohamed Seedat. The 2016 conference was the first ICCP to be hosted on African soil and as such carries a special significance for us. Indeed, radical forms of community psychology in South Africa have historically been at the forefront of critical engagements with mainstream psychology, raising questions about the discipline's relevance to understand and address the challenges of a (post) apartheid/colonial context, but also questioning the role of psychology as a discipline in legitimising forms of inferiorisation and control. The year 2016 was also significant as it marked the second year of widespread protests on university campuses in South Africa under the banner of #FeesMustFall calling for a free and decolonised education. These protests led to a renewed interest in decolonial theorising on the continent and the diaspora, and renewed synergies with decolonial movements in the rest of the world. In this new wave of decolonial thinking, psychologists have once again drawn on community psychology praxes to expand on their theorising of power, identity and knowledge, and contribute to enactments of resistance, transformation and social change.

S. Kessi (✉)
Department of Psychology, University of Cape Town, Cape Town, South Africa
e-mail: shose.kessi@uct.ac.za

S. Suffla · M. Seedat
Institute for Social and Health Sciences, University of South Africa, Johannesburg, South Africa

South African Medical Research Council-University of South Africa Masculinity and Health Research Unit, Cape Town, South Africa
e-mail: suffls@unisa.ac.za; seedama@unisa.ac.za

S. Kessi et al. (eds.), *Decolonial Enactments in Community Psychology*, Community Psychology, https://doi.org/10.1007/978-3-030-75201-9_1

The writings in this collection are contributions from scholar-activists living and/or 'in practice' across the world: in Australia, Palestine, the Philippines, the USA, New-Zealand, Chile, and South Africa. In their contributions, they advance conceptions and praxes for community psychology grounded within a decolonial framework and bring to life the overt and covert, intricate and invisibilising ways in which coloniality operates. It is important to acknowledge the South American influence on what we see as a decolonial turn in much contemporary scholarship in the global south and the diasporas. Quijano's (2007) articulation of coloniality, as the continuation of the hierarchical organisation of society following the end of formal colonial rule, provides a framework to bring into sharp focus the continued western imperialist agenda of the global political economy, and the Eurocentric patterns of social and cultural life. This global lens on contemporary life, and particularly on the struggles of people and communities on the margins of society, is precisely what has been absent in much community psychology. This volume heeds the call for a generation of approaches to community psychology that link local struggles to broader questions of power, identity, and knowledge production. Deconstructing colonial ways of knowing (Mignolo, 2009), being (Maldonado-Torres, 2007), and doing (Ramugondo, 2015) necessitates a commitment to *disobedient praxis*. Bringing together these examples of praxes from different contexts is a political project of visibilising indigenous struggles toward self-determination and their interconnectedness within a global understanding of power and the need for a common platform for action and resistance.

The chapters in this volume embody a decolonial agenda for community psychology that foregrounds social justice; the lives and knowledges of the marginalised and oppressed; epistemic disobedience and transdisciplinarity; and decolonial aesthetics.

A Fundamental Quest for Social Justice

Social justice has been a key principle of critical forms of community psychology and often associated with conceptions of freedom and liberation. In that sense, social justice is the articulation of resisting and overcoming the violences of coloniality, violence being understood as the physical, psychological (and embodied), cultural, and structural legacies of colonial power (Malotane-Henkeman, this volume). A community psychology orientation to social justice is useful as it highlights both the communal (or collective) as well as the psychological imperatives for deconstructing and resisting coloniality. Social justice is a mutual pursuit as it involves a collective consciousness (Freire, 1970) of the historical dimensions of colonial violence and conjoint action towards change; it is simultaneously an individual pursuit to undo internalised violence (Fanon, 1970), and build self-esteem and empowerment. A decolonial lens on this individual-collective interface, as exemplified in many of the chapters, locates both individual and community struggles in a global context (Campbell, 2006) and one that is historicised in colonial relations of power (Mignolo, 2009).

A Critical Focus on the Marginalised and Oppressed

A decolonial community psychology is premised on centering the knowledge and capabilities of those who are on the margins of society as the result of wars, conflicts, or disenfranchisement; and whose values and beliefs are devalued, misrecognised and misrepresented in dominant sectors of society. Denying an individual's or a collective's right to their beliefs and values is undermining their agency and self-determination to exist. Some key conceptual values for centering marginalised knowledges are community participation, empowerment, and action. A key argument in this book, however, is that the notion of the 'marginalised' or 'oppressed' needs further engagement and deconstruction, and that naming marginalisation and oppression should not be equated with victimhood (Dutta, this volume). Community psychology research has shown us that power is multidimensional. We know, for instance, that communities are not homogenous and that power struggles exist within them (Jovchelovitch, 2007). Our understandings of community must therefore also be revisited. In contemporary transracial and transcultural societies, power fluctuates in intricate and complex ways. Questions of identity and belonging are often the result of intersecting dimensions of race, class, gender, nationality, and other markers of difference. A decolonial approach is perhaps useful in finding points of convergence and solidarity amongst those who have endured the legacies of colonisation. We have seen, for example, in recent times how social movements such as the Occupy movement, Rhodes Must Fall and Black Lives Matter have gathered solidarity and led to productive and concrete actions across identities, cultures, and geographical spaces.

Epistemic Disobedience and Transdisciplinarity

Almost every chapter in this volume emphasises the need for transdisciplinarity in academic knowledge production. Disciplinary-focused scholarship falls short of a critical examination of the hegemonic structure of knowledge production and the linkages between rationality, modernity and coloniality (Mignolo, 2007). Community psychology is not exempt from this history. The focus on the modern individual and rational thought has often led psychologists to come to erroneous assumptions and conclusions about the marginalised and the oppressed, and to reinforce victim-blaming projects and deficit models to explain difference and inequality. The reductionism of a disciplinary lens also confines us to methodological boundaries which, for disciplines like psychology, espouse experimental methods that find their roots in eugenics, phrenology, and race-science (Richards, 1997). This epistemological violence (Teo, 2014) contradicts the project of social justice as it reproduces ideas, attitudes, and beliefs that undermine, inferiorise and exclude, and has therefore led community psychologists to engage in epistemic disobedience (Mignolo, 2007). A decolonial community psychology must engage seriously with epistemic disobedience and transdisciplinarity as an orientation that embraces "… multiple disciplines and methods … new methodological categories, discursive forms, pedagogical and political practices, and institutional spaces that seek to expand a movement for emancipation, liberation, and the decolonisation of power, being, and knowledge" (Maldonado-Torres, 2019, p. 234).

Embracing A Decolonial Aesthetic

A common thread in these decolonial conceptualisations of community psychology praxes is the role of affect, embodiment, and spirituality in social change efforts, what others have referred to as a decolonial aesthetics (Mignolo & Vazquez, 2013). At the heart of the colonial project is the dehumanisation of people who do not fit the norm of the white, male, hetero-patriarchal individual. Being outside of the norm is akin to being in the zone of non-being (Grosfoguel, 2016) and thus being denied rights and material resources, but also the subjective resources for identity construction, epistemic and cognitive justice (Ndlovu-Gatsheni, 2018), and the recognition of religion and spirituality. Hence, the coloniality of being has affective, embodied, and spiritual elements beyond the materiality of existence. South African artist Khanyisile Mbongwa, in conversation with Nomusa Makhubu (Makhubu & Mbongwa, 2019), speaks to the shortfalls of certain forms of protest activity, such as public marches, as traumatic rather than inspiring healing and radical changes to deep-seated beliefs in black inferiority. This is not to say that marches are not effective forms of resistance as they can lead to concrete material changes, but rather that they do not necessarily address the affective and spiritual transformations needed for psychosocial healing and wellbeing, or "the historical denial of self-love through slavery, migrant labour systems, segregation policies and racist judicial systems which ruptured and continue to sever black family life" (Makhubu & Mbongwa, 2019, p. 11). Many of the chapters in this volume affirm this dimension as critical to the decolonial project. They speak to how intergenerational trauma, ancestral knowledge, and other forms of pain and anguish, but also love resides at the core of their political consciousness and action; and emphasise alternative forms of praxis through the arts: poetry, music, photography, theatre, storytelling, and more that tap into our senses in ways that bring together the intellectual, the experiential, and the soulful. Placing decolonial love at the center of community psychology praxes and centering the "needs of the spirit" (Makhubu & Mbongwa, 2019, p. 11) is perhaps a most significant and revolutionary contribution to a discipline such as psychology.

The quadruple focus on social justice; knowledges of the marginalised and oppressed; epistemic disobedience and transdisciplinarity; and decolonial aesthetics resonates with the radical traditions of community psychology manifest in different regions of the global south (see Seedat & Suffla, 2017), which aim to overturn the onto-epistemological and material manifestations of the coloniality of knowledge and obtain groundedness through an immersion in everyday struggles.

A Critical Assembly of Theoretical, Methodological and Praxical Resources

The book is divided into two parts. Part I, on *Conceptions of Engagement for Community Psychology*, delves into the conceptual framework for a decolonial community psychology, whilst Part II, *Modes of Enactments and Praxes for Community Psychology,* builds on these theoretical advancements through examples of praxis in different contexts.

Conceptions of Engagement for Community Psychology

In Chapter 2, Kopano Ratele and Nick Malherbe consider a liberatory, Africa(n)-centered community psychology located in decolonial thought. They animate this by drawing on the work of South African anti-apartheid activist and scholar Neville Alexander. Ratele and Malherbe's decolonial vision brings into question the problem of disciplinary boundaries in academic research and praxes coupled with challenging the reductionist idea of Africa as a monolithic space for theorising the psychological. Rather, they claim that Africa(n)-centered scholarship is premised on an orientation that is *from, by, with and for Africa* which opens up possibilities for collaborative research beyond the constraints of essentialised ideas of identity differences or the confines of disciplinary traditions: "Africa(n)-centred critical community psychologists who are guided by the new historical community are urged to take notions of difference seriously by working with people to draw out points of connection within and between seemingly unconnected struggles". This paves the way for a generation of critical community psychologists to take on the premise of transdisciplinary scholarship and activisms in addressing Africa's multiple, complex, and intersecting challenges as they are simultaneously located in African and global historical contexts; and grounded in a clear emancipatory agenda and vision of justice and resistance to localised and global forms of oppression. Neville Alexander's work is critical in this endeavor and allows the authors to prioritise community and human connectedness for an African-centered decolonial community psychology.

Building on these ideas, Jesica Siham Fernández, in Chap. 3, introduces participatory action research (PAR) as a paradigm for 'doing' decolonial community psychology. She argues that PAR can challenge hegemonic knowledges when located in a decolonial paradigm and proposes 10 axioms for a decolonial PAR praxis. Resonating with Ratele and Malherbe, she refers to decolonial community psychology as a "collective of theoretical and epistemological praxes that are not located within one specific discipline or field of inquiry but are rather multidisciplinary and transnational in their critique of coloniality and colonialism". For Fernández, community psychology must engage seriously with coloniality if it is to become relevant in the global south and draw on indigenous and local histories, cultures, and knowledges. The chapter focuses on how PAR can undo systems of oppression through critical perspectives on participation, action, and research. Through the 10 axioms that she proposes, Fernández introduces a number of themes that appear consistently throughout the chapters in this collection as ethical principles towards reimagining ways of knowing, being and doing.

Continuing with the theme of disrupting disciplinarity, Urmitapa Dutta, in Chapter 4, troubles the metanarrative of 'community' and its historical rootedness in colonial relations of power. She argues that in dominant North American conceptions of community psychology, community is always pathologised and written as deficient. Such essentialised notions reproduce power relations in the research process and exacerbate hierarchies between researcher and participant with the risk of further marginalising groups. This, Dutta argues, emerges from notions of 'helping' others that are deeply embedded in clinical psychology. A decolonial community psychology needs to therefore disrupt mainstream ideas of community, away from a "damage-centered framework" and question the boundaries of community that are focused on place and belonging. She proposes alternative concepts drawn from feminist decolonial thinkers, such as communality, diversality, and survivance as starting points for an epistemic repositioning of the field where sovereignty is considered the epistemological basis that would enable a radical move away from colonial stasis; and decolonial love as the generator of social movements for change.

Modes of Enactment and Praxes for Community Psychology

Part II begins with reflections on radical love and rebellion. In Chapter 5, Devin G. Atallah expands on Dutta's articulations of decolonial feminisms and decolonial love through his own decolonial community clinical practice in multiple sites of engagement in Chile, Palestine and the USA. These examples of praxis in different contexts capture the meaning of building alliances and solidarity across communities through transnational forms of resistance and rebellion. Atallah speaks of decolonial solidarity as radical love: "This revolutionary relationality that we build together, and that we build against, the colonial 'system of compartments', is grounded on love – decolonial love". Illustrated through powerful examples of brotherhood, motherhood, marriage, family, and friendship on the backdrop of brutal violence, Atallah invites us on his own journey of self-discovery and of becoming human through 'communal flourishing'. Atallah brings to life what community means in a global fight for liberation and how finding love in these spaces generates a mature and profound vision for liberation.

Chapter 6 presents community-based arts and cultural practices as decolonial enactments in community-based research. Christopher C. Sonn, Amy F. Quayle, and Pilar Kasat reflect on participatory cultural activities with Aboriginal Noongar communities in the Wheatbelt region of Western Australia. They highlight the role of storytelling through different artforms as instrumental in fostering dialogue, reflection and relationship building in minoritised and marginalised communities.

Drawing on liberatory praxis, the authors show how human connectedness and healing is nurtured through pleasure and skills-building activities in participatory settings. A central concern for these communities is the recognition of their cultural knowledges and the need to find ways of sustaining this knowledge for current and future generations, echoing Atallah's focus on the importance of ancestral knowledge in becoming human. Similar to Fernández, the authors emphasise the imperative for participatory research to "adopt a decolonial attitude" and not fall into the trap of reproducing knowledge hierarchies.

Building on participatory artistic methods for decolonial research, Siew Fang Law, Malual Deng, Diego Cifuentes and Richard Barber, in Chapter 7, examine the role of dialogue through Dialogue Theatre. Also based in Australia, the authors reflect on a Dialogue Theatre project, South Sudanese Voices, about the experiences of Sudanese migrants living in Melbourne. A method developed in Thailand and used in this project by a mixed group of activists (artist, community leader, documentalist-photographer, and scholar) from vastly different cultural backgrounds shows the transcultural and transnational possibilities and necessities of doing decolonial work in contemporary settings. In this example of Dialogue Theatre, a most fascinating aspect of the project was the disruption of the oppressor-oppressed dichotomy by looking at how oppressions and privilege circulate in complex ways. The participatory aspect is an essential element in disrupting these power dynamics and lies in the co-creation of the stories being staged: the characters, the script, and the performance, as well as the subsequent engagement with the audience through critical dialogue. The intention of Dialogue Theatre is to create a safe space for people to engage with one another around questions and challenges that affect them. Through role-playing, characters on stage enact and disrupt the hierarchies that exist in everyday life and ask the audience to reflect on these dynamics: "The process mediated a discussion that opened to thought-provoking, complex, and solutions-oriented engagement, stimulated by the lived experience of community actors." This framework presents possibilities for meaningful engagement with the potential to conscientise, shift attitudes, as well as foster a global interconnectedness between individuals and communities from different walks of life.

Chapters 8 and 9 are further examples of arts-based participatory methods through the use of Photovoice. Simone Peters, Shose Kessi, and Floretta Boonzaier explore the lived experiences of 'coloured' communities in Bishop Lavis, a township in Cape Town, South Africa. The authors use Photovoice as a way of disrupting colonial images of young men who identify as 'coloured' and provide a more nuanced narrative of racial and place identities. The participants in the project used their photographs to tell stories about their lives, their dreams and aspirations and in this process countered the "hegemonic northern modes of theorising men in global southern contexts". Similarly, Ruth Urson, Shose Kessi, and Shari Daya comment on a Photovoice project held with members of Reclaim the City, a grassroots housing movement in Cape Town. Through Photovoice, participants shared stories about their experiences of displacement and located these experiences in a historical analysis of colonial conquest and land dispossession. Drawing on critical and feminist geographies, the authors reflect on the concept of the "colonial spatial imaginary" to explore ideas of identity, belonging, place, and citizenship in the aftermath of apartheid and colonisation. They claim that building alternative spatial imaginaries is possible by centering the lived experiences of displaced peoples combined with active resistance through conscientisation and physical occupations. These actions are needed to expand our imagination of identities in space. Similar to Sonn et al. and Law et al., these chapters highlight how participatory arts can advance a decolonial community psychology by combining participation, affect, consciousness and social action within a liberatory agenda.

In Chapter 10, Elelwani L. Ramugondo and Isla Emery-Whittington explore a decolonial approach to health promotion. Health promotion being a key focus in much of community psychology, this chapter contributes to reframing the "individualistic, apolitical and ahistorical" dominance of biomedicine in community and health psychology, building on social determinants of health approaches in contrast to biomedical ones. Occupational therapists Ramugondo and Emery-Whittington introduce the concept of occupational consciousness—that is, becoming aware of the ideological dimension of everyday professional practices and expanding on decolonial theory through the notion of the *coloniality of doing*. Bringing this concept to bear on scholarship on health promotion challenges fundamental assumptions of ill-health, the definitions and distribution of disease, and the academic responses to health in a neocolonial system that "drives the commodification of health". They illustrate the necessity of disobedient practice through examples from their experiences as occupational therapists in South Africa and New Zealand. Their sharp analysis points to how "scholarship as embodied performance within the academy, once co-opted, is no longer scholarship, but commodified mimicry." These novel concepts have much to offer for a decolonial community psychology, focusing on working with marginalised and oppressed communities to reclaim agency and build critical consciousness for "diagnosing problems that lead to ill-health".

In Chapter 11, Yvonne Clark and Tanja Hirvonen engage with the role of aboriginal psychologists in decolonising Australian psychology, in particular challenging the shortcomings of mainstream psychology in addressing the wellbeing and healing of Aboriginal and Torres Strait Islander communities. In this account of the formation of the Aboriginal and Torres Strait Islander People and Psychology (ATSIPP) interest group, the Australian Indigenous Psychologists Association (AIPA) and the Australian Indigenous Psychology Education Project (AIPEP) formed through the Australian Psychological Society (APS), the authors describe the role of indigenous psychologists in advancing professional psychological services for Aboriginal and Torres Strait Islander people and their contributions to furthering national debates on reconciliation. Clark and Hirvonen highlight the importance of paying attention to the institutional context in which psychologists practice and also offer a historical account of the trajectory of Aboriginal psychologists in Australia. They conclude with their own experiences and work as indigenous psychologists in Australia.

In Chapter 12, Sarah Malotane-Henkeman discusses a conceptual approach to everyday violence and the role of denial in reproducing everyday violences. She draws on her own experiences as a conflict resolution practitioner in South Africa to offer a model for disrupting denial. Her narration is a powerful description of the role of transhistorical trauma and how trauma can surface, sometimes unexpectedly, in reaction to everyday violence. This embodied knowledge, she argues, is often missing from scholarship on violence, surfacing the need for the oppressed to centre their knowledge as "counterpoint to decentre coloniality". The model is a practical conceptual presentation of different forms of violence: structural, cultural, psychological and physical, also described as visible/invisible violence. Building self-determination, agency and consciousness are highlighted as the tools to disrupt transhistorical and transnational violence.

Rosa Cordillera A. Castillo's reflections on historical violence in, Chapter 13, engage with memory work as decolonial praxis. Illustrated through the struggles of the Moro Islamic Liberation Front in the Cotabato region in Southern Philippines, Castillo theorises memory as the basis for recovery and resistance and the imagining of alternatives futures, liberation, or what she refers to as "re-existence". The Muslim communities she describes evoke memories of violence, in particular the massacres of Muslims under the Marcos regime, to help "shape their struggle and right to self-determination" against dehumanisation and violence. These memories also open up avenues for peace: "… imagining the other's experience with violence and suffering that has been transmitted through memory-making practices can link individuals with each other, thereby helping constitute (but also challenge) the collective, and provides a framework for understanding resistance and peace-making." Once again, the concept of survivance is introduced as it is memories of survivance rather than victimhood that form the basis for liberation from oppression. These memories drive people to act for social change and build a consciousness of the experiences of exclusion of Filipino Muslims as rooted in coloniality.

From understandings of colonial violence to enactments of decolonial love, the chapters in this edited collection provide scholars and activists interested in community psychology (broadly speaking) with innovative concepts, practical tools, and inspiration to continue engaging in struggles for social justice. These examples of decolonial enactments in community psychology have brought to the fore the "inextricable entanglements of ontological extractivism, material extractivism, and epistemic extractivism as colonial/imperial technologies of subjectification" (Ndlovu-Gatsheni, 2017, p. xi) and thus the need for disobedient praxes in our quest for humanity across the globe. What is clear is that understanding human connectedness starts with those who are marginalised and oppressed by colonial violence and that through decolonial practices we can begin to channel deeply engrained attitudes, beliefs, and traumas in productive and generative ways.

The framework presented in this volume authenticates social justice, knowledges of the marginalised and oppressed, epistemic disobedience and transdisciplinarity, and decolonial aesthetics as the axes for decolonial praxes and privileges the import of affect, love, and spirituality in reclaiming our humanness. Notwithstanding the vastness of the decolonial project and the extent and reach of violent articulations of coloniality across the globe, the contributors suggest that locating our efforts in local struggles and finding solidarity in everyday contexts can lead to meaningful and sustained efforts towards social justice. Local enactments may constitute the building blocks for transnational forms of decolonial resistance and change.

References

Campbell, C. (2006). HIV/AIDS: Politics and inter-group relations. In K. Ratele (Ed.), *Intergroup relations: South African perspectives* (pp. 171–192). Juta.

Fanon, F. (1970). *Black skin, white masks*. Grove Press.

Freire, P. (1970). *Pedagogy of the oppressed*. Continuum.

Grosfoguel, R. (2016). What is racism? *Journal of World-System Research, 22*(1), 9–15. https://doi.org/10.5195/jwsr.2016.609

Jovchelovitch, S. (2007). *Knowledge in context: Representations, community and culture*. Routledge.

Makhubu, N., & Mbongwa, K. (2019). Radical love as decolonial philosophy: In conversation with Khanyisile Mbongwa. *Journal of Decolonising Disciplines, 1*(1), 10–26. https://doi.org/10.35293/2664-3405/2019/v1n1a2

Maldonado-Torres, N. (2007). On the coloniality of being. *Cultural Studies, 21*(2–3), 240–270. https://doi.org/10.1080/09502380601162548

Maldonado-Torres, N. (2019). Ethnic studies as decolonial transdisciplinarity. *Ethnic Studies Review, 42*(2), 232–244.

Mignolo, W. D. (2007). Coloniality of power and decolonial thinking. *Cultural Studies, 21*(2), 155–167. https://doi.org/10.1080/09502380601162498

Mignolo, W. D. (2009). Epistemic disobedience, independent thought and de-colonial freedom. *Theory, Culture, and Society, 26*(7–8), 1–23. https://doi.org/10.1177/0263276409349275

Mignolo, W. D., & Vazquez, R. (2013). Decolonial aestheSis: Colonial wounds/decolonial healings. *Social Text*. https://socialtextjournal.org/periscope_article/decolonial-aesthesis-colonial-woundsdecolonial-healings/

Ndlovu-Gatsheni, S. J. (2017). Foreword: The case for a decolonised/Africanised Africa. In V. Msila (Ed.), *Decolonising knowledge for Africa's renewal* (pp. x–xv). Kr Publishing.

Ndlovu-Gatsheni, S. J. (2018). The dynamics of epistemological decolonisation in the 21st century: Towards epistemic freedom. *Strategic Review for Southern Africa, 40*(1), 16–45.

Quijano, A. (2007). Coloniality and modernity/rationality. *Cultural Studies, 21*(2), 168–178. https://doi.org/10.1080/09502380601164353

Ramugondo, E. L. (2015). Occupational consciousness. *Journal of Occupational Science, 22*(4), 488–501. https://doi.org/10.1080/14427591.2015.1042516

Richards, G. (1997). *Race, racism, and psychology: Towards a reflexive history*. Routledge.

Seedat, M., & Suffla, S. (2017). Community psychology and its (dis)contents, archival legacies and decolonisation. Special Issue: Liberatory and critical voices in decolonising community psychology. *South Africa Journal of Psychology, 47*(4), 421–431.

Teo, T. (2014). Epistemological violence. In T. Teo (Ed.), *Encyclopedia of critical psychology* (pp. 593–596). Springer.

Liberatory Africa(n)-Centred Community Psychology of Psychosocial Change

Kopano Ratele and Nick Malherbe

Introduction

Psychosocial change-making demands of the Africa(n)-centred critical community psychologist a historical, activist, local, global and disciplinary awareness. It is in this that we can begin to draw on community to critically (re)form the self in relation to others as well as broader society (see Martín-Baró, 1994). Action thus occurs with and through critical reflection (see Freire, 1970), meaning that in fostering psychosocial well-being, we endeavour to make visible, understand and fight against oppression as it is experienced directly, epistemologically, culturally, materially, and systemically in communities throughout Africa and the world.

In speaking to the notion of psychosocial change, that is, as necessitating transformation from within and from without (see Frosh, 2003), the two of us—along with a number of colleagues—attempted to carve out a space for thinking through the kinds of contradictions, politics, problematics and emancipatory potentialities that emerge from a particular kind of collaborative and potentially decolonising psychological work, what we came to call liberatory Africa(n)-centred work, which in this chapter we turn in the direction of community psychology. Within this collaborative space, we did not seek to offer a definitive definitional outline of such a psychology, but rather to draw on a multitude of disciplines, skill-sets, biographies, and experiences as a means of fleshing out the look, feel and utility of such a

K. Ratele (✉)
Psychology Department, Stellenbosch University, Stellenbosch, South Africa
e-mail: kratele@sun.ac.za

N. Malherbe
Institute for Social and Health Sciences, University of South Africa, Johannesburg, South Africa

South African Medical Research Council-University of South Africa Masculinity and Health Research Unit, Cape Town, South Africa
e-mail: malhenr@unisa.ac.za

S. Kessi et al. (eds.), *Decolonial Enactments in Community Psychology*, Community Psychology, https://doi.org/10.1007/978-3-030-75201-9_2

psychology (or perhaps more accurately, psychologies). Therefore, the space itself came to represent an institutionally-located site of psychosocial change.

A number of collective writing projects concerning Africa(n)-centred psychologies have been conceptualised in this space. Although our colleagues did not join us in writing this chapter for a number of reasons, we nonetheless present this piece as having emerged from this space. Appropriately then, we begin the chapter with an extended consideration of the role that collective writing-thinking-questioning has played in our attempts to think through and enact issues related to psychosocial change, including decolonising psychosocial change, within community psychology. While we realise that this section on collaborative writing may appear somewhat disconnected to the case we are making for a liberatory Africa(n)-centred community psychology of psychosocial change, it is in fact integral to our conceptualisation of such a psychology. This is to say, we consider collaborative writing as a form that is inextricable to the content of liberating and decolonising African psychologies.

We then briefly consider the emancipatory contours, as well as the change-making limitations and capabilities, of community psychology in, from, for and with Africa. While it is not possible to harness a complete, definitive or stable image of Africa, nor are we interested in such an image, we are nevertheless urged to think through problems of colonial as well as anticolonial and decolonialist histories, economic and social injustice, structural and group inequalities, and psychosocial ill-health from our different positions within Africa. To look at the world and ourselves in and from Africa is an issue of actual, imagined, ideological and politico-economic geography, hence we consider how critical psychologists working in communities might situate Africa—with its myriad and often contradictory fantasies, historiographies, politics, economies, ideologies and cultures—within their work, and how this might re-write the disciplinary boundaries of community psychology.

In the chapter we argue for the transdisciplinary impulse of liberatory Africa(n)-centred community psychology and its decolonising potentialities. This is certainly not a novel idea within community psychology praxis. Indeed, Kimberlé Crenshaw's (1991) intersectional feminism, Paulo Freire's (1970) critical pedagogy, Steve Biko's (1987) Black Consciousness philosophy, Linda Tuhiwai Smith's (2012) work on decolonising methodologies, and Orlando Fals-Borda's (1987) participatory action sociology have all been useful in developing emancipatory modes of community psychology from outside of the psy-disciplines. However, these thinkers, along with a number of others, have come to represent a fairly rigid interdisciplinary, even post-disciplinary, 'canon' of sorts within critical community psychologies. Accordingly, in our collective writing-thinking-questioning we have sought to enrich, trouble and stretch critical community psychology's non-psychological, anticanonical ancestry. Elsewhere (Ratele et al., Forthcoming), we have sought to do so through the work of Walter Rodney (1972), Edward Said (1978) and the Combahee River Collective (2000). In this chapter, informed by the same goal, we turn to the work of the revolutionary scholar and activist Neville Alexander. Although Alexander's work had no formal ties to community psychology and is rarely—if ever—drawn on by community psychologists, we show how he offers to psychologists a number of invaluable tools for interrogating problems of history, change, psychology, society, ideology, and identity within and beyond Africa.

We conclude the chapter by considering some of the challenges faced by positioning the psychological within social change efforts, and what this means for liberating psychologies of community within and beyond Africa. We emphasise that formalised, decolonising, and spontaneous political activism must constitute part of the emancipatory agenda of a liberatory Africa(n)-centred psychology and cannot merely be drawn on as a rhetorical or optional adjunctive.

A Note on A Method of Thinking and Writing Together

In 2019, a group of students and professors began developing, trialling and putting into practice a collaborative method of writing and thinking. Many of us had done other forms of collaborative writing and thinking, naturally. Many had participated in different community projects in which they were collaborators with each other, with other researchers, with other groups or individuals, and with communities not linked to universities or research centres. However, this particular mode of writing and thinking *together* followed a demanding track with possibilities for an Africa(n)-centred, liberatory, feminist, antiracist, decolonial or otherwise critical community psychology. The two of us were part of this group.

As a response to the academic machine where publishing is, ultimately, what counts, our loose group came together with the intention to go beyond publishing for its own sake. We wanted to move away from mercenary work and toward more meaningful ends; to have a different relationship with our words. A famous book has the title, *I write what I like* (Biko, 1987). We wanted to write what we like. But we also wanted to like what we wrote. And before the writing, we wanted to like what we thought.

The writing and thinking that we intended to do sought to achieve several goals, besides publication. One goal was to challenge ourselves and, hopefully, others with regards to *how* to do intellectual work; in other words, to decolonise our intellectual engagement and knowledge making. Intellectual work is as much personal as it is political. How we do this work is as important, and often enough not more so, than its end result. There are always, of course, power differentials that colour the process of collaborative writing and thinking. An obvious example is in cases where professors write with students and where the former's voices are assumed to have more weight than the latter's simply because of occupational status or experience. The strength of the collective's contribution is in this sense made secondary to individual habitus. In another example, because political and economic structures give unearned advantage to one language group over others (see Alexander, 2013), the result is that some members of a collective of authors—such as ours—have a much better command of the language used to write than other members. As such, our loose collective effort is put into undermining hierarchies and complicating differences around who knows more between students and professors—or between first-language English speakers and other speakers, between genders, or between academics and community activists—so that we might profoundly learn from each other. There are members of our research institute who already effect this difference-blurring move through writing with community members who are not academics or professional researchers (see, e.g., Cutts et al., 2016; Isobell et al., 2018). Thus, we wish to put in place a different way of understanding power, of relating to each other, of thinking of ourselves as creators of knowledge, and of the character and ownership of the knowledge work that we do as students and professors, community psychologists and community activists—old and young, black and white—who are interested in resolving various social problems and working on diverse topics from different academic disciplines and locations.

Another goal of our writing was to generate a greater sense of collective voice within the text, whereby none of us could readily identify who said what. Indeed, edited books, journal articles and book chapters oftentimes reveal the trace of individual contributors even when they are co-authored (e.g., Bond et al., 2017; Hook, 2004; Ratele et al., 2004; Reich et al., 2007; Seedat et al., 2001). Hence, in contradistinction to an individual authorial voice, we wanted the outcome of our thinking to be a radically collective voice; a voice that is at once voices. Achieving such voice is much harder than we can convey. We do not think we will ever be able to perfect this. Traces of the individual are hard to erase. This difficulty will become clearer soon.

It was with these goals and possibilities in our minds that we began to enact a certain kind of collaborative writing and thinking as a method of knowledge production. We wanted to begin, or perhaps intensify, psychosocial and relational shifts in and for the self and the other.

The method begins with a question by an individual. Sometimes the question is not readily obvious or it may come to one of the group as an invitation for a contribution. Indeed, for this particular chapter, the question was not self-evident and had to be extracted from the invitation to contribute a chapter on psycho-social change and Africa-centred community psychology.

The person who brings the question (or invitation to the others) is not, we have come to appreciate, asking the question with the expectation of receiving a definitive answer. Rather, asking becomes a way into group discussion, pondering, contestation, and reformulations. Thus, the individual's question, once it is asked, becomes the collective's. The question can never be taken as is. There is always another question or other questions enveloped within the initial question, another potential problem or problems behind the solution with which we are presented. And so the individual must allow the others in the group (that is, allow the self before others are allowed) to think of the question from their own socio-cultural situatedness, or indeed material situation—both of which denote an objective, concrete situation, rather than a theoretical one (see Freire, 1970).

It is this questioning of the question that generates the others' interest, possibly through animating the desire to change not just the initial question, but the eventual answer. More broadly, in enabling the dialectics of intersubjective questioning, we may not only generate interest in the problem but, potentially, change the situation.

With each question that is brought to the group, a series of other, implicit questions follow, such as: is this the question we want to work on? What is the problem at stake in this question? This implicit questioning encourages each group member to think about the relevance of the initial question, what it is meant to problematise, its hidden or unarticulated meanings, the potential impact of our responses, the ethics of publishing, how the question might connect to other groups members, how the piece of writing connects to institutional requirements, whether one wants to be part of a group that thinks about this question, and whether one has the time to do so. We can attempt to engage this initiating moment of explicitly and implicit questioning in the language of research methods, that is, as an active teaching and learning of the practice of crafting a clear and concise research question, while also thinking through—with others—the politics, sociology, and ethics of research questions.

Throughout the process of questioning, the initial question changes as it is tested, opened up, critiqued. The question usually gives birth to several other questions as it is broken down to its constituent elements, as was done in this chapter. The breaking-apart entails different members of the group formulating different questions related to the initiating question. They, in turn, ask other members of the collective to review, challenge, reflect on, change, and reformulate new questions before they are incorporated into the writing piece.

The next significant moment begins when the question(s) have been decided and the collective agrees on how different group members will respond to the question(s). For example, in the present case, the question we decided on was, as indicated already, what might a liberatory Africa(n)-centred community psychology in contexts of decoloniality look and feel like, and what might the consequences of such a praxis for psychosocial change be. After an approximate number of words with which to respond to different question(s) has been allocated to each group member, the writing begins. Initially, an individual author will offer their, her, or his first response to a question, which is then sent to another of the collaborators. As with the (re)formulation of the question(s), an individual's answer to a question becomes that of the group's as soon as it is received. The collaborators are encouraged to add, change and rewrite another's work, and not merely to edit the work. This active group engagement can result in feelings of defensiveness from the original author. However, this confrontation with knowledge ownership allows for lessons into how ideas are shared. In other words, disagreement of this sort can give rise to the conditions necessary for critical and liberating learning. While this interaction is generative, constitutive and re-creative, psychological resistance can raise its head, and strain in the group can become visible. Nevertheless, the development of the manuscript does proceed and it is thus taken through several iterations, changes and corrections before submission to external peer review.

We should say something about three other elements of this collaborative questioning-writing-thinking method. The first, and the most obvious—yet one that nonetheless remains overlooked in discussions on writing—is the need to be in the same room. Being in the same room has come to include making time to eat together and talk about other things unconnected to the writing itself. While being in the same room is not always possible, such enmeshment within the lives of others makes for a "thick" mode of collaboration that is, we believe, best facilitated through interpersonal, physical interaction. Indeed, email and other online kinds of "thin" collaborative modalities, do not always work as well when attempting to explain why a sentence reads the way it does, or understand a co-author's bodily experience.

Second, stories. In addition to the question and response that we work on iteratively and collaboratively, our method of collaborative writing and thinking incorporates real-life stories. It soon becomes clear that it was vital to find stories that illustrate as much as possible the epistemological, political, economic, theoretical and social problems to which we wish to respond. Additionally, the stories must be personal. Drawing from and critically analysing our own lived experiences is, we believe, essential if we wish to make our answers relevant for readers, and to engage in a mode of subjective and psychological decolonisation.

It turns out that finding stories is vital, but the skilful narration of those stories is as essential. Telling stories well is something that we teach to and learn from one another. Similarly, listening becomes a skill that is actively worked on. Therefore, it is necessary to instigate a dialogue concerning the stories that we tell and hear from each other, which is what fosters the third element of our group writing process: facilitation. We can listen to stories passively, but as we listen to them, they tend to stimulate thoughts, feelings, memories, stories and questions of our own. Learning to question others with consideration and respect, sitting with discomfort, careful interruption—and we encourage interruption—and hearing each other well are things that we learn in the writing process. It is this respectful and uncomfortable questioning which points to the need for competent facilitation. Notably, any one of the collective is able to take on facilitation; it is not necessarily a role that is designated to the eldest or most experienced member of the group.

This collaborative process of working is psychologically demanding. It discomposes individuals and taxes one's emotional capacities as well as their intellectual faculties, both of which can give rise to anxieties and guardedness. Yet, at the same time, collaboration of this sort can open up potentially decolonising possibilities for thinking and writing, and for being-with-others in new and uncommon ways. As such, it can make possible a decolonising mode of knowledge production.

This collaborative writing method continues to be elaborated. We have grown increasingly appreciative of what we are doing as we put the method into action. It is through the transdisciplinary and actional orientation of such writing-thinking-questioning that we now turn to the question of psychosocial change within a liberatory Africa(n)-centred community psychology—and the other questions to which it gave birth as we kept turning it around.

How to Situate Africa Within Community Psychology to Re-write the Discipline?

The notion of geography may appear somewhat irrelevant to a community psychology whose disciplinary historiography tends to prioritise intellectual—rather than geographic—currents (Reich et al., 2017). Yet, as critical and radical human geographers have long realised (see, e.g., Blomley, 2006), the histories of political, cultural, epistemological and social struggle are locked into particular geographies. Questions of geography are therefore crucial in understanding the present moment and its emancipatory requirements. Of relevance for those of us attempting to situate Africa within critical community psychology praxis, Heidi Hudson's (2016) conception of meta-geographies is especially germane. Colonial meta-geographies, she argues, cast an idealised vision of where and how something should (and also can) look and feel. Thus, the Global North becomes a fixed, cosmopolitan and aspirational ideological entity. Similarly, and by contrast, the Global South emerges as an objectified, fundamentally lacking, deviant and helpless place. For community psychology, meta-geographies delineate what the discipline is (e.g. birthed in Swampscott, Massachusetts in 1965; institutionally-bound; faithful to psychological dictum; rhetorically, rather than politically committed to radical social movements), and what it cannot be (politicising; decolonial; African; anti-psychological). Understanding the meta-geographic constitution of community psychology indicates to us why and how centring Africa within community psychology is able to re-write the discipline, and even *bend* it in decolonising directions.

Yet, centring Africa in critical community psychology presents a similar set of problems facing those who have sought to centre community within critical psychology (see Dutta, 2018; Kagan et al., 2011). Africa—like community—contains a universe of meanings, histories, and materialities which can be drawn upon, essentialised and deployed for political purposes. For instance, both the colonial trope of "Africa as savage" and the capitalist one of "Africa rising" employ very different images of Africa towards (differently constituted) exploitative ends. And just as a monolithically determined set of values, languages, religions, conceptual vocabularies, races or principles do not exist in Africa (Appiah, 1992), community psychology in Africa sees a wide range of political and epistemological orientations that incorporate a mix of Afro-centric, Euro-America(n)-centric and localised frameworks (see Lazarus et al., 2006). It would seem then that Africa as a construct has no definitional centre, which makes difficult the centring of it within any discipline, let alone one as politically diverse as community psychology.

We wish to argue, however, that it is precisely because of Africa's fundamental unrepresentability (see Hall, 1990) that an emancipatory Africa(n)-centred community psychology need not concern itself with delineating a singular way of doing community psychology. Rather, Africa(n)-centred critical community psychology points to an orientation—specifically, a global orientation that is always from, by, with, and for Africa—which is to be assumed by community psychologists (see Ratele, 2019). Africa(n)-centred critical community psychology, accordingly, may serve to guide how we enact, as well as decolonise, psychologies of community from within Africa. These need not look nor feel like formalised community psychology, which is often—quite legitimately—taken as an object of suspicion in a number of African countries (see Lazarus et al., 2006). Much of the most liberatory, decolonising, and socially useful kinds of community psychology work in Africa and elsewhere are undertaken by community organisers, activists, social movements, artists, radical policy-makers, cultural workers and community leaders who have little to no association with community psychology (see Dutta,

2018; Frosh, 2003; Lazarus et al., 2006; Watkins & Shulman, 2008). Thus, for our purposes, it is perhaps suitable to engage Africa(n)-centred critical community psychology not as a professional discipline and organisation, but as a set of assumptions and values which are able to resist restricting adherences to disciplinary dogma (see Fryer, 2008).

Africa(n)-centred critical community psychology should conceive of both Africa and community psychology as epistemologically open. Neither concept can be fixed if they are to be used together for the purposes of psychosocial justice. However, advancing a socially just project of this kind requires that Africa(n)-centred critical community psychologists assume a political stance that is transparent in its vision of justice, yet willing to adapt this vision to various emancipatory requirements. We do well to remember that just as justice has been conceptualised in liberalised, radical and pseudo-progressive ways from within community psychology (see Gokani & Walsh, 2017), there have been brutally oppressive colonial visions of Africa that have been imposed by people like Ian Smith, King Léopold II, Cecil Rhodes and Louis Faidherbe as well as decolonising, African-centred ones advanced by people like Thomas Sankara, Albertina Sisulu, Frantz Fanon, Funmilayo Ransome-Kuti and countless others. Therefore, while a liberatory Africa(n)-centred community psychology resists precise definition, its potential for advancing decolonising and politically progressive praxes rests on its engagement with emancipatory traditions within Africa and—usually—outside of community psychology (see Mignolo, 2009). As the work of Stuart Hall (1990) reminds us, liberatory politics requires that we look for new ways of relating to the past so that we can discover new places from which to speak and act in the present. Those concerned with psychosocial liberation should, in every instance, submit Africa(n)-centred critical community psychology to the exigencies of decolonial politics.

Finally, in addition to being a political and, in many ways, an anti-disciplinary project, the task of centring Africa within critical community psychology is one characterised by epistemic disobedience (see Mignolo, 2009). This is to say, a liberatory Africa(n)-centred community psychology rejects colonial epistemes that seek to justify the inferiorised and/or essentially different *place* of Africa and Africans in the world by harnessing marginalised knowledge systems and histories for the purpose of (re)asserting the dignity and humanity of Africa (Mignolo, 2009; Suffla & Seedat, 2015). This focus on agency and subjecthood rejects the "cult of victimisation" (Mbembe & Rendall, 2002, p. 244) that is so often apparent within so-called modes of "African self-writing". And still, a focus on epistemic disobedience should not displace internationalised and solidarity-driven impulses from within Africa(n)-centred critical community psychology. It should instead signal to us that it is often those at the margins who are best equipped to study the coloniality of power, being, and knowledge as they exist at the margins as well as at the centre (Ratele, 2018).

In What Ways Might Neville Alexander Be Useful in Unearthing Socially Just Africa(n)-Centred Critical Community Psychologies?

Neville Alexander was born in South Africa's Eastern Cape in 1936. His political consciousness was sharpened as a university student in Cape Town, where he eventually embraced radical, anti-colonial socialism that would guide all of his subsequent thought and activism. After receiving a PhD in Germany at the age of 26 (Vally, 2014), he returned to South Africa in 1961 and became a high school teacher committed to critical pedagogy (see Freire, 1970). He was also, at this time, involved in the fiercely anti-apartheid Black Consciousness Movement and the Non-European Unity Movement (see Alexander, 2013). As a result of his activism, in 1963 he was sentenced to prison on Robben Island, where he engaged in intensive prisoner-run education programmes. After a decade on Robben Island, he spent another 5 years under house arrest (Magnien, 2012; Vally, 2014). He stayed politically active in post-apartheid South Africa and, right up until his death in 2012, remained bitingly critical of the country's embrace of neoliberal capitalism (see Alexander, 2002, 2010), as well as neo-imperialist policy (Alexander, 2013).

As noted already, Alexander was not a community psychologist in any formalised sense, and his work is almost never considered within community psychology. However, we argue that throughout his life as a community-oriented activist scholar he practiced the kind of Africa(n)-centred, liberatory, decolonising and community-situated psychology which has been conceptualised in this chapter and elsewhere (e.g. Kurtiş & Adams, 2015; Ratele, 2019). Certainly, Alexander's three primary intellectual concerns—political economy, education and sociolinguistics (see Alexander, 2013)—were always grounded in, and linked back to, notions of Africa, psychology and community. Indeed, a palpable sense of hope and community appears to have undergirded Alexander's work (Vally, 2014). His was a politics of humanism that sought to interrogate the psycho-material social nexus while, at the same time, striving to bring this into the orbit of life and human freedom.

Aligning with the Marxian sentiment that history proper begins only once all people are liberated from exploitation and oppression (see Malherbe, 2018), Alexander (2013) proposed that community activists should engage state apparatuses (while always retaining sovereignty and control) as a means of creating a *new historical community* where colonial reasoning cannot prosper. Although dispelling class reductionism and race essentialism in his analysis (Vally, 2014), Alexander believed that race must be given space and openly discussed within public and in social movements if people and communities are to be reimagined in ways that centralise their humanity. This brings to mind Paul Gilroy's (2000) well-known statement that "action against racial hierarchies can proceed more effectively when it has been purged of any lingering respect for the idea of 'race'" (p. 13). The idea of a new historical community presents Africa(n)-centred community psychologists with an evocative and inspiring mode of politics. Certainly, we should develop ways of engaging the psycho-political consequences of identity categories which have been weaponised to cataclysmic effect and yet, at the same time, have formed a basis of pride, community organising, solidarity and resistance (see Alexander, 1995, 2013). We cannot, however, forget the mytho-ideological basis upon which these categories are founded. It is important, as Alexander (2013) emphasised, to retain a critical sense of differentness from each other (socially constructed, materially experienced and psychologically felt) while not capitulating in every instance to social or political fragmentation, colonial ontologies, and identitarian essentialism. We must nurture an appreciation for what can be done with differences while retaining the psychosocial and politico-economic necessity of human connectedness. Africa(n)-centred critical community psychologists who are guided by the new historical community are urged to take notions of difference seriously by working with people to draw out points of connection within and between seemingly unconnected struggles. The underexplored psychosocial contours of solidarity are therefore paramount to a liberatory, decolonial, radical or otherwise critical psychology that looks to centre Africa within community-engaged praxes.

In speaking out against the narrative of exceptionalism that often surrounds the 'miracle' of South Africa's transition from apartheid capitalism to liberal democratic capitalism, Alexander (2002) frequently referred to South Africa as an "ordinary country". By this he meant that a legitimate radical politics must advance from the premise that South Africa is a product of a particular set of global and localised socio-economic forces. He emphasised that South Africa, like other countries on the continent, had indeed experienced a political revolution of sorts, however the socio-economic revolution required for psychological liberation remained to be seen (Alexander, 2013). Echoing the work of Fanon (1967), Alexander always engaged psychological pathology as fundamentally social in constitution. Accordingly, the work of Africa(n)-centred critical community psychologists is always situated within an ordinary country, whose situation has been made by systemic historical forces and can thus be remade in the image of decolonial liberation. It is only by understanding how power works through history to structure ordinary, daily life that we can take it back (see Alexander, 2002, 2013; Frosh, 2003).

While Marxist theory undergirds much of Alexander's thinking, he was especially taken by the work of Antonio Gramsci, the Italian communist revolutionary. Of particular interest was Gramsci's notion of the war of position, which denotes how counter-hegemonic activity is able to challenge oppressive hegemons from within oppressive social structures, slowly building a critical mass that is necessary for revolutionary change, or what Gramsci referred to as the war of manoeuvre (see Gramsci, 1971). Alexander understood the war of position as a kind of ethics of hope which sought to harness people's creative and experimental capacities in disseminating humanistic ways of organising society (Alexander, 2010; Vally, 2014). He noted how the community development programmes carried out by the Black Consciousness Movement during the apartheid era represented a war of position which, along with other movements, eventually toppled the apartheid regime (see Alexander, 2013). These programmes, as Seedat and Lazarus (2011) argue, were also a precursor to critical community psychology in South Africa. For Africa(n)-centred critical community psychologists, the war of position is a reminder to refuse limited decolonial horizons (see Alexander, 2010; Dutta, 2018), while always facing our work toward the enormous task of liberating subjectivities, relations with each other, communities and the world. Community psychology work attuned to the war of position is not an end in and of itself, but seeks to clear space for more transformative psychological, community and global change. Small victories within and beyond communities can begin to bring into focus the kind of life that people deserve, one devoid of systemic social suffering. Indeed, victories of this sort can function to bring about other victories that, together, move the needle towards decolonised psychosocial change. In this regard, the war of position seeks to build alternative and demonstrable possibilities for decolonisation (Vally, 2014).

Another Gramscian notion that informed Alexander's thinking and activism was that of the organic intellectual. In line with Gramsci's (1971) thinking, intellectuals, Alexander (1995) argued, are not confined to the academy and instead can be found in communities. These organic intellectuals constitute organisers and activists who work with people to construct emancipatory ideologies and attempt to understand the world in order to take an informed political stance in trying to change it (Alexander, 2013). In the context of liberatory community psychology, the organic intellectual is conceptualised as one who must embody a critical mode of reflexivity that interrogates the self with respect to its emancipatory utility within psychological work (see Malherbe, 2018). Africa(n)-centred community psychologists should always be guided by decolonising struggles as they are articulated and understood by communities themselves (see Dutta, 2018; Ratele, 2019). The task of liberatory Africa(n)-centred community psychology is to allow community-based organic intellectuals to determine how community psychology is to be used, if at all, within and for particular struggles for decolonial justice.

For Alexander, one of the ways by which organic intellectuals can enter into the war of position is through their engagement with 'common sense'. He passionately lamented about "the entrepreneurial, individualistic whiz-kids of the neo-liberal epoch" (see Alexander, 2013, p. 42) whose 'commonsensical' economic rationalism acts to normalise systems of inequality and obscure the possibility of socially just human relations (Alexander, 2010). In turn, he argued for an ethic of radical humanism that rejects the individualising meaning-making capacities which characterise the capitalist hermeneutic. In other words, organic intellectuals who involve themselves in the war of position have a duty to engage political programmes that oppose, rather than emulate, the 'common sense' constraints that are placed on our collective emancipatory imagination (Vally, 2014). Alexander's conception of 'common sense' presents a useful means of constructing counter-hegemonies which, for those community psychologists concerned with liberation, can inform processes of de-ideologisation, denaturalisation and critical consciousness-raising (see Malherbe, 2018). By rejecting, in the first instance, all common sense-making which denigrates and/or fetishises African cultures and societies, a project of decolonial (re)humanisation can begin to take place through political activism, critical pedagogy and cultural production (see Dutta, 2018). Therefore, if it is to be politically useful, Africa(n)-centred critical community psychology must involve itself in such activities, seeking always to make commonsensical Alexander's (2013) notion of the new historical community.

Alexander (1995) stressed that within a political resistance strategy, we are required to manoeuvre between creativity and various political, economic and cultural structures. In this way, we should remain as utopian as we are analytic. Alexander exemplified an ethos and an ethico-political paradigm that is able to guide the enactment and theorisation of Africa(n)-centred critical community psychology in ways that prioritise the decolonial needs of communities over any kind of disciplinary allegiance (see also Kurtiş & Adams, 2015; Fryer, 2008; Martín-Baró, 1994; Ratele, 2019). We thus argue that Alexander's work represents an especially emancipatory and relevant iteration of critical community psychology.

How Can Africa(n)-Centring Psychologists Within Communities Locate the Psychological Within Social Change?

Social change presents a somewhat spurious case within community psychology. Despite social justice being included as a value and an aim in the Vision of the Society for Community Research and Action, Section 27 of the American Psychological Association, (Evans et al., 2014), mainstream community psychology has, historically, attended to coloniality inadequately, including issues related to class inequality (Gokani & Walsh, 2017), racial oppression (Carolissen et al., 2010), Euro-centrism (Suffla & Seedat, 2015) and gender-based violence (Kurtiş & Adams, 2015). In short, in struggling to break from psychology's focus on the individual, community psychologists regularly work to adjust people to social inequities rather than altering oppressive and exploitative social conditions at the root (Moane, 2003). In speaking to this problematic, decolonial psychosocial work looks to address liberalised modalities of psychology that separate the psychological from the social. Stephen Frosh (2003) argues that such work advances from the thesis that human subjectivity is always constituted through social forces, meaning that neither the psychological nor the social contain an 'essence' that is formed independently from the other.

Ignacio Martín-Baró (1994) highlights how oppression, in addition to its material components, presents a fundamentally psychosocial character. He notes that the psychosocial subject is dialectical in character, and that trauma is determined by a number of intersecting currents that are, themselves, individual and social. Such trauma, Martín-Baró (1989) argues—as did Frantz Fanon (1967) before him—is written into the fabric of an unequal and oppressive (neo)colonial society. Thus, in attempting to heal psychosocial traumas, psychological interventions are urged to, similarly, locate the psychological within decolonising society change-making efforts, most of which occurs at a considerable distance from the discipline of psychology (Dutta, 2018). The point becomes not to settle the fractured and/or traumatised self into existing, fracturing social relations, but to create new ways of self-styling (Mbembe & Rendall, 2002) that are geared toward fostering Alexander's (2013) notion of the new historical community.

For Africa(n)-centred critical community psychology, psychosocial healing requires an engagement with the socio-historical and economic contours of trauma within the continent which, because of the colonial encounter, necessitates a global, as well as a local, purview. How, in other words, are people's present-day psychological conditions linked to their historical positionality on, against and with Africa? Further, how do these historical currents reach into and shape suffering in the present? In answering these questions, Africa(n)-centred critical community psychologists should not understand themselves as healers in any definitive sense, but rather as facilitators working with people to nurture critical and psychological consciousness, and to connect this with the kinds of decolonising social action required for psychosocial healing. Indeed, social activism is able to, and should, attend to the ways by which the wounds of coloniality are experienced consciously as well as unconsciously (Mbembe & Rendall, 2002). At the same time, such activism (and its inevitable ongoing setbacks) undoubtedly incur a psychological toll, to which Africa(n)-centred critical community psychologists must also attend. When we consider trauma as expansively as possible, our healing processes take on a similarly ambitious formation.

Concluding Thoughts

In speaking to both the form and content of a community psychology that looks to place itself within the orbit of Africa(n)-centred decolonising liberation, we have attempted in this chapter to engage issues of psychosocial change through a series of related questions on geography, ideology, the limits of psychology, transdisciplinarity, and political activism. We have also taken the opportunity to delineate the method of joint thinking and writing that we increasingly use in our work. We introduced to community psychology the revolutionary scholar and activist Neville Alexander. Although he is a well-known scholar in his chosen areas of work, his inspiring combination of activism and scholarship is able to offer to community psychology particular insights with respect to decolonial psychosocial praxis. Indeed, despite, or maybe because, Alexander had no formal links with community psychology, his work presents us with a useful set of resources for articulating a liberatory Africa(n)-centred community psychology that extends beyond institutional confines. Ultimately, we have sought in this chapter to place an accent on the boundary-blurring that is required when formulating liberatory, decolonial, anti-racist, feminist or other kinds of critical community psychology. This work is, however, nascent, emerging and incomplete. We encourage others to work with one another to bring into their psychosocial community work a mode of political action that is not disciplinary bound, and that seeks to engage liberation in a necessarily ongoing, decolonising, ambitious, expansive, radical and imaginative fashion.

References

Alexander, N. (1995). "The moment of manoeuvre": "Race," ethnicity and nation in post-apartheid South Africa. *Comparative Studies of South Asia, Africa and the Middle East, 15*(1), 5–11.
Alexander, N. (2002). *An ordinary country: Issues in the transition from apartheid to democracy in South Africa*. University of Natal Press.
Alexander, N. (2010, May). *South Africa—Unfinished revolution*. Paper Presented at the Fourth Strini Moodley Annual Memorial Lecture, University of KwaZulu-Natal. http://links.org.au/1693
Alexander, N. (2013). *Thoughts on the new South Africa*. Jacana.
Appiah, A. K. (1992). *In my father's house: Africa in the philosophy of culture*. Methuen.
Biko, S. (1987). *I write what I like: Steve Biko. A selection of his writings*. Heinemann.
Blomley, N. (2006). Uncritical critical geography? *Progress in Human Geography, 30*(1), 87–94. https://doi.org/10.1191/0309132506ph593pr
Bond, M. A., Serrano-García, I., & Keys, C. (Eds.). (2017). *Handbook of community psychology (Vol. 1): Theoretical foundations, core concepts, and emerging challenges*. American Psychological Association Press.
Carolissen, R., Rohleder, P., Bozalek, V., Swartz, L., & Leibowitz, B. (2010). "Community psychology is for poor, black people": Pedagogy and teaching of community psychology in South Africa. *Equity & Excellence in Education, 43*(4), 495–510.
Combahee River Collective. (2000). The Combahee river collective statement. In B. Smith (Ed.), *Home girls: A feminist anthology* (pp. 264–274). Rutgers University Press.
Crenshaw, K. (1991). Mapping the margins: Intersectionality, identity politics, and violence against women of color. *Stanford Law Review, 43*(6), 1241–1299.
Cutts, T., Olivier, J., Lazarus, S., Taliep, N., Cochrane, J. R., Seedat, M., van Reenan, R., Hendricks, C., & Carelse, H. (2016). Community asset mapping for violence prevention: A comparison of views in Erijaville, South Africa and Memphis, USA. *African Safety Promotion: A Journal of Injury and Violence Prevention, 14*(1), 1–25.
Dutta, U. (2018). Decolonising "community" in community psychology. *American Journal of Community Psychology, 62*(3–4), 272–282. https://doi.org/10.1002/ajcp.12281
Evans, S. D., Rosen, A., & Nelson, G. (2014). Community psychology and social justice. In C. V. Johnson, H. L. Friedman, J. Diaz, Z. Franco, & B. K. Nastasi (Eds.), *Praeger handbook of social justice and psychology* (pp. 143–163). Praeger.
Fals-Borda, O. (1987). The application of participatory action-research in Latin America. *International Sociology, 2*(4), 329–347.
Fanon, F. (1967). *Black skin, white masks*. Grove Press.
Freire, P. (1970). *Pedagogy of the oppressed*. Herder and Herder.
Frosh, S. (2003). Psychosocial studies and psychology: Is a critical approach emerging? *Human Relations, 56*(12), 1545–1567. https://doi.org/10.1177/00187267035612005
Fryer, D. (2008). Some questions about "The history of community psychology". *Journal of Community Psychology, 36*(5), 572–586. https://doi.org/10.1002/jcop.20240
Gilroy, P. (2000). *Against race*. Harvard University Press.
Gokani, R., & Walsh, R. T. (2017). On the historical and conceptual foundations of a community psychology of social transformation. *American Journal of Community Psychology, 59*(3–4), 284–294. https://doi.org/10.1002/ajcp.12141
Gramsci, A. (1971). *Selections from the prison notebooks*. Lawrence & Wishart.
Hall, S. (1990). Cultural identity and diaspora. In J. Rutherford (Ed.), *Identity: Community, culture difference* (pp. 222–237). Lawrence and Wishart.
Hook, D. (Ed.). (2004). *Critical psychology*. University of Cape Town Press.
Hudson, H. (2016). Decolonising gender and peacebuilding: Feminist frontiers and border thinking in Africa. *Peacebuilding, 4*, 194–209. https://doi.org/10.1080/21647259.2016.1192242
Isobell, D., Taliep, N., Lazarus, S., Seedat, M., Toerien, E., & James, A. (2018). Reflections on the development and utility of a participatory community violence surveillance methodology. *African Safety Promotion: A Journal of Injury and Violence Prevention, 16*(1), 20–37.

Kagan, C. M., Burton, M., Duckett, P. S., Lawthom, R., & Siddiquee, A. (2011). *Critical community psychology*. Wiley.

Kurtiş, T., & Adams, G. (2015). Decolonising liberation: Toward a transnational feminist psychology. *Journal of Social and Political Psychology, 3*(1), 388–413. https://doi.org/10.5964/jspp.v3i1.326

Lazarus, S., Bojuwoye, O., Chireshe, R., Myambo, K., Akotia, C., Mogaji, A., & Tchombe, T. (2006). Community psychology in Africa: Views from across the continent. *Journal of Psychology in Africa, 16*(2), 147–160. https://doi.org/10.1080/14330237.2006.10820116

Magnien, N. (2012). Dr. Neville Edward Alexander. *South African History Online*. https://www.sahistory.org.za/people/dr-neville-edward-alexander

Malherbe, N. (2018). Expanding conceptions of liberation: Holding Marxisms with liberation psychology. *Theory & Psychology, 28*(3), 340–357. https://doi.org/10.1177/0959354318767757

Martín-Baró, I. (1989). Political violence and war as causes of psychosocial trauma in El Salvador. *International Journal of Mental Health, 18*(1), 3–20.

Martín-Baró, I. (1994). *Writings for a liberation psychology*. Harvard University Press.

Mbembe, J. A., & Rendall, S. (2002). African modes of self-writing. *Public Culture, 14*(1), 239–273.

Mignolo, W. D. (2009). Epistemic disobedience, independent thought and decolonial freedom. *Theory, Culture & Society, 26*(7–8), 159–181. https://doi.org/10.1177/0263276409349275

Moane, G. (2003). Bridging the personal and the political: Practices for a liberation psychology. *American Journal of Community Psychology, 31*(1–2), 91–101.

Ratele, K. (2018). Toward cultural (African) psychology. In G. Jovanović, L. Allolio-Näcke, & C. Ratner (Eds.), *The challenges of cultural psychology: Historical legacies and future responsibilities* (pp. 250–267). Routledge.

Ratele, K. (2019). *The world looks this from here: Thoughts on African Psychology*. Wits University Press.

Ratele, K., Duncan, N., Hook, D., Mkhize, N., Kiguwa, P., & Collins, A. (Eds.). (2004). *Self, community and psychology*. University of Cape Town Press.

Ratele, K., Stevens, G., & Malherbe, N. (Forthcoming). An orienting conversation on Africa(n)-centred decolonial community psychologies. In G. Stevens & C. Sonn (Eds.), *Decoloniality, knowledge production and epistemic justice in contemporary community psychology*. Springer.

Reich, S., Bishop, B., Carolissen, R., Dzidic, P., Portillo, N., Sasao, T., & Stark, W. (2017). Catalysis and connections: The (brief) history of community psychology throughout the world. In M. A. Bond, I. Serrano-García, & C. Keys (Eds.), *Handbook of community psychology volume 1: Theoretical foundations, core concepts, and emerging challenges* (pp. 21–66). American Psychological Association Press.

Reich, S. M., Riemer, M., Prilleltensky, I., & Montero, M. (Eds.). (2007). *International community psychology: History and theories*. Springer.

Rodney, W. (1972). *How Europe underdeveloped Africa*. BogleL'Ouverture Publications.

Said, E. W. (1978). *Orientalism*. Pantheon Books.

Seedat, M., Duncan, N., & Lazarus, S. (2001). *Community psychology: Theory, method and practice*. Oxford University Press.

Seedat, M., & Lazarus, S. (2011). Community psychology in South Africa: Origins, developments, and manifestations. *Journal of Community Psychology, 39*(3), 241–257. https://doi.org/10.1002/jcop.20429

Smith, L. T. (2012). *Decolonising methodologies: Research and indigenous peoples* (2nd ed.). Zed Books.

Suffla, S., & Seedat, M. (2015). Reflexivity, positionality, context, and representation in African and Arab enactments of community psychology. *Journal of Community Psychology, 43*(1), 4–8. https://doi.org/10.1002/jcop.21716

Vally, S. (2014). The battle for the truth–youth resistance, neoliberalism and an appreciation of Neville Alexander. *Critical Arts, 28*(1), 69–75. https://doi.org/10.1080/02560046.2014.883690

Watkins, M., & Shulman, H. (2008). *Toward psychologies of liberation*. Palgrave Macmillan.

Decolonising Participatory Action Research in Community Psychology

Jesica Siham Fernández

Introduction

> It is also imperative that we not lose sight of the methods of the oppressed that were developed under previous modes of colonisation, conquest, enslavement, and domination, for these are the guides necessary for establishing effective forms of resistance under contemporary global conditions: they are key to the imagination of "postcoloniality" in its most utopic sense.—Chela Sandoval (2000, p. 9)

Community psychology is deeply committed to the empowerment, liberation and thriving of communities who are socially, politically, economically and institutionally marginalised. Across the West and the Global South the discipline has evolved to address the pressing needs of communities affected by interlocking systems of power and oppression. These evolutions, or waves of change within the discipline have been referred to as "agitations" (Langhout, 2016), and calls for "decentering" the field (Dutta, 2016), thereby making community psychology theory, research, and practice better rooted in its values and commitments to liberation and social change. Empowerment, like liberation, has been a key tenant of community psychology. Both empowerment and liberation seek to foment transformational social change, albeit through distinct processes. Seemingly missing from these concepts and from the discipline more generally is an explicit intentional interrogation of coloniality; how it surfaces and is maintained in community psychology theory, research, and practice. Specifically, how the logics of coloniality and the mechanisms of imperialism—and empiricism in research, for example—make their way into the field through epistemologies and methodologies that inform the production of knowledge and power.

Understanding how community psychology engages in decoloniality and anticoloniality is imperative. In general, within psychology some scholarship has

---

J. S. Fernández (✉)
Santa Clara University, Santa Clara, CA, USA
e-mail: jsfernandez@scu.edu

S. Kessi et al. (eds.), *Decolonial Enactments in Community Psychology*, Community Psychology, https://doi.org/10.1007/978-3-030-75201-9_3

engaged decoloniality, decolonisation and anti-coloniality (Adams, 2014; Adams et al., 2015; Bhatia, 2017; Bulhan, 2004; Fanon, 1970; Walsh, 2009, 2014; Wendt & Gone, 2012). Yet more is needed, specifically within community psychology (for exceptions see Reyes Cruz & Sonn, 2011; Kessi, 2017; Lykes, 2001; Lykes & Moane, 2009; Seedat & Suffla, 2017; Sonn et al., 2013; Watts & Serrano-García, 2003). The implication of leaving unexamined community psychology's disciplinary ontologies, epistemologies and methodologies is inherently problematic. Furthermore, how hegemonic knowledge is reinforced and reproduced often unknowingly by well-intentioned researchers, practitioners, and even scholar-activists, lies in the ability and capacity to engage decoloniality as a praxis, especially when engaging with issues of power, systemic intersectional oppression, and racialised inequities.

From a community psychology disciplinary standpoint, one paradigm within the field that has engaged in a decolonial praxis, as well as the democratisation of knowledge and epistemic justice, is participatory action research (PAR). Characterised by strategic methodological interventions that seek to disrupt and transform systemic structural inequities, PAR is aligned with a liberatory decolonial research justice framework (Ayala et al., 2018; Jolivétte, 2015; Smith, 1999). By explicitly centering community voices, lived experiences, collective agency and sociopolitical leadership, PAR works to deconstruct the structures of knowledge and power in research. In keeping with the decolonial process, PAR allows for the disruption of hegemonic knowledge production by facilitating opportunities for critical consciousness, self and collective determination, shared decision-making, and transformational social change as defined and determined by communities, who draw from local knowledge or Indigenous cosmologies (Ayala et al., 2018; Torre, 2009; Torre et al., 2012; Walsh, 2009, 2014).

Over the past two decades community psychology has embraced decoloniality and PAR as valuable praxes to engage in theory, research, and practice with communities and institutions as a means to facilitate conditions toward liberation. Special issues in flagship community psychology journals, such as the *South African Journal of Psychology* (volume 47, issue 4, 2017), *American Journal of Community Psychology* (volume 62, issue 3–4, 2018), and *Journal of Social & Political Psychology* (volume 3, issue 1, 2015) for example, have made space for decolonial scholarship. Yet as Tuck and Yang (2012) warn, academic structures have not been prepared to engage decolonisation and decoloniality. Disciplines, like community psychology, may well be constrained by the foundations of psychology, scientific inquiry, and academe that reify the disciplinary logics of the field and its conventions. Pursuing trajectories toward the decolonisation of research in the (re)production of knowledge can be perceived as transgressive or antithetical to a field that was once rooted in positivist paradigms.

The need for a decolonial community psychology, specifically a decolonial PAR praxis toward the transformational liberation and sovereignty of communities, builds on the work of Tuck and Yang (2012) who pointedly claim that "decolonisation is not a metaphor:"

> Decolonise (a verb) and decolonisation (a noun) cannot be easily grafted onto preexisting discourses/frameworks, even if they are critical, even if they are anti-racist, even if they are justice frameworks (p. 3).

Unequivocally, decolonisation and decoloniality cannot substitute and are not synonymous terms with anti-oppression, liberation or social justice. However, it must be acknowledged that PAR as a tool can be strategically used to disrupt colonial relations of power. In arguing for the importance of decolonising PAR, I emphasise the necessity for community psychology as a discipline to engage in a praxis of "decolonial doing" that is characterised by the de-linking and disrupting/interrupting of hegemonic colonial structures and systems of oppression.

A decolonial praxis that goes beyond merely interrogating our research subjectivities in PAR is crucial, otherwise we risk reproducing colonial relations. Of most significance, however, is decolonising the epistemological and methodological orientations that inform our theory, research, and practice. Given the implications of Tuck and Yang's (2012) cautionary statement, what does it mean to decolonise PAR within community psychology? What are key elements toward the development of a decolonial PAR praxis? Pursuing these questions requires the disruption of particular logics within the discipline, and centering decoloniality within liberatory PAR paradigms rooted in the Global South and Indigenous cosmologies.

Before proceeding with this chapter, a reflection of my subjectivities is in order for it shapes the theoretical and epistemological underpinnings of my work, including my writings. As a U.S. trained social-community psychologist, and community-engaged researcher with specific subaltern realities as a *mestiza* in a settler colonial nation-state, I am critical of my subjectivities and how these operate to subvert or re-inscribe the logics of coloniality within my work. Therefore, a goal for this chapter is to trouble and engage with elements of decoloniality and PAR in order to work toward a decolonial PAR praxis in community psychology.

Inspired by Maldonado-Torres' (2016) *Outline of Ten Theses on Coloniality and Decoloniality*, which builds on the work of Fanon (1963, 1970), in this chapter. I propose a set of interconnected mutually constitutive values described as axioms for decolonising PAR in community psychology. Axiom is defined as a starting point for conversation and reasoning. The word is also used to denote an agreement over a claim considered to be self-evident to those who share a common perspective. Thus, my use of the word axiom is intentional in inviting reflection and dialogue among those of us committed to anti-colonial/decolonial/decolonising movements. I invite us to consider if and how PAR can serve as a tool toward decoloniality, as well as be(come) decolonised. The goal of decolonising PAR is indeed incomplete and ongoing as it constitutes part of a larger movement within and beyond the discipline of community psychology toward the *decolonial turn*.

Decolonial Thinking and the Decolonial Turn

> Decolonisation never takes place unnoticed, for it influences individuals and modifies them fundamentally.—Frantz Fanon (1963, p. 36)

Decolonisation is a process, not an end goal. Decolonial thinking is characterised by modes of thought, action, relationships and practices that do not maintain colonial systems of racialised imperial power, oppression and epistemic violence. Quijano (2001) characterises the coloniality of power as the continuous colonial thinking processes of the oppressed; processes that stem from cycles of oppression and violence upon differential ways of knowing, thinking and being. Coloniality, therefore, is characterised by relations of power that produce colonial race and gender structures, locally and transnationally, and in connection with capitalistic systems of neoliberal oppression and dispossession (Lugones, 2008; Quijano, 2001). Because the *coloniality of power* (re)produces colonial thinking, an essential task of a decolonial PAR praxis must be to decolonise knowledge and disrupt hegemonic power in all of its forms. Decolonial thought can cultivate and support the radical imagination of oppressed people, and produce distinct patterns of expression and thought that can challenge or refuse Western epistemologies associated with the colonial gaze (Smith, 1999) and disembodied research objectivity (Tuck & Yang, 2014). To contribute to the project of the *decolonial turn*, as proposed by Maldonado-Torres (2007), PAR in community psychology must be interrogated and transformed into a praxis that allows for the disruption of the *coloniality of power*, which informs knowledge and being.

Colonial disruption is a central goal to the project of the *decolonial turn*. Grosfoguel (2006, 2011) credits the term the *decolonial turn* to Puerto Rican critical philosopher Nelson Maldonado-Torres. Maldonado-Torres (2007) characterises the *decolonial turn* as a collective of theoretical and epistemological praxes that are not located within one specific discipline or field of inquiry, but are rather multidisciplinary and transnational in their critique of coloniality and colonialism. Collectively, these transnational multidisciplinary praxes identify "coloniality as a fundamental problem in the modern (as well as postmodern and information) age, and decolonisation or decoloniality as a necessary task that remains unfinished" (Maldonado-Torres, 2011, p. 3). In line with a critique of coloniality and modernity, Grosfoguel (2006) makes an explicit differentiation between the terms colonialism and coloniality, which he argues are not the same, yet are often problematically equated. The former refers to manifestations of colonial relations, or the coloniality of power, rooted in histories of racialised imperialism. According to Grosfoguel (2006), the latter describes:

> the continuity of colonial forms of domination after the end of colonial administrations, produced by colonial cultures and structures in the modern/colonial capitalist/patriarchal world-system (p. 175).

Thus, coloniality remains well after the existence of colonialism, and is strengthened by the systemic and systematic organisation of epistemic, material and cultural products that reify the imperial project of modernity, or new modes of colonial power.

The *decolonial turn* is therefore an epistemic, as well as methodological intervention, which calls upon all committed to decoloniality to pursue anti-colonial/decolonial actions that unearth and dismantle the effects of colonisation. A key feature of the *decolonial turn* is its explicit commitment to contesting the *status quo* and challenging taken for granted "truths," whilst generating knowledge, building coalitions and supporting movements that seek to undo the *coloniality of power* in theory, research and practice. When informed through the lens of the *decolonial turn*, PAR can be used as a tool toward this purpose of decoloniality/decolonisation. Community psychology must therefore engage in this project of the *decolonial turn*, and respond to Maldonado-Torres's (2007) call to action, who writes:

> Whereas colonialism constituted an act of violence whereby an imperial power invaded another country and imposed an economic and administrative order on the invaded group in order to extract wealth from them for the benefit of the invading group's power base in the colonising metropole – an act closely associated with the advent and development of modernity – coloniality suggests that after the removal of this invading power, the power structures they put in place still have an influence on the invaded subjects. In other words, coloniality outlives colonialism (p. 343).

The power structures that informed Euro-American research, which are superimposed upon people and communities as "research subjects/objects," remain. Yet a critical decolonial PAR praxis rooted in local knowledge or Indigenous cosmologies alters or, at the very least, creates fissures toward epistemic justice. The process of undoing the *coloniality of power* through PAR characterises one approach toward the *decolonial turn*.

The decolonisation of PAR is also a part of that process of undoing, however. Decolonising PAR entails generating critical theories, knowledge, methods and critiques that are counter-hegemonic, and that center the voices, agency and power of those who are affected by and resist systems of oppression. The aim of decoloniality and decolonising is to re-center histories and perspectives that have been devalued in knowledge production through opportunities for unlearning, disrupting and transgressing disciplinary boundaries, whilst grounding knowledge in histories and relationships of mutual co-existence and accountability. Critical community psychologists Reyes Cruz and Sonn (2011) call upon us to develop

> ways of being, knowing and doing that contribute to decolonisation and liberation [and opening] the way to transformation as we recognise different ways of knowing that value the lived experiences and voices of the marginalised" (p. 207).

To decolonise the discipline also implicates us; it requires that we decolonise our thinking in order to engage in a process that aligns with the *decolonial turn*'s aims to disrupt the coloniality of knowledge, power and being. When we commit ourselves to a decolonial praxis we thread against currents that can drown our body, soul, and mind—and yet it is this experience of drowning that can led to rising and renewal for radical decolonial liberation.

The Roots of PAR

PAR is typically used as a broad term to describe various approaches that share core values of inclusive collaboration, participation and transformative action *by* and *with* community members engaged in every phase of the research process. Approaches within PAR include, but are not limited to, the following: community-based participatory research, action research, empowerment evaluation, participatory rural appraisal, and community-engaged research. With origins in the social sciences, specifically in education, psychology and sociology, as well as the health sciences, such as public health, PAR developed in the West and the Global South. The lineages that informed a Western approach to PAR are distinct from those of the Global South, which are rooted in specific historical and sociopolitical contexts. Although there are some similarities, a discussion on the roots of PAR is important for the decolonising of PAR.

The Western Approach

In the West, the aftermath of World War II compelled social scientists to ask deeper and more critical questions about the human condition, social structures and contexts. Thus, in the early 1940s Kurt Lewin (1946, 1947), often credited as the father of action research, introduced it as an approach to promote organisational change. Through continuous iterative cycles of inquiry, action and evaluation led by and with social groups who engaged as active research participants, Lewin advanced action research in the U.S. and in Europe.

Building on the legacies of Lewin (1946, 1947), specifically action research processes of planning, acting, observing and reflecting, Reason (1994; Reason & Bradbury, 2001) developed a participative inquiry approach that involved the democratisation of research through cooperative investigation and shared power. Instead of viewing research participants as passive subjects, this approach engaged people as active agents and co-researchers. A participative inquiry approach is characterised by people developing and conducting research relevant to their own lives, rather than doing research on people. Both action research (Lewin, 1946, 1947) and participative inquiry (Reason, 1994) approaches reflect Dewey's (1916/2004) emphasis on experiential learning as a foundation for the democratisation of knowledge. Learning from individual and collective experiences is therefore a key feature of PAR in the West.

By creating conditions where people can have access to greater control over the production and application of knowledge, including the research methods that produced it, democratisation is achieved and social change is possible. Democratisation, like social change, as Tuck and Yang (2012) remind us, is not equal nor the same as a decolonising/decolonial praxis. Henceforth, PAR in the Global South aligns more with decolonial liberatory perspectives.

Perspectives from the Global South

Central to the development of decolonial epistemologies and knowledge, which characterise paradigms rooted a PAR from the Global South, is the work of Frantz Fanon (1963, 1970). In *Black Skin, White Masks*, Fanon (1970) makes visible, through vignettes and case studies, the subjectivities and realities of Black people, whose lives and communities are imbued in systems of power and oppression stemming from coloniality. Fanon's work was critical to the formation of new forms of inquiry that centered the person and the political, as well as the historic, the cultural and the sociopolitical. Research inquiry that begins "from below," from within and through the experiences of people's bodies, lives and social conditions, as illustrated by Fanon, and how these are subsequently manifested in the catalysing of social movements, mark the beginnings of decolonial research paradigms reflected in PAR.

Thus, unlike the West, PAR in the Global South has roots in social movements, as well as Indigenous cosmologies. Distinct from the cycles of inquiry and action that Lewin (1947) purported, Freire (1970) emphasised the importance of fostering liberatory transformation by interrogating social realities through dialogic critical reflection and organising to bring about socially just reforms (Gaventa, 1993). Through the implementation of community-based critical literacy projects with working-poor rural communities in South America, Freire (1970) conceived the notion of praxis as critical reflection and action toward what he described as a process of *conscientização* (conscentization). Within this PAR tradition, Freire's work was key to the development of a critical praxis *of* and *for* liberation. Freire's contributions helped inform the work of Jesuit social psychologist Ignacio Martín-Baró (1994) in El Salvador, Colombian sociologist Fals-Borda (1979), and critical decolonial pedagogue Catherine Walsh (2009, 2014) with Afro-Andean communities. According to these critical scholars, the aim of research must be to redress systemic injustices from the perspective of communities most afflicted by oppression.

Martín-Baró (1994), often credited as the founder of liberation psychology in Latin America, advocated for the development a critical social psychology that examined the historic, sociopolitical and cultural contexts of communities on the margins of society. Extending Freire's (1970) theorising of conscientizacion, Martín-Baró (1994) proposed the concept of de-ideologisation to describe the process of becoming aware of the sources, practices, and interconnected systems of oppression in order to deconstruct and transform these. Like Martín-Baró (1994), Fals-Borda (1979, 2006) understood the significance of praxis as a process centered on deconstructing social conditions with the aim of fomenting liberatory transformation. A central feature of Fals-Borda's (1979) work is the concept of *vivencia,* for which there is no adequate English translation other than perhaps the words "lived experience." Unique to PAR is the notion of *vivencia*, which Fals-Borda (1979) defined as a full embodied experience of living, feeling and being with the oppressed that leads a researcher to engage communities with compassion, solidarity and *sentirpensar* (feeling and thinking). By combining theory and practice, within sociocultural community contexts of *vivencia*, a more focused understanding of injustice is gained in order to create social change.

In line with this relational way of thinking with communities is Gloria E. Anzaldúa's theorising (1987, 2002), specifically two interconnected concepts: *nos-otras* and spiritual activism. For the former, Anzaldúa (1987, 2002) posits the following:

> The *nos* is the subject 'we,' that is the people who were in power and the colonised others. The *otras* is the 'other,' the colonised group. Then there is also the dash, the divide between us. However, what is happening, after years of colonisation, is that all of the divides disappear a little bit because the coloniser, in his or her interaction with the coloniser takes on a lot of their attributes. And, of course, the person who is colonising leaks into our stuff. So we are neither one nor the other; we are really both. There is not a pure other; there is not a pure subject and not a pure object. We are implicated in each other's lives. (*Borderlands/La Frontera: The New Mestiza,* 1987, p. 243)

Applying this concept to PAR, Maria Elena Torre (2009) uses *nos-otras* to describe the intersections of everyday lived experiences and social relations that lead to contact zones of mutual implication in forms of oppression, as well as projects for liberation. Although located in a U.S. context, Anzaldúa (1987, 2002), like Torre (2009), produced writings grounded in ontological and epistemological theories of the *borderlands* and Global South. Theories that prioritise undoing the *coloniality of power* in theory, research, and practice, including decoloniality in ways of thinking, being and doing with others. Additionally, PAR must be anchored in a praxis of spiritual activism, which is the latter concept proposed by Anzaldúa (1987, 2002). This anchoring must strive to reconcile the spiritual—mind, body, and soul—with the theoretical and political dimensions of addressing real world issues entrenched in histories of colonial violence, morphed into scientific objectivity and disembodied research.

PAR's roots in the South are thus comprised of the following elements: liberation, multidisciplinarity, community-centered engagement, collective action, and a reclamation of Indigenous histories, cultures and knowledges. The role of the researcher must not be solely to gain and produce knowledge through research collaborations. Instead, the goal must be to decolonise research epistemologies and methodologies to produce knowledge, relationships, and ways of being with the self and others that lead to emancipatory decolonial transformation. In accordance with Indigenous teachings, this means we must unlearn in order to know (*desaprender para concocer*) and deeply understand in order to learn (*comprender para aprender*) (Paiva, 1992). These are interconnected processes toward the decolonisation of Western logics that (re)produce hegemony in the field, and researcher-community relations. The struggle for liberation with oppressed peoples within the research process begins with the decolonisation of a researcher's subjectivities, and the ontological and epistemological foundations that underlie the process, purpose and relationship to the research and "researched" communities it implicates. A PAR from the Global South posits that knowledge must be produced *by* and *with* communities. Through the co-creation of knowledge, from academic research to lived experience and shared social conditions, decolonisation unfolds and is driven by a commitment toward the decolonial/anti-colonial liberation of oppressed communities.

Decolonising Participation, Action and Research in PAR

> Decolonisation, however, does not mean and has not meant a total rejection of all theory or research or Western knowledge. Rather, it is about centering our concerns and world views and then coming to know and understand theory and research from our own perspectives and for our own purposes.—Linda Tuhiwai Smith (1999, p. 41)

PAR as tool toward decoloniality has garnered significant value over the past decades. PAR is not a methodology, however. As an iterative process of reflection, dialogue, decision-making and collective action, PAR seeks to disrupt relations of power that contribute to the inequitable production of research through the essentialising of communities (Torre et al., 2012). PAR is a research paradigm that challenges conventional research approaches by developing relationships of meaningful collaboration between researchers and community members in the process of conducting action research (Cahill, 2007; Torre & Ayala, 2009; Tuck, 2009). The paradigm seeks to restructure researcher-participant/community relations by centering Indigenous or local knowledge, experiences and community stories, whilst undoing systems of oppression that foreclose possibilities for social change and decolonial/anti-colonial liberation.

PAR does not follow a standard procedure. Yet generally there are four interconnected processes that characterise it. These include the following: (1) working together with community members, or research participants, to identify an issue or social problem affecting the community; (2) determining a method for information gathering, and collecting and analysing that information (e.g., data) to discern the root cause of the problem; (3) collectively developing an action, or an intervention, to address the problem; and (4) assessing and evaluating the impact of the action on redressing the problem. PAR strives to problematise researcher-participant/community dynamics by fomenting values of transparency and shared decision-making, whilst emphasising collaboration (Torre et al., 2012). Through these practices, researchers and community members collectively identify a problem to address and agree upon the best approach to bring about social change.

Decoloniality is therefore imperative within PAR. In partnership with communities who are most affected by systems of oppression, PAR strives for a decolonial liberatory approach to theory, research, and practice that includes decoloniality/anti-colonial action. PAR centers the community's voice, as well as their histories, culture and experiential knowledge in order to record and reproduce knowledge, whilst resisting paradigms that do not work to de-like or undo the entanglements of coloniality. Yet PAR also requires decolonising. Despite all good intentions, PAR can re-inscribe systems of power and oppression among disenfranchised groups. No theory, method or research paradigm is uncontaminated by a colonial history, or the colonial binaries of scientific inquiry. As PAR becomes more popular among social scientists and practitioners, it runs the risk of it becoming co-opted by colonial hegemonic projects that do not align with the *decolonial turn*. Therefore, in what follows three elements of PAR—participation, action and research—are interrogated through examples of PAR projects that engaged a decolonial praxis.

Participation Not Tokenism

PAR is predicated on the notion of participation, not tokenism. In PAR, communities are often included or involved in the process, yet rarely are they participants with power, agency, and ownership of the theories, data and framing of the outcomes produced from the research (Ayala et al., 2018; McTaggart, 1991; Segalo et al., 2015; Smith, 1999; Torre et al., 2012). Furthermore, participation can often be falsely equated with involvement or inclusion, which does not lead to equitable, authentic and meaningful engagement. Participation is defined as an action, an active practice that involves taking part in a process. Involvement, on the other hand, means to be included but not necessarily actively engaged, which can lead to tokenism. A distinction between participation and involvement is therefore necessary in order to facilitate conditions for authentic participation. Characterised by processes of shared decision-making, engagement, collaboration and accountability, authentic participation leads to collective ownership of the research process and outcomes. This is premised on the understanding that communities most affected by social issues must participate as collaborators, contributors and co-researchers in every step of the process if the research is to be truly committed to decolonising toward liberatory social change (Jolivétte, 2015; Smith, 1999).

Authentic participation can be more difficult to achieve, however. Researchers must be culturally attuned and intentional in their approaches to engage communities as active participants. In a PAR project with Aboriginal Health Workers (AHW), "participatory focus groups" were formed to assist with the development and analysis of research centered on documenting the experiences of AHW working in urban Aboriginal health service areas (Genat, 2009). Participatory focus groups were constituted by a group of diverse AHW, as well as professionals in the health services, such as nurses, doctors, welfare workers and clients. Through the formation of two participatory focus groups, one centered on best practices and the other on increasing the status of AHW, strategies for developing resources, forms of support, and increasing visibility of AHW were discerned. Similarly, drawing from her work with Maya women in Guatemala, Brinton Lykes (2001) has written extensively about the importance of facilitating opportunities for participation that are complementary and compatible with the customs, practices, traditions and local knowledge of the community.

Institutions marked by violence, disenfranchisement and oppression, must conceive of and advocate for creative modes of participation. Fine et al. (2003), for example, engaged eight women as co-researchers, contributors, and collaborators on a PAR project to identify the impact of college on incarcerated and formerly imprisoned women. Women with experience in the carceral system joined the project through a college course that offered them opportunities to learn and gain college credit, whilst develop essential research skills in PAR.

In a different, yet significantly structured context, Cammarota and Romero (2009) developed the Mexican American Studies program at a public high school within the Tucson Unified School District to engage predominantly Latinx students in Mexican American Studies and Ethnic Studies curricula. Via opportunities for critical learning, Latinx students engaged PAR as a tool for political action that led them to document their education experiences of inequity and disenfranchisement. As active co-researchers and participants, Latinx students discerned the impact and value of Ethnic Studies in promoting their critical consciousness, political identities, academic thriving, and educational aspirations. Both PAR projects—one in the prison system and the other on inequitable schooling opportunities—demonstrate that even in the most institutionally oppressive settings, authentic participation in PAR can be fostered. Researchers must commit themselves to intentional, explicit and creative modes of community inclusion and meaningful engagement that go beyond mere superficial involvement as tokenised performance.

Action-Oriented Transformation

Action must beget action-oriented transformation along multiple settings, systems and institutions. Kurt Lewin (1946, 1947) claimed, "No research without action, no action without research" (as cited in Adelman, 1997, p. 90). These words are imperative to a PAR process, especially given the paradigm's commitment to changing or disrupting power systems that oppress and produce hegemony within group relations, settings and institutions. Action must not be of any kind, however. In fact, it must be decolonial/anti-colonial social justice-oriented. Action must work toward the transformation of oppressive systems in all of its manifestations.

One mode of engendering action that is community determined and driven is by incorporating opportunities for creativity, arts, and performance, including the use of photovoice methodologies (Kessi, 2017; Langhout et al., 2015; Seedat et al., 2015; Suffla et al., 2015). Opportunities such as these can foster the creation of decolonial spaces where local knowledge, practices and stories can surface without censorship, shame or essentialised commodification. Cultivating decolonial spaces that allow for innovative decolonial methodologies, whilst centering the voices and subjectivities of communities wounded by the legacies of colonialism reflected in contexts of systemic violence and apartheid, are imperative to decoloniality (Segalo et al., 2015). The work of Indigenous and critical race theory scholar Eve Tuck (2009) with the Collective of Researchers on Educational Disappointment and Desire (CREED) and the Youth Researchers for a New Education System (YRNES), is an example of the PAR coalitions that can be formed to conceptualise action-oriented work. With support of CREED, YRNS developed a PAR project to document the experiences of youth who were denied educational opportunities. Through the use of theatre, poetry and collective free-writes (e.g., slam books), the YRNS produced powerful counter-stories that "spoke back" to the hegemonic narratives ascribed to students who "dropped out" of school. The production of counter-hegemonic knowledge, in this case, served to support educational campaigns led by youth/co-researchers against educational disenfranchisement in New York City public schools.

PAR is inherently a relational and political process, as well as a personal one that requires taking action rather than passively observing. By engaging with communities and stakeholders, PAR can facilitate the creation of transformational change and the redressing of social issues. Rooted in community experiences, informed by Indigenous cosmologies, the work of Torre (2009) with the Opportunity Gap Project demonstrates the importance of engaging in decolonial action-oriented PAR that aims to disrupt relations of power. Action-oriented transformation was thus characterised by the interruption of power and hierarchical structures. The formation of new methodologies, relationships and actions that can support and contribute to amplifying the voices of communities who are institutionally, socially and politically on the margins, is key to a decolonising PAR praxis.

Research Justice for Decoloniality

When defined, determined, and driven by communities on their own terms, PAR can yield decolonial social change. According to Jolivétte (2015) a research justice framework is premised on three interconnected fundamental practices. These include the development of (1) opportunities for knowledge co-construction via dialogic reflection, agency and collective determination; (2) approaches for grassroots community mobilisation and coalition building; and (3) strategies for social transformation and policy reforms via direct actions, creative products (e.g., public performances, art), and data dissemination. Research justice is most enduring when enacted by Indigenous people and disenfranchised communities. Furthermore, when communities who are most affected by the systems of oppression are acknowledged and engaged as experts, they hold the right to (re)claim, own, and wield all forms of knowledge, resources and power to produce actions and political influence over conditions in their lives that can lead to decolonial social change.

A decolonising PAR praxis combines various forms of knowledge, experience and relationships to produce information that can shape institutional change, public policy, and legislation rooted in redressing conditions of epistemic injustice, social inequity, and disenfranchisement. For example, in an undocumented student-led PAR project, Cahill (2007) describes the importance of engaging an "ethic of care" to foster and support the creation of conditions where social change can be determined and owned by the community of undocumented students. By producing a video documentary about their lives and experiences on silence and invisibility, young people engaged in the Equal Access to Higher Education For All PAR project, which disrupted anti-immigrant discourses that positioned them as undeserving of an education and human rights. Similar to Tuck's (2009) and Torre's (2009) process, undocumented students' actions were not necessarily intent at gathering data. Instead, their actions were aimed at disrupting narratives, practices and policies constructed to reinforce colonial ways of thinking that produce conditions of disenfranchisement.

A decolonial PAR praxis that fosters social justice cannot begin with specific research objectives or agendas tied to academe. On the contrary, PAR must begin with relationships, lived experiences, and intentionally listening to communities and people's lived experiences. To decolonise institutional spaces, like higher education for example (Reyes Cruz & Sonn, 2011; Kessi, 2017), epistemic modes of knowledge production and ways of being that researchers are conditioned to follow must be decolonised and interrogated. Decoloniality as a praxis takes fold when PAR is informed by being, doing and learning with others. This allows for a process of de-linking away from notions of modernity, including research objectivity and scientific racism (Richards, 1997/2012), which reify the *coloniality of power*. A decolonial PAR praxis is thus committed to justice, liberation and transformation.

Ten Axioms for a Decolonial PAR Praxis

> To acquire new psychological knowledge it is not enough to place ourselves in the perspective of the people; it is necessary to involve ourselves in a new praxis, an act of transforming reality that will let us know not only about what is, but also about what is not, and by which we may try to orient ourselves toward what ought to be.—Ignacio Martín-Baró (1994, pp. 28–29)

In the spirit of a decolonial PAR praxis rooted in Global South and Indigenous cosmologies, the following ten axioms are meant to be used as ethical principles to reimagine and decolonise ways of knowing, being and doing research. Presented in a non-hierarchical, non-specific order, these axioms are characterised as starting points for thinking about approaches toward decolonising PAR; they are not to be taken as absolutes, however. Instead these are to be considered as values toward a decolonial PAR praxis. Motivated by Maldonado-Torres' (2016) writings on coloniality and decoloniality, the axioms outlined below align with decolonial epistemologies. The ten axioms must be mindfully considered as tools toward a decolonializing PAR praxis, as well as an orientation toward decolonising participation, action and research.

Participatory Relational Engagement

Facilitating opportunities toward participatory relational engagement is key to the development and engagement of PAR. Cultivating relationships of mutual reciprocity, accountability and trust, fosters the authentic participation and contributions of communities as co-researchers and co-creators of knowledge. Practitioners and researchers engaged in PAR must be intentional in their efforts to collaborate with communities from diverse lived experiences, positionalities and relationships with colonialism and decoloniality. In doing so, they must respect the dignity and integrity of the community to determine the conditions of their engagement, participation, and refusal (Tuck & Yang, 2014). The formation of diverse participatory relational opportunities for meaningful engagement can facilitate the development of critiques of research conducted via the colonial gaze (Smith, 1999).

Socio-historical Inquiry

PAR must be grounded in decolonial/anti-colonial, anti-oppressive, intersectional feminisms and critical race theories that examine the socio-historical, cultural, political and environmental contexts of people's lives. When researchers explicitly and intentionally engage with the colonial legacies and socio-historical circumstances of communities, as well as the contemporary issues that impinge upon their liberation, only then can decolonial liberation take place (Segalo et al., 2015). Relatedly, when inquiry, problem-posing and knowledge production takes place outside of academe, and is legitimately recognised and valued as knowledge, only then will epistemic decolonial justice be realised. This, however, begins with a critical socio-historical inquiry approach that decenters the Euro-American master narrative of a past marked by the *coloniality of power* that leads to permutations of systemic violence.

Interrupting Power

Power is best understood as a system. As a structure that works to alter, delimit and constrain community actions, power has the capacity to determine what is possible. Power can influence the imagination, and thus forestall radical possibilities. PAR can contribute to the disruption and interruption of power by setting up structures for participation and engagement that include people who are on the margins. Because power moves and circulates through relationships, practices and modes engagement, it is intrinsically tied to coloniality. When deployed to oppress people and communities, power must be interrupted. Power is created to suppress the agency and determination of communities, whose habits of being, thinking and doing are constrained by coloniality. Developing decolonial praxes like PAR are therefore essential for interrupting the *coloniality of power* in theory, research, and practice. A decolonial PAR praxis must include a critical socio-historical analysis of the relationship between power and knowledge production, which can make salient the intersections of oppression embedded within historic and contemporary structures.

Critical Reflexivity

Deep introspective self-inquiry characterises the practice of critical reflexivity. A PAR collective must critically reflect upon their research process at every step; they must also reflect upon their relationships to one another, the research context and the research process. When critical reflexivity unfolds in relation to the research and the "researched," genuine trusting relationships of accountability, transparency, and mutuality among constituents in the project can develop. Critical reflexivity examines and deconstructs how power surfaces within practices and relationships in PAR collaboration (Torre, 2009). PAR is predicated on the capacity for people to engage in a critical analysis of a situation, or the circumstances wherein they work, how they work and with whom. Critical reflexivity in PAR consistently interrogates the *coloniality of power* at the intersections (Ayala et al., 2018; Torre et al., 2012). Thus it is imperative that a decolonial PAR praxis engage critical reflexivity that interrogates the positionalities and role of the researcher, along with their relationship to collaborators, the community, and literatures or knowledges in conversation with the work.

Sovereignty

Sovereignty is a demand to be treated with dignity, respect and integrity, and to exist without colonial influence. Quechua activist-scholar Sandy Grande (2015) purports that the sovereignty of communities, especially Indigenous peoples, is a necessary criterion for democracy. The significance of sovereignty is elaborated by Tuck (2009), who defines it as "a call for recognition and full realisation of rights to social, cultural and spiritual (tribal) identities and to our own envisioned political development" (p. 56). Claims to sovereignty must not to be limited to or equated with notions of autonomy and self-determination, however. Failure to uphold the sovereignty of Indigenous peoples, as well as communities resisting various forms of intersecting oppression, will never lead societies to the fulfillment of decolonisation, liberation and healing. Thus, a decolonial PAR praxis must aim to de-center the power of the nation-state and its institutions for the realisation of decolonisation, which as Segalo et al. (2015) write "honor local language and practices as they theorise the affects, embodiments, and consciousness through color, gender, ethnicity, class and religion" (p. 361). Sovereignty, therefore, is about imagining new relations of being that center the autonomous collectivity of oppressed communities with the agency to express and create decolonial/anti-colonial worlds—cosmologies of knowing, doing and being in alignment with the *decolonial turn*.

Tensegrity as Balance

Tensegrity is a term that combines features of pressure and balance. The equilibrium between these two processes leads to the experience of tension with integrity. These interconnected processes can lead to an experience of contention, which can be understood as a mechanism toward catalysing change when embraced for its possibility of shedding light on moments of discomfort that require a restructuring of power relationships. Contention, or the process of working to reconcile experiences of tension, integrity and synergy, can then facilitate a semblance of balance, as well as non-hierarchical relations. There is no formula for PAR; however, the key is to strive for balance, or tensegrity, in the development of action toward decolonial social change. Grande (2015) notes responsibility and power can never be equally shared nor should they. Within Indigenous communities, the elderly hold a very different position of responsibility and power that differs from that of other members of the community. Yet when considered as a whole, different members within these communities act in balance with synergy in relation to one another. Balance is necessary for equitable responsibility, power and knowledge co-construction. Tension with integrity is imperative to a decolonial process, premised on the *decolonial turn* project of liberation from the *coloniality of power*.

Co-constructing Knowledge

Knowledge that informs action is produced in collaboration with communities where researchers and "the researched" become knowledge-producers/change agents. Kincheloe (2009) writes:

> Knowledges constructed by colonised peoples and Indigenous peoples around the planet reflect what might be called the colonial divergence; that is the impact colonialism made on seeing the world and being in the world. Peoples who have been subjugated and classified as inferior by colonial, political, economic and epistemological systems of classification will produce different knowledges than those who come from societies implicated in the colonisation process (p. 119).

Given the positionalities and experiences with coloniality and colonialism that have marked people's lives, communities will have different relationships to and approaches to decoloniality. Therefore, knowledge produced from within these subjectivities becomes essential to decolonising PAR praxis. The production of knowledge and research is a relational, social, and political process that is deeply seeded in histories of coloniality. Thus, it is necessary that within PAR, knowledge be co-constructed and co-produced with communities with whom differences and similarities can be understood beyond binary colonial ways of thinking. Opportunities for unlearning and co-learning require a degree of humility on the part of the researcher, who must attend to the local knowledge, histories and social conditions of communities.

Community Determined Applied Action

Actions that emerge from a decolonial PAR praxis must be community determined and applied. Communities who have experienced the wounds of colonial violence and oppression hold deep understanding and knowledge about their own and their communities' lived experiences. Thus, it is communities affected by conditions of oppression who must determine the issues, questions, methods, tools for interpretations and data analysis, as well as modalities of dissemination and knowledge sharing, which will ultimately contribute to their decolonial liberation. Approaches for disseminating, presenting or implementing an action that results from the data must be informed by the cultural practices and contexts of the community. The production of material resources or interventions must be presented in creative and innovative ways that can be accessed by communities.

Emancipatory Transformational Praxes

A decolonial PAR praxis is a process. It is part of a movement toward the project of the *decolonial turn*. Central to this process is the development of a critical consciousness that aligns with social justice-oriented action and decolonial liberation. Researchers and practitioners have a social, ethical and political responsibility to serve as allies and advocates toward the production of decolonial paradigms and epistemic justice. As Kincheloe (2009) purports:

> Without a critical epistemology and transformative politics of knowledge, the story of the colonised will continue to be told by the colonisers. Reclaiming these stories and this history is a central task of a critical PAR and for the general process of decolonisation (p. 119).

Thus, it is crucial to examine how PAR is engaged by communities, and *if* and *how* it disrupts hegemonic knowledge that prevents the realisation of decoloniality. Praxes that center social justice are not confined to academe or scholarly products. On the contrary, stories and experiences, what some keenly refer to as "data," serve as wisdom and pedagogical tools to connect, build and speak back to hegemonic colonial power. The Zapatistas refer to this experiential knowledge as "*piegogia*" (Amrah Salomón, 2015), which in Spanish is a play on words that translates to "foot-ology," and is defined as the learning that happens when different groups of people come together to learn from one another, often literally by walking with each other. Actions via shared experiences that led communities to co-create knowledge are rooted in walking/co-being, asking and unlearning/re-learning together (Walsh, 2009, 2014). As Anzaldúa (1987) reminds us "caminante, no hay puentes, se hace puentes al andar" (voyager, there are no bridges, one builds them as one walks). By experimenting with different modes of collective existence toward action-oriented decolonial PAR, emancipatory transformational praxes unfold.

Decolonial Love

Women of color feminist scholars have written extensively about the value and importance of centering love in movements for social justice and liberation (Anzaldúa, 1987, 2002; Collins, 2002; hooks, 2018; Lorde, 2012; Moraga & Anzaldúa, 2015). Love can also claim a space in contexts of research, especially a decolonial PAR praxis. In writing about a "methodology of the oppressed" centered on decoloniality and decolonial movements, Sandoval (2000) describes decolonial love "as a social movement [that] is enacted by revolutionary, mobile, and global conditions of citizen-activists who are allied through the apparatus of emancipation" (p. 184). Love, as a decolonial embodiment, demands a deep recognition of our humanity and mutual implacability in undoing colonial relations of power and oppression that lead to indifference, contempt and dehumanisation. These conditions allow for the permanence of the *coloniality of power*.

Decolonial love begins from within, and from below or through the subjectivities of communities who have risen and resisted colonial forces, as they pursue modes of healing and liberation. Conceptualising love as a decolonial, radical, political and relational act is deeply rooted in Indigenous cosmologies akin to the Chicanx expression of *in lak'ech* (I am you, you are me). Decolonial love strives to foment a sense of accountability to communities engaged in PAR collaborations. To put it clearly, it means answering sincerely the question of where we are, or how have we contributed to the community in ways that honor the integrity of who they are, whilst strengthening their voice, agency and determination to lead the decolonial liberatory change they want to see. Decolonial love answers these questions, including the deeper question that invites PAR practitioners to consider their role in disrupting the *coloniality of power*.

Conclusion

Over the past decade the discipline of community psychology has gained interest in decolonial/anti-colonial praxes, epistemologies, and methodologies centered explicitly on the decolonisation of psychology and allied disciplines. Yet these discussions have remained rather sparse or limited in their scope (for exceptions see

Reyes Cruz & Sonn, 2011; Kessi, 2017; Seedat & Suffla, 2017; Sonn et al., 2013; Watts & Serrano-García, 2003). The pursuit of decolonisation/decoloniality is indeed an entangling and disquieting endeavor because it requires the researcher to implicate the self and their work in a questioning of the epistemologies that informed their work. Furthermore, researchers must examine if and how their work maintains or (re)produces the *coloniality of power*, or how it co-opts the language of decolonisation without centering it in communities' local knowledge or Indigenous cosmologies from the Global South.

The neoliberalization of academe inherently operates to maintain the *coloniality of power* characterised by racial, social, political (Quijano, 2001) and gender (Lugones, 2008) hierarchical structures that prescribe value to certain forms of knowledge, people and cultural practices or products. The Euro-American paradigms of knowledge production thereby leave the racialised and gendered neoliberalization of certain disciplines like psychology intact. Practices of research objectivity, positivism and empiricism, rooted in centuries of Western colonial hegemony, are thus re-inscribed in research, theory, and practice, and therefore how communities are approached in the name of research.

PAR is not immune to these forms of colonial hegemony, however. In spite of its potential to decolonise and transform research paradigms toward more humanising liberatory epistemologies, PAR comes with its unique set of challenges and limitations. These merit a lengthier discussion, yet a few are offered for consideration. First, PAR with social groups who maintain positions of power or social privilege is not decolonial. Unless, however, it is intentionally explicit in working toward reparations for communities who have been displaced, dehumanised and exploited as a result of the *coloniality of power*. Socio-historical inquiry, interrupting power, sovereignty, community determined and applied action, as well as emancipatory transformational praxes are some axioms that could facilitate how, when and why privileged social groups engage in decolonial/anti-colonial PAR collaborations. Second, even when rooted in local or Indigenous knowledge and practices, PAR can serve to maintain exclusionary or oppressive interests, especially when it is done in a way that does not critically attend to the intersections of oppression. Hence, participatory relational engagement, critical reflexivity, tensegrity, co-constructing knowledge and decolonial love can be helpful axioms for ensuring that PAR aligns with a decolonial orientation grounded in cultivating and amplifying the decolonial/anti-colonial liberation of communities.

The ten axioms outlined in this chapter are meant to support the development of a decolonial PAR praxis within community psychology. These axioms, however, are not fixed nor are they absolutes. On the contrary, these should be considered as starting points, principles to strive for, and practices to develop toward a decolonial PAR. Decolonial liberation cannot be realised if it does not expand to include a critique of community psychology and its paradigms. As PAR practitioners aligned with decolonial/anti-colonial work, our histories, experiences, identities and positionalities are situated in relation to power and oppression, just as the experiences of the communities with whom we engage. Yet we must remain mindful and cautious of the colonial gaze through which we might view the experiences of communities.

Interrogating our positionalities through the lens of decoloniality can allow us to understand our place as learners, allies and facilitators toward the production of decolonial/anti-colonial knowledge that seeks to undo—or at the very least interrupt—the *coloniality of power* in theory, research, and practice within and beyond community psychology.

A decolonial PAR praxis aims to disrupt the *coloniality of power* that manifests as epistemic injustice, and through this process contribute to the *decolonial turn*. PAR paradigms oriented toward decoloniality, deconstruct the hegemonic theories and practices that perpetuate violence against Indigenous peoples and oppressed communities whose identities, experiences, and social conditions have been essentialised. Communities marked by coloniality, as well as difference on account of their race, gender, sexuality, class, ability and other categories, have been and continue to be misrepresented, silenced and/or objectified through imperialist/empirical research. A decolonial PAR praxis, rooted in local community knowledge and Indigenous cosmologies, can lead to emancipatory transformation. Socio-historical, cultural, and relational ways of knowing and being with others through a decolonial PAR praxis can cultivate seeds of knowledge and action for decolonial/anti-colonial liberation.

Author Note
The author wishes to thank the anonymous reviewers who offered constructive commentary on an earlier version of this chapter.

References

Adams, G. E., Dobles, I., Gómez, L. H., Kurtiş, T., & Molina, L. E. (2015). Decolonising psychological science: Introduction to the special thematic section. *Journal of Social & Political Psychology, 3*(1), 213–238. https://doi.org/10.5964/jspp.v3i1.564

Adams, G. (2014). Decolonising methods: African studies and qualitative research. *Journal of Social and Personal Relationships, 31*(4), 467–474.

Adelman, C. (1997). Action research and the problem of participation. In R. McTaggart (Ed.), *Participatory action research: International contexts and consequences* (pp. 79–106). State University of New York Press.

Amrah Salomón, J. (2015). Telling to reclaim, not to sell: Resistance narratives and the marketing of justice. In A. J. Jolivétte (Ed.), *Research justice: Methodologies for social change (pp. 185–198)*. Policy Press.

Anzaldúa, G. (1987). *Borderlands: La Frontera*. Aunt Lute.

Anzaldúa, G. (2002). Preface: (Un)natural bridges, (un)safe spaces. In G. E. Anzaldúa & A. Keating (Eds.), *This bridge we call home: Radical visions for transformation* (pp. 1–3). Routledge.

Ayala, J., Cammarota, J., Rivera, M., Rodriguez, L., Berta, Avila, M., & Torre, M. E. (2018). *PAR EntreMundos: A pedagogy of the Americas*. Peter Lang Publishing.

Bhatia, S. (2017). *Decolonising psychology: Globalisation, social justice, and Indian youth identities*. Oxford University Press.

Bulhan, H. A. (2004). *Frantz Fanon and the psychology of oppression*. Springer.

Cahill, C. (2007). Repositioning ethical commitments: Participatory action research as a relational praxis of social change. *ACME: An International Journal for Critical Geographies, 6*(3), 360–373.

Cammarota, J., & Romero, A. F. (2009). A social justice epistemology and pedagogy for Latina/o students: Transforming public education with participatory action research. *New Directions for Youth Development*, 53–65.

Collins, P. H. (2002). *Black feminist thought: Knowledge, consciousness, and the politics of empowerment*. Routledge.

Dewey, J. (1916/2004). *Education and Democracy*. Palgrave Macmillan.

Dutta, U. (2016). Prioritising the local in an era of globalisation: A proposal for decentering community psychology. *American Journal of Community Psychology, 58*(3–4), 329–338. https://doi.org/10.1002/ajcp.12047

Fals-Borda, O. (1979). Investigating the reality in order to transform it: The Colombian experience. *Dialectical Anthropology, 4*, 33–55.

Fals-Borda, O. (2006). The north-south convergence: A 30 year first person assessment of PAR. *Action Research, 4*(3), 351–358. https://doi.org/10.1177/1476750306066806

Fanon, F. (1963). *The Wretched of the earth*. Grove Press.

Fanon, F. (1970). *Black skin, white masks*. Grove press.

Fine, M., Torre, M. E., Boudin, K., Bowen, I., Clark, J., Hylton, D., Martinez, M., Roberts, R. A., Smart, P., & Upegui, D. (2003). Participatory action research: From within and beyond prison bars. In P. M. Camic, J. E. Rhodes, & L. Yardley (Eds.), *Qualitative research in psychology: Expanding perspectives in methodology and design* (pp. 173–198). American Psychological Association.

Freire, P. (1970). *Pedagogy of the oppressed*. Continuum.

Gaventa, J. (1993). The powerful, the powerless, and the experts: Knowledge struggles in an information age. In P. Park, M. Brydon Miller, B. Hall, & T. Jackson (Eds.), *Voices of change: Participatory research in the United States and Canada* (pp. 21–40). OISE Press.

Genat, B. (2009). Building emergent situated knowledges in participatory action research. *Action Research, 7*(1), 101–115. https://doi.org/10.1177/1476750308099600

Grande, S. (2015). *Red pedagogy: Native American social and political thought*. Rowman & Littlefield.

Grosfoguel, R. (2006). World-systems analysis in the context of transmodernity, border thinking, and global coloniality. *Review (Fernand Braudel Center), 29*(2), 167–187.

Grosfoguel, R. (2011). Decolonising post-colonial studies and paradigms of political-economy: Transmodernity, decolonial thinking, and global coloniality. *Transmodernity: Journal of Peripheral Cultural Production of the Luso-Hispanic World, 1*(1), 1–38.

hooks, B. (2018). *All about love: New visions*. William Morrow Paperbacks.

Jolivétte, J. A. (2015). *Research justice: Methodologies for social change*. Policy Press.

Kessi, S. (2017). Community social psychologies for decoloniality: An African perspective on epistemic justice in higher education. *South Africa Journal of Psychology, 47*(4), 506–516.

Kincheloe, J. L. (2009). Critical complexity and participatory action research: Decolonising "democratic" knowledge production. In D. Kapoor & S. Jordan (Eds.), *Education, participatory action research, and social change* (pp. 101–121). Palgrave Macmillan.

Langhout, R. D., Fernández, J. S., Wyldbore, D., & Savala, J. (2015). Photovoice and house meetings as tools within participatory action research. In L. A. Jason & D. S. Glenwick (Eds.), *Handbook of methodological approaches to community-based research: Qualitative, quantitative, and mixed methods* (pp. 81–92). Oxford University Press.

Langhout, R. D. (2016). This is not a history lesson; this is agitation: A call for a methodology of diffraction in US-based community psychology. *American Journal of Community Psychology, 58*(3–4), 322–328. https://doi.org/10.1002/ajcp.12039

Lewin, K. (1946). Action research and minority problems. *Journal of Social Issues, 2*(4), 34–46.

Lewin, K. (1947). Frontiers in group dynamics: Concept, method and reality in social science; social equilibria and social change. *Human Relations, 1*(1), 5–41.

Lorde, A. (2012). *Sister outsider: Essays and speeches*. Crossing Press.

Lugones, M. (2008). Coloniality and gender. *Tabula Rasa, 9*, 73–102.

Lykes, M. B., & Moane, G. (2009). Editors' introduction: Whither feminist liberation psychology? Critical explorations of feminist and liberation psychologies for a globalising world. *Feminism & Psychology, 19*(3), 283–297. https://doi.org/10.1177/0959353509105620

Lykes, M. B. (2001). Activist participatory research and the arts with rural Mayan women: Interculturality and situated meaning making. In D. L. Tolman & M. Brydon-Miller (Eds.), *Qualitative studies in psychology. From subjects to subjectivities: A handbook of interpretive and participatory methods* (pp. 183–199). New York University Press.

Maldonado-Torres, N. (2007). On the coloniality of being: Contributions to the development of a concept. *Cultural Studies, 21*(2–3), 240–270. https://doi.org/10.1080/09502380601162548

Maldonado-Torres, N. (2011). Thinking through the decolonial turn: Post-continental interventions in theory, philosophy, and critique – An introduction. *Transmodernity: Journal of Peripheral Cultural Production of the Luso-Hispanic World, 1*(2), 1–15.

Maldonado-Torres, N. (2016). *Outline of ten theses on coloniality and decoloniality.* http://caribbeanstudiesassociation.org/docs/Maldonado-Torres_Outline_Ten_Theses-10.23.16.pdf

Martín-Baró, I. (1994). *Writings for a liberation psychology.* Harvard University Press.

McTaggart, R. (1991). Principles for participatory action research. *Adult Education Quarterly, 41*(3), 168–187.

Moraga, C., & Anzaldúa, G. (Eds.). (2015). *This bridge called my back: Writings by radical women of color.* SUNY Press.

Paiva, A. P. (1992). *Y…El anciano hablo [And … The elder spoke].* J. C. editors.

Quijano, A. (2001). Colonialidad del poder, cultura y conocimiento en América Latina. In W. Mignolo (Ed.), *Capitalismo y geopolítica del conocimiento. El eurocentrismo y la filosofía de la liberación en el debate intelectual contemporáneo.* Signo.

Reason, P. (1994). Three approaches to participative inquiry. In N. K. Denzin & Y. S. Lincoln (Eds.), *Handbook of qualitative research* (pp. 324–339). Sage.

Reason, P., & Bradbury, H. (Eds.). (2001). *Handbook of action research: Participative inquiry and practice.* Sage.

Reyes Cruz, M., & Sonn, C. C. (2011). (De) colonising culture in community psychology: Reflections from critical social science. *American Journal of Community Psychology, 47*(1–2), 203–214. https://doi.org/10.1007/s10464-010-9378-x

Richards, G. (1997/2012). *Race, racism and psychology: Towards a reflexive history.* Routledge.

Sandoval, C. (2000). *Methodology of the oppressed.* University of Minnesota Press.

Seedat, M., Suffla, S., Bawa, U. (2015). Photovoice as emancipatory praxis: A visual methodology toward critical consciousness and social action. In Bretherton, D., Law, S. F. (Eds.), Methodologies in peace psychology: Peace research by peaceful means (pp. 309–324). Cham, Switzerland: Springer International Publishing.

Seedat, M., & Suffla, S. (2017). Community psychology and its (dis) contents, archival legacies and decolonisation. *South Africa Journal of Psychology, 47*(4), 421–431.

Segalo, P., Manoff, E., & Fine, M. (2015). Working with embroideries and counter-maps: Engaging memory and imagination within decolonising frameworks. *Journal of Social and Political Psychology, 3*(1), 342–364.

Smith, L. T. (1999). *Decolonising methodologies: Research and indigenous peoples.* Zed Books.

Sonn, C., Stevens, G., & Duncan, N. (2013). Decolonisation, critical methodologies and why stories matter. In G. Stevens, N. Duncan, & D. Hook (Eds.), *Race, memory and the apartheid archive* (pp. 295–314). Palgrave Macmillan.

Suffla, S., Seedat, M., & Bawa, U. (2015). Reflexivity as enactment of critical community psychologies: Dilemmas of voice and positonality in a multi-country photovoice study. *Journal of Community Psychology, 43*(1), 9–21. https://doi.org/10.1002/jcop.21691

Torre, M. E., & Ayala, J. (2009). Envisioning participatory action research entremundos. *Feminism & Psychology, 19*(3), 387–393. https://doi.org/10.1177/0959353509105630

Torre, M. E. (2009). Participatory action research and critical race theory: Fueling spaces for nos-otras to research. *The Urban Review, 41*(1), 106–120. https://doi.org/10.1007/s11256-008-0097-7

Torre, M. E., Fine, M., Stoudt, B. G., & Fox, M. (2012). Critical participatory action research as public science. In H. Cooper, P. M. Camic, D. L. Long, A. T. Panter, D. Rindskopf, & K. J. Sher (Eds.), *APA handbook of research methods in psychology (Vol. 2). Research designs: Quantitative, qualitative, neuropsychological, and biological* (pp. 171–184). American Psychological Association.

Tuck, E. (2009). Re-visioning action: Participatory action research and Indigenous theories of change. *The Urban Review, 41*(1), 47–65. https://doi.org/10.1007/s11256-008-0094-x

Tuck, E., & Yang, K. W. (2012). Decolonisation is not a metaphor. *Decolonization: Indigeneity, Education & Society, 1*(1), 1–40.

Tuck, E., & Yang, K. W. (2014). Unbecoming claims: Pedagogies of refusal in qualitative research. *Qualitative Inquiry, 20*(6), 811–818. https://doi.org/10.1177/1077800414530265

Walsh, C. (2009). Interculturalidad crítica y pedagogía de-colonial: Apuestas (des)de el in-surgir, re-existir y re-vivir. *UMSA Revista (entre palabras), 3.*

Walsh, C. (2014). Pedagogías decoloniales caminando y preguntando: Notas a Paulo Freire desde Abya Yala. *Entramados: Educación y sociedad, 1*, 17–30.

Wendt, D. C., & Gone, J. P. (2012). Decolonizing psychological inquiry in American Indian communities: The promise of qualitative methods. In D. K. Nagata, L. KohnWood, & L. A. Suzuki (Eds.), *Qualitative strategies for ethnocultural research* (pp. 161–178). American Psychological Association.

Watts, R. J., & Serrano-García, I. (2003). The quest for a liberating community psychology: An overview. *American Journal of Community Psychology, 31*(1–2), 73–78.

Decentering "Community" in Community Psychology: Towards Radical Relationality and Resistance

Urmitapa Dutta

The decolonial turn encompasses diverse worldviews and theoretical positions that share the view of coloniality as a fundamental problem in the contemporary world (Lugones, 2010; Maldonado-Torres, 2011). Decoloniality involves the production of counternarratives, counterknowledges, and counterpractices that seek to dismantle the oppressive patterns of coloniality that operate through interlocking systems of domination such as racism and racialisation, capitalist exploitation, neoliberalism, patriarchy, and heterosexism (Maldonado-Torres, 2016). Decoloniality entails a fundamental transformation of the terms of being, relating and producing knowledge that disrupt the imagined temporal movement from barbarism to modern civilisation embedded in colonial progress narratives (Khatun, 2018). A decolonial lens positions colonised peoples as "questioner, thinker, theorist, writer, and communicator"—a fundamental reorientation in frames of reference that dismantles the normativity or self-justified status of white Euro-American/colonising groups (Fanon, 2004; Gordon, 2014; Maldonado-Torres, 2016). Importantly, the decolonial impulse has always been part of many people's movements and struggles for sovereignty and collective self-determination long before the decolonial turn of the early twentieth century.

Decoloniality does not constitute a single theoretical position nor can it be captured in a single discipline or unit of study; it is inherently unfinished, being continually made, remade, and reconfigured (Maldonado-Torres, 2011). As such, decolonial community psychology is not another sub-field of psychology whose boundaries have to be determined, defined, and legitimated as one of many. This sort of "inclusion" of decoloniality—adjudicating its boundaries vis-à-vis other critical traditions in psychology—acts to maintains colonial hierarchies, compartmentalisation, and disciplinary solipsism in ways that are antithetical to decolonial thought (Arvin et al., 2013; Maldonado-Torres, 2011). In fact, decolonial theorist

U. Dutta (✉)
Department of Psychology, University of Massachusetts Lowell, Lowell, MA, USA
e-mail: Urmitapa_Dutta@uml.edu

S. Kessi et al. (eds.), *Decolonial Enactments in Community Psychology*, Community Psychology, https://doi.org/10.1007/978-3-030-75201-9_4

Lewis Gordon (2014) has called for "teleological suspension of disciplinarity" as critical to epistemic decolonisation. Thus, the construction of a complementary decolonial model among different iterations of community psychology, would compromise radical possibilities of decoloniality. When we center communities-in-resistance and struggle, we have to reject both theoretical and disciplinary orthodoxy. As Palestinian scholar Rita Giacaman (2019) recently noted, "people do not live in disciplines, they live in communities."

In this chapter, I bring decolonial perspectives to bear on the metanarrative of "community" in community psychology. I scrutinise the construct of "community"—underlying assumptions and how it is conceptualised/studied—and its implications for the maintenance of colonial hierarchies. This interrogation is framed by a close attention to epistemic (in)justice, for example: Who produces knowledge and from what space/location? Whose/what knowledge is considered legitimate and generalisable? What remains silent and what are the implications of this silence? I argue that we should trouble the architecture of community as a core concept and examine colonial legacies tied to it. I do so with the humble recognition that knowledge production is one of many interconnected sites of decolonial struggles.

Decolonial enactments in community psychology are not only about disrupting and dismantling normativity of the Euro-American vantage point; it is also about nurturing capacities to imagine and embody alternatives that are not predicated upon racialised colonial hierarchies. Toward this end, the second part of the chapter explores decolonial possibilities that disrupt the coloniality inherent in the metanarrative of *community.* I argue that a decolonial community psychology needs to reframe its relational engagement to communities-in-struggle, centering the experiences and struggles of those at the frontlines of diverse struggles for justice (Shalhoub-Kevorkian, 2009). Such a stance disrupts the coloniality inherent in the assumed knowability of oppressed or disenfranchised groups and denaturalises the role of research so that it is one of many possible avenues for action. Thus, I propose that we disrupt the (colonial) disciplinary gaze from community as object-of-inquiry, committing instead to lived struggles—focusing on mutuality, solidarity, reciprocity, and justice, along with the exploration of specific praxes that could nurture this shift. In making these arguments, I am inspired by decolonial social movements unfolding across the Global South[1] as well as the work of decolonial, Indigenous, and transnational feminists (Ahmed, 2012; Anzaldúa, 2002; hooks, 1994; Lugones, 2010; Mohanty, 2003; Morales, 1998; Pérez, 1999; Shalhoub-Kevorkian, 2009; Smith, 2013; Tuck & McKenzie, 2015). I also draw upon my own journeys accompanying people's struggles in Northeast India and navigating colonial, neoliberal terrains of the academy as someone from the Global South.

[1] I use the term 'Global South' to refer to communities and contexts dominated by systematic and unjust human suffering produced by global capitalism, colonialism, and patriarchy. The term is inclusive of communities and groups in the geographic North (e.g., Indigenous communities, people of color) who continue to experience marginalisation and oppression (Esteva & Prakash, 2014).

The Metanarrative of "Community" in Community Psychology

The concept of community is virtually synonymous with our sense of what community psychology is about. As a consequence, "community" tends to acquire the invisibility of the obvious. I interrogate the metanarrative of *community* primarily within U.S. based community psychology because of the latter's hegemonic power and privileged access to knowledge production (e.g., Carolissen et al., 2017; Dutta, 2016). We cannot remain oblivious to the fact that theoretical and methodological trends generated in mainstream US community psychology enjoy a certain universality and transportability precisely because it is American. This critique does not undermine the importance of southern theory and South-South dialogues in decolonial endeavors. Nevertheless, the recognition and contestation of the hegemonic power of United Statesian community psychology is integral to decolonial enactments.

One of the salient features of the Swampscott conference—often regarded as the genesis of community psychology in the US—represents a marked departure from laboratory-based research accompanied by a focus on the broader contexts. Yet, positivist assumptions were tacit in members' conceptualisation of community-based research. As an example:

> Today's community is a veritable laboratory uniquely suited to the study of certain crucial problems of man's behavior in his natural habitat. Participation in such community activities by qualified research-minded observers enables the investigator to observe and study certain phenomena not otherwise available to his inspection. (sic) (Swampscott participant, as cited in Anderson et al., 1966, pp. 22–23)

Although Swampscott participants strongly underscored the need to bridge the dichotomy between "university" and "field stations" in community-based research, the underlying epistemological position that dichotomised the researcher (university) and researched (field/community) remained largely unquestioned (see Sonn, 2016). Fast forward to the present, here is an excerpt from Society for Community Research and Action (APA Division 27) describing the role of community psychologists (http://www.scra27.org):

> We seek to expand "helping" beyond traditional psychotherapy to promote wellness; We consult with and provide tools to organisations to build capacity to address social problems such as exploitation and victimisation; We fight oppression, work to reduce social inequalities, and work with marginalised people toward their empowerment.

The articulation of community psychology in the US has certainly evolved over the years.

Community psychology in the U.S. is historically rooted in clinical psychology and the discourse of an expanded notion of "helping" has become part of its professional self-definition. Yet this distinction is not simply one of unit of analysis, but an ideological one. Historically, the so called "helping" professions (e.g., social work, social welfare, psychology) have had a close and reciprocal relationship with colonialism and racism—justifying, consolidating, and furthering minoritising and marginalising particular groups (Bulhan, 1985; Lavallée & Poole, 2010). The categorical divide between community psychologists and the communities they serve is a key aspect of coloniality, the exclusion of majority world peoples as knowledge producers.

In the recently published *APA Handbook of Community Psychology*, Krause and Montenegro (2017) conceptualised community as a multi-level construct with seven core dimensions: psychological-emotional, relational, action, time, space, motivational, cultural, and social identity. They contend that understanding these dimensions facilitate "the understanding of communities in its complexity and diversity, and can be relevant for practice in diagnostic studies, intervention, and evaluation" and that it "allows for a multidimensional diagnosis that may clarify the strengths and weaknesses of the specific community" (p. 291). Although they emphasise the need to study transformations and new forms of community, the broader approach continues to mirror a curative community psychology, one that views communities as sites of intervention (Butchart & Seedat, 1990; Carolissen et al., 2010, 2017; Dutta, 2018). To the extent that the *APA Handbook of Community Psychology* is taken to reflect the state of the field, it has epistemic power with enduring implications for majority world peoples. Despite expressed commitment to relationality and social justice, the predominant approach to research and theory in mainstream community psychology has not been extricated from the coloniality of establishment psychology (see Bell, 2017). This epistemological inconsistency is mirrored in persistent efforts to identify and measure various dimensions of community, which are then used to predict participation, leadership, cohesion, healthy development, well-being, etc. (e.g., Brodsky et al., 1999; McMillan & Chavis, 1986; Neal & Neal, 2014). Thus, any decolonial engagement in community psychology has to begin with a confrontation of its deep ties to colonial foundations of establishment (Euro-American) psychology (Bell, 2017; Bulhan, 1985; Kessi & Boonzaier, 2018). Against this backdrop, I excavate some of the specific colonial processes embedded in the metanarrative of *community*.

Community as Other: The Academy: Community Binary

In the last few decades, there has been considerable research and theorising of community, most notably through extensive conceptualisations of (psychological) sense of community (PSOC)—often referred to as the conceptual centre or organising concept in community psychology (e.g., Brodsky, 2009; Chavis & Wandersman, 1990; Hill, 1996; McMillan & Chavis, 1986; Neal & Neal, 2014; Sonn et al., 1999; Townley et al., 2011). One of the most extensively cited work in this area has been that of McMillan and Chavis (1986), who formulated sense of community in terms of four components—membership or sense of belonging, mutual influence, fulfillment of needs, and shared emotional connection. Interwoven in dominant conceptualisations of community in community psychology—both explicit and tacit—is a persistent othering: the academy as self and community as the socially, politically, and economically disadvantaged other (Carolissen et al., 2010; Dutta, 2016, 2018; Butchart & Seedat, 1990). The vast majority of research on psychological sense of community does not consider, for example, how communities being *studied* might speak back to the literature, instead, legitimising disciplinary powers and assumed expertise, to speak on behalf of oppressed peoples. This is not to say that conceptualisations of PSOC were devoid of critique; in fact, the concept has since been expanded, modified, and critiqued (e.g., Brodsky, 2009; Hill, 1996; Sonn & Fisher, 1998; Wiesenfeld, 1996). But the point I am making is that a decolonial community psychology demands an ongoing, systematic, and critical engagement with the kinds of theories that are prioritised in the field and their colonial legacies. Otherwise we stand to misinterpret and neglect aspirations, desires, as well as dissidence that challenge social collectivities, especially those carried out in the Global South.

In outlining the role of community psychologists vis-à-vis PSOC, McMillan and Chavis (1986, p. 19) stated: "A clear and empirically validated understanding of sense of community can provide the foundation for lawmakers and planners to develop programs that meet their stated goals by strengthening and preserving community;" the *programs* being primarily government assistance and treatment programs. Along similar lines, Ira Iscoe (1974) noted:

> Characteristic of high-risk populations such as the poor, minorities, and the aged is a high degree of community disorganisation and the lack of effective coping strategies. How a community is to be made more competent is an enormously complex matter, and no prescription presently exists. (p. 608)

Thus conceptualised, *communities* are where problems exist, whether they are attributed to individual deficits or social structural issues; they are the ones without power or competence. This is an example of what Tuck (2009a) calls damage-centered research—well-intentioned social scientific research that privileges voices from the margins, but only to the extent that they convey pain and suffering; for example, portraying poor urban communities, Indigenous communities, and communities of Color as sites of endemic disinvestment and dispossession. This kind of research is socially and historically situated in that so-called poor outcomes are linked to conditions of poverty, racism, and exploitation. The suppression of heterogeneity and production of monolithic narratives of damage, however unintended, serves to maintain and reproduce relations of structural domination. It also robs people of their complex personhood.[2] The discursive use of community as

[2] Avery Gordon (2008) characterises complex personhood as follows: "Complex personhood means that all people (albeit in specific forms whose specificity is sometimes everything) remember and forget, are beset by contradiction, and recognise and misrecognise themselves and others. Complex personhood means that people suffer graciously and selfishly too, get stuck in the symptoms of their troubles, and also transform themselves. Complex personhood means that even those called "Other" are never never that. Complex personhood means that the stories people tell about themselves, about their troubles, about their social worlds, and about their society's problems are entangled and weave between what is immediately available as a story and what their imaginations are reaching toward... At the very least, complex personhood is about conferring the respect on

predominantly sites of suffering or intervention undermines the nature, scope, and possible relationships between community psychologists and communities, reproducing instead, hierarchical relationships that render people and contexts as subservient to our disciplinary interest as community psychologists. It thwarts our ability to develop and imagine multiple configurations of relationships, voices, actors, and knowledge that are possible (Segalo & Kiguwa, 2015). Even within more critical traditions that emphasise accompaniment and solidarity, they still do so within a binary context—the "relationship between disadvantaged community members and CP professionals" (Prilleltensky & Nelson, 2009, p. 128). As a consequence, the idea of "helping" or "giving voice to" marginalised communities are the dominant modes of community engagement.

The epistemic divide between academy and community has another major set of implications: What happens when we are part of the marginalised and oppressed communities we are expected to study, when our complex histories cannot be disciplined into a linear researcher-researched binary? Epistemic injustice is not limited to the content of research, but also pertains to questions of who possess the power to produce knowledge and to determine what constitutes valid knowledge? What is at stake here is not simply the rhetoric of voice; the more crucial question concerns the sort of voice one comes to have as the result of one's racialised, gendered, classed political locations. A decolonial project is more than a mere "inclusion" or "addition" of subaltern voices or perspectives. Rather, it requires a fundamental restructuring and even defiance of disciplinary norms to reimagine research as an activity undertaken by frontliners, grassroots intellectuals, and scholar activists in ways that do not draw strict boundaries between academy and community. At the core of decolonising community lies a radical re-envisioning of epistemic parameters that disrupt essentialist notions of community.

Sense of Community: Troubling Colonial Legacies in Theory and Applications

In this section I trouble some of the colonial legacies entrenched in the theorisation and applications of psychological sense of community. Specifically, I problematise how notions of belonging and diversity have been engaged within this theoretical framework. My goal here is to illuminate some of the conditions and mechanisms through which these discursive practices perpetuate and maintain coloniality and structural injustice. Conceptualisations of sense of community tend to rely a great deal on individuals' "sense of belonging" to particular groups (McMillan & Chavis, 1986). Yet belonging is not a reified, static entity but a dynamic process configured and adjudicated by particular patterns of hegemonic power relations (Yuval-Davis,

---

others that comes from presuming that life and people's lives are simultaneously straightforward and full of enormously subtle meaning." (pp. 4–5).

2006). As we witness unprecedented displacement, with more and more people across the globe leading a precarious existence, it is imperative to grasp belonging as an existential and political project rather than a social-cognitive project. When delineated boundaries are impermeable or exclusionary, contesting the values that govern their maintenance constitutes a mechanism for shifting parameters of what constitutes community and who can belong; as such, we need to center contestations, collective struggles, and resistance in how we think about belonging. Failing to do so reproduces the colonial and racialised center-periphery dichotomy as ossified in the language of accommodation, acculturation, adaptation, and assimilation (Dutta, 2016). An emphasis on the politics of belonging allows for more active engagement with tensions, contradictions, and conflict, but also troubles adjustment and acquiescence as inherently desirable (Adams et al., 2018; Martín-Baró, 1996). Thus, impediments to belonging that produce collective discontent along with desire to transform boundaries that police belonging need to be centered in the present moment.

While concepts such as territoriality, locality, and to some extent place have been considered in conceptualising sense of community, they have been done so within dominant white Western frames of reference. These conceptions of place tend to include physical dimensions of specific locations, the meanings and emotions that people assign to them, and their implications for civic democratic and participatory processes (e.g., Devine-Wright, 2009; Manzo & Perkins, 2006; Thomas et al., 2015). Notably, these are predominantly white Western conceptualisations of place. There is a fundamental incommensurability between Western and indigenous conceptions of place, with the former rooted in processes of settler colonialism. The accumulation of place and the structural denial of indigenous sovereignty were integral to the establishment of settler colonies (Arvin et al., 2013; Pratt, 1992; Tuck & McKenzie, 2015). In the process, land was remade into capital that could be sold, exchanged, and purloined; indigenous relationships to place restructured into those of individual ownership. Dominant psychological conceptions of place and placemaking fail to consider these complex, contested, and historicised relationships to place in the broader context of settler colonialism, globalisation, and neoliberalism.

Yet another key instance of coloniality embedded in the sense of community literature is its perceived tension with diversity, commonly referred to as the community–diversity debate, where the two are seen as somewhat opposing constructs (e.g., Neal & Neal, 2014; Neal, 2017; Stivala et al., 2016; Townley et al., 2011). The fundamental issue in the debate is this: if sense of community is about homogeneity/ cohesion while diversity is about heterogeneity /difference, then how are these two core values to be balanced, reconciled, and/or fostered simultaneously. Townley et al. (2011), for example, proposed a dialectical approach to this apparent tension while Brodsky (2017) and Stivala et al. (2016) underscored the need to expand context specific understandings of both constructs while emphasising the need to consider superordinate values that tie both diversity and community. This *debate* underscores the need for a decolonial community psychology to scrutinise *diversity*, specifically the ways in which its usage propagates epistemologies of domination. As a "core value," *diversity* has become part of the common parlance of mainstream community psychology, often taken to pertain to "the other"—non-dominant identities and non-hegemonic ways of being. Although diversity can be understood as a way of repairing the injustice wrought by pervasive and normalised practices of discrimination and social exclusion, its everyday usage obfuscates systemic inequalities under the banner of difference. Untethered from histories of oppression and a commitment to redistributive justice, diversity is reduced to an exercise in public relations that protects rather than disrupts whiteness (Ahmed, 2007a, b, 2012; Mitchell, 2018). Bound to Enlightenment knowledge formations, diversity can serve to maintain coloniality.

Notably, Wiesenfeld (1996) preempted the perceived tension between community and diversity long before the debate took shape, characterising it an artifact of epistemological inconsistencies in community psychology. She argued that although relationality and diversity were theoretical goals, the practical reliance on dualism and the tendency to undermine intersubjective complexities led to categorical thinking that rendered community and diversity as oppositional. More recently, Watts (2017) challenged approaches that equate diversity with demographic dispersion (e.g., Neal & Neal, 2014). In particular, he argued for the need to broaden our understanding of superordinate group identities, contending that a superordinate group identity does not replace "us and them" but complicates those categories in ways that identity boundaries overlap or become porous (Watts, 2017, p. 282). Decoloniality then demands more than diversity as categorical representation, it requires altering power relations and fundamentally changing the structure of social relations. Following Andreotti (2011), Carolissen et al. (2017) have argued for prioritising *diversality* as opposed to diversity. Diversality is predicated upon the expression of non-western cosmologies and epistemes—a renegotiation of epistemic privilege, which is better aligned with a decolonial agenda than diversity. Diversality represents an antidote to dualism and categorical thinking; it values "dissensus" or the "ability to hold paradoxes and not be overwhelmed by complexity, ambiguity, conflict, uncertainty, and difference" (Andreotti, 2011, p. 395). Crucially, diversality accords people "the right to be different because they are equals" (Mignolo, 2000, p. 311), which is radically distinct from the idea of treating people as equal *despite* difference (e.g., in relation to whiteness, heteronormativity, west)—something that is inherent in the notion of diversity.

In this section, I have reviewed some key concepts, themes, and moments in understandings of community within community psychology to illustrate how coloniality circulates in mainstream United Statesian community psychology. In doing so, I argue that decolonial enactments in community psychology entails a radical shift in the terms of the conversation and not merely its content. This interrogation does not assume or imply that community psychology is a monolithic entity, nor that it is without critical or dissenting voices. Rather, the point is to name colonial complicities embedded in core principles of mainstream US community psychology given its privileged location within hegemonic knowledge structures. If we do not interrogate these hierarchies produced and maintained through theory, we risk reducing "voice" to a rhetorical value. As Gordon (2014, p. 88) reminds us: "It is not that colonised groups fail to speak. It is that their speaking lacks appearance or mediation; it is not transformed into speech."

Decolonial Possibilities for Community Psychology: Towards Resistance, Desire and Radical Relationality

Decoloniality is more than disruption of colonial hierarchies and practices, it also involves concerted efforts to produce counterknowledges and counterpractices that are rooted in the lived struggles against legacies of coloniality. In the context of the present chapter, I argue that decoloniality demands a radical transformation in how we think about *community* in community psychology—namely, abandoning notions of community as phenomenon of interest or site of inquiry/intervention to focus instead on a radical relationality capable of dismantling or rendering extraneous colonial patterns of power and relationality. I look toward the contributions of transnational, Indigenous, women of Color, and decolonial feminists in thinking through alternatives to current ways of approaching community in the field. Mobilising knowledges of majority world peoples, these epistemic standpoints are based on the premise that oppression is always intersectional and contingent—shaped by specific historical and institutional realities. They help us reclaim *community* as political, as resistance, in contrast to essentialist (biological, cultural, or sociological) conceptions of community, while underscoring the criticality of lived engagement with struggles. In that spirit, I discuss forms of radical relationality as decolonial possibilities that would help us escape the trappings of binary oppositions characteristic of epistemic coloniality. I do so with the recognition that these cosmologies and knowledge relations are anchored in concrete historical experiences and struggles for justice.

An Orientation Towards Desire

An orientation towards desire is paramount as we strive to step into a different kind of knowledge relations and construct fuller, ethical, and compassionate representations of people and communities (see Anzaldúa, 1987; Tuck, 2009a). Tuck (2009a) has proposed desire-based framework as an alternative to damage-centered research that rely on monolithic portrayals of oppressed communities as damaged and suffering. It is important to note that desire-based frameworks do not negate the lived experiences of pain or the wisdom that comes from suffering (Tuck, 2009a; Tuck & Yang, 2014). Instead, a desire-based framework entails "working inside a more complex and dynamic understanding of what one, or a community, comes to know in (a) lived life" (Tuck & Yang, 2014, p. 231). Desire is always intergenerational and rooted in the collective.

The failure to interrogate the idea of presumably damaged community as objects or inquiry serves to maintain the privileged status of social science research and its rights to know. Positivist "rigor" then takes precedence over issues such as collective or representational harm (Tuck & Yang, 2014; also see Bulhan, 1985). Desire is fundamental to decolonial enactments in community psychology as it interrupts the colonial gaze, the metanarrative of damaged oppressed communities, and of social scientific research as rescue or progress. Desire-based frameworks trouble binary oppositions and disrupt categorical thinking—a revival from the epistemic death wrought by marginalisation of colonised peoples' knowledge/ knowledge systems as well as colonial practices of classification, categorisation, and codification (Chilisa, 2005; Kim, 2018; Lugones, 2010). Crucially, desire-based frameworks open up epistemic possibilities for those of us from non-hegemonic and subaltern standpoints. It allows for the complexities and contradictions inherent in those standpoints instead of slotting us into mutually exclusive categories such as the colonising researcher or oppressed/uncivilised communities (e.g., Reyes Cruz, 2008; Dutta, 2017).

Desire-based frameworks should not be conflated with strengths-based approaches in community psychology. In fact, damage vs. desire-based frameworks is not analogous to deficit vs. strengths-based approaches. Rather than focusing on deficits or pathology characteristic of mainstream psychology, community psychology values the strengths and capacities of people/communities dealing with adverse and oppressive conditions. Although this move was guided by the recognition that a deficits approach furthers subjugation of oppressed groups, it did not go so far as to trouble the idea of communities as sites of intervention. Community strengths are largely considered to the extent that it can guide intervention. Thus, it falls short of a decolonial epistemic shift where colonised and oppressed subjects emerge as "questioner, thinker, theorist, writer, and communicator" (Maldonado-Torres, 2016); or as feminist scholar Rajan (1993, p. 138) put it, "their relationship to the social structure is marked as external and adversarial rather than as integral and ambivalent." A desire-based framework eschews social scientific (and inherently ideological) categories that tend to reproduce the victim-savior trope or rescue narratives that were and continue to be used to justify colonial domination.

The Indigenous concept of survivance (Vizenor, 1994) is an important instantiation of a desire-based framework. Unsettling categorical representations, survivance is about creating "spaces of synthesis and renewal" (Vizenor, 1994, 9. 15). Rooted in Indigenous cosmologies, survivance is predicated on "intergenerational connection to an individual and collective sense of presence and resistance in personal experience and the word, or language, and particularly through stories" (Vizenor et al., 2014, p. 107). Along similar lines, decolonial feminist scholar Maria Lugones, argued for a shift from *community* to *communality* as an ontological and epistemological alternative to Eurocentrism. She described communality as follows:

> I am interested in communality as a way of being in the world of the oppressed, the colonised, the enslaved. I think of communality as we resist oppressions that constantly deteriorate our very being through the disintegration of our communality itself. I do not mean the being together somewhere we can call "community," but in the communal wanting, imagining, visioning, intending and acting together. When intending is communal, the self that intends is communal. (Lugones, 2015)

The concepts of survivance and communality offer powerful examples of what Tuck (2009a) means by a desire-based framework—both center the standpoint of colonised and oppressed peoples, their lived experiences of suffering, of collective resilience and wisdom, as well as their resistance to oppression. They offer possibilities for engaging militant utopian desires for decolonial action—utopian desires that demand a break from the shackles of colonial categories, that center the personal and communal in relation to the social in order to act on the desire to change the world (Bell et al., 2020; Morales, 1998).

Epistemic (Re)positionings, Centering Relational Ethics: Reconfiguring the Role of Community Psychologists

A desire-based framework requires a radical unsettling of the subject-object binary, which lies at the core of exogenous colonialism, settler colonialism, and coloniality. As such, epistemic repositionings, specifically those that recognise the centrality of relationships and relational ethics, are critical to decoloniality. This is different from approaches that view relationships as transactional, as key to "gaining entry" into communities or producing "accurate" analyses. It means that we honor and absorb the wisdom of cosmologies that prioritise the interconnectedness and interdependence of human lives; re-embed research in relationships, relationality, and collectivity as central rather than instrumental/peripheral concerns, thereby establishing radically different forms of accountability than those in mainstream social science research (Rowe & Tuck, 2017; Wilson, 2001). Psychosocial accompaniment represents one such epistemic stance (Lykes & Crosby, 2015; Watkins, 2015). With its roots in liberation theology, psychosocial accompaniment involves walking alongside those who "desire listening, witnessing, advocacy, space to develop critical inquiry and research, and joint imagination and action" to address oppression and suffering (Watkins, 2015, p. 324). Psychosocial accompaniment centers relational understandings that include a structural analysis of oppression as well as well as embodied and affective meanings of suffering. Maria Lugones (2010) speaks of this as being "creatively in coalition," which involves mutual learning without harming each other, contending with the "deep betrayals" of the everyday, and crossing without taking over (p. 755).

Torre and Ayala (2009) offer a powerful form of decolonial epistemic positioning through the notion of *nos-otras.* They use the term *nos-otras* to mean mutual implication, which underscores the ways in which the oppressor and the oppressed "'leak' into each others' lives" (p. 390, citing Anzaldúa, 1987). Drawing upon Anzaldúa's conceptualisation of a Mestiza consciousness that rejects simple dualisms (e.g. insider/outsider), they mobilise the notion of *nos-otras* to advocate for an alternate "in between" psychological and spiritual space that hold complexity, situatedness, and/or contradiction in all their ambiguity. Importantly, the notion of *nos-otras* does not presume a sameness. It renders salient rather than flatten oppressive histories and demands that we contend with the ways in which idealised notions of solidarity-as-sameness can reinscribe settler colonial logics. As Tuck and Yang (2012) remind us in "Decolonisation Is Not a Metaphor," colonisation and decolonisation are time-specific and land-specific; as such, the conflation of decolonisation with social justice and human rights projects serve to reproduce the erasure of Indigenous genocide and conquest. They argue:

> Because settler colonialism is built upon an entangled triad structure of settler-native-slave, the decolonial desires of white, nonwhite, immigrant, postcolonial, and oppressed people, can similarly be entangled in resettlement, reoccupation, and reinhabitation that actually further settler colonialism. (Tuck & Yang, 2012, p. 1)

As a radical departure from colonial epistemologies, Tuck (2009b) argues for *sovereignty as epistemology*:

> Sovereignty encapsulates what I know about knowing, where knowing comes from and goes to, how knowledge stretches and rises, even and because of the punching down. *It is not just internal however, not a state of mind, but a real thing, a lived thing that through the treaties, through the apologies, through our survivance, through the reconfigurations, removals, and repatriations is still unrecognised*...Broadly, sovereignty is a call for recognition and full realisation of rights to social, cultural, and spiritual (tribal) identities and to our own envisioned political development. It is a call for respect for our integrity as whole, significant, contemporary civilisations with long histories (some, like mine, spanning 10,000 years) and even longer futures. (p. 56; Emphasis added)

This has important implications for those of us—migrants, people of Color—who live and work on stolen and unceded native lands; we have to confront our complicity in ongoing colonisation of indigenous lands, resources, and lives. Along these lines, Tuck and Yang (2012) have advocated for contingent collaborations as an alternative to false promises of solidarity that do not oppose settler colonial logic. They define contingent collaborations as those temporary partnerships that are organised around an ethic of incommensurability (rather than sameness). In the context of settler colonialism, for instance, contingent collaborations recognise the significance of sovereignty and self-determination as distinct from other human and civil rights based social justice projects. It allows for the possibility that some social justice projects may not be aligned or allied but does not view incommensurability as an impediment to collective action. This idea is particularly relevant for decolonial enactments in community psychology as we reclaim, recenter, imagine, and transform renderings of knowledge and knowledge relations that exceed coloniality and conquest.

Another powerful example of decolonial epistemic movements can be found in *Australianama; The South Asian Odyssey in Australia,* where decolonial feminist historian Samia Khatun steps outside of a colonialist multicultural migrant narrative to seam together narratives from different knowledge traditions—particularly Muslim historiography and Aboriginal knowledge traditions—to render visible rich imaginative landscapes that push back against colonial assumptions of linear historical and progress narratives (see Khatun, 2018). Importantly, she does not argue for the mere inclusion of texts in non-English, colonised peoples' tongues, but she demonstrates how doing so "can transform the very grounds from which we view the past, present and future" (Khatun, 2018, p. 4). Of particular relevance to the present chapter is how Khatun reads *Ishq* or decolonial love in the *Kasasol Ambia* (Stories of the Prophets), a Bengali-language Sufi songbook, a copy that was published in Calcutta in the nineteenth century and, which Khatun located in a mosque in Broken Hill, in the Australian desserts. Ishq, she argues, becomes a metaphor for knowledge relations, as site of encounters with difference, of sophisticated ways of narrating and theorising relationships between self and other. This resonates powerfully with the politicised, social, and revolutionary love in Chela Sandoval's *Methodology of the Oppressed*—love as intricate modes of survival, resistance, knowing, and being, love as social movement (Sandoval, 2013). All of these examples afford glimpses into ways of being and relating (including knowledge relations) beyond the colonial difference.

Many indigenous and non-Western epistemes of relationships are concerned with aural traditions where the emphasis is on listening and hearing rather than the more common understanding of it as oral tradition that focuses on speaking (Alfred, 2005; Khatun, 2018). In many South Asian indigenous traditions for example, śruti (listening) and smriti (collective memory) rooted in lived experience have been critical intergenerational modes of knowledge transmission. To listen or to hear then is as an engaged, embodied, and affective process rather than a muted or passive activity; it is based on reciprocity and entails an authentic striving to comprehend, to grasp, to feel, and to be moved by what is heard. Bringing these ways of being and relating to knowledge relations means approaching research "from within collectivity," with "a commitment to complex personhood, with a responsibility toward preparedness, listening, reflection, and reparation" (Tuck, 2009b, p. 48). These epistemic reorientations dismantle colonist ideologies that hold the individual a priori to the group and drives home the need to respect that not everyone and everything is or has to be knowable. As such, they have the potential to interrupt dominant research trends that reduce complex, multifaceted social and psychic lives of individuals and communities to their suffering and to their role as research participants; and in doing so, it allows for alternatives to damage-centered research that reify disenfranchised and colonised communities.

Toward a Decolonial Education and Curriculum: Reading Against the "Canons"

Educational systems, as we know too well, function as tools for colonisation and coloniality. One of the important ways in which education operates to maintain coloniality is by disregarding, marginalising and suppressing colonised peoples' knowledge systems and ways of knowing—relegating those to otherness, as uncivilised, barbaric and superstitious. Chilisa (2005) for example excavates the trajectory of how Maori knowledge was brought into the university—under the aegis of anthropology and linguistics—such that Maori knowledge systems are rendered knowable only from Western epistemologies. Thus, the coloniser is always constructed as the one who knows and the colonised as inherently knowable. This is a form of epistemic death, a deadening of colonised people's knowledges either as a historical artifact or by reading, classifying and categorising it in ways that destroys its essence. Education systems, including contemporary neoliberal systems, that are founded on this essential binary continue to produce psychic alienation of colonised people by splitting the self from knowledge and understanding of itself (Bell, 2018; Chilisa, 2005). Thus, as much as education is touted as a liberatory force, self-knowledge and journeys toward collective healing remain elusive in neoliberal institutions of higher education (Bell, 2018). Furthermore, the preponderance of damage centered research (discussed earlier in this chapter) limit students' abilities to imagine and develop multiple configurations of relationships, voices, actors, and knowledge—gravitating to the idea of "helping" or "giving voice to" marginalised communities as the dominant mode of engaging with communities (e.g., Dutta, 2018).

Coloniality is inextricably intertwined with and perpetuates itself through whiteness, which is rooted in European colonisation that established the superiority of white identity through interlinked processes of material exploitation and cultural domination (Fanon, 2004; Quijano, 2000). As such, it is as imperative to interrogate whiteness—as a structural advantage, as a worldview, and as a set of unmarked, unnamed cultural practices that inflect how people 'take up' space and what they 'can do' (e.g., Ahmed, 2007a; Frankenberg, 1993). Turning the gaze on to whiteness represents a marked shift from an almost exclusive focus on people of color and associated disadvantage or disinvestment, a move that is fundamental to decentering the notion of community-as-Other.

A decolonial curriculum is not only about disrupting and dismantling normativity of the Euro-American vantage point; it is also about nurturing capacities to imagine alternatives that are not predicated upon hierarchies of difference such as local-global, researcher-researched, and university-community. Decolonising community psychology curriculum requires us to recenter knowledge from standpoints of historically marginalised groups. The decolonial imperative cannot be met through mere inclusion of readings that focus on historically marginalised groups, as long as the white Euro-American centric knowledge continues to be the preferred vantage point. We have to the set the stage for students to question fundamental assumptions about knowledge and their connections to power, normativity, and representation; engaging with questions such as what counts as knowledge, who produces knowledge and how, and what/who are absent from this enterprise (e.g., Dutta, 2016; Gegeo & Watson-Gegeo, 2001; Khatun, 2018; Montero, 2002; Smith, 2000). This is an example of what it means to read "against" canonical texts—deconstructing the perceived normalcy of dominant knowledge, exposing its historically and culturally specific character and inherent exclusions. Second, these critical framings must be foregrounded in indigenous and other marginalised ontologies and knowledge systems that induct students into a plurality of cosmologies and narrative pathways (e.g., Battiste, Bell, & Findlay, 2002; Bulhan, 1985; Khatun, 2018; Reyes Cruz, 2008; Smith, 2013; Tuck, 2009b). We have to ask: what do different modes of discovery and knowledge transmission look like beyond Enlightenment? And we have to be courageous, to take risks—to (re)imagine the past and present, and to write a new future into existence (Carby, 2019; Khatun, 2018; Pérez, 2010).

Concluding Remarks

Much of the present chapter has been about scrutinising how colonial patterns map on to knowledge relations and research relations in mainstream community psychology. Specifically, I have interrogated the metanarrative of *community* though a decolonial lens. I examined tacit assumptions inherent in discourses of *community* in U.S. based community psychology, dominant trends of theorising *community* via psychological sense of community, and naming theories and standpoints that are both elevated and silenced. In doing so, I illuminated some of the ways in which knowledge production is implicated in the production and maintenance of colonial, racialised, classed, and gendered hierarchies. I then presented some possibilities for a decolonial community psychology—possibilities that strive to center the lived experiences of subjugated groups, in all their complexity, but also turn the gaze back on power. While the issues raised in the process of deconstructing the metanarrative of *community* are historically contingent, there are certain crosscutting themes that are fundamental to decolonial and decolonising community psychology.

First, decolonial engagements in community psychology require an essential recentering of majority world peoples, histories, knowledges, ways of being, standpoints, and modes of knowledge production that are marginalised. Crucially, this is not simply an additive process of "including" excluded perspectives but a fundamental shift that repositions marginalised and oppressed peoples—traditionally "objects of inquiry"—as epistemic subjects. At the same time, it is about turning the gaze back upon power, specially the colonial modalities of knowledge production and regimes of representation.

Second, we have to adopt a stance of epistemic skepticism defined by the interrogation of mainstream knowledge claims including what we think of as core values or accepted methods in our field. This means perpetually questioning the centers and boundaries of the field, critically considering its "turf" and the kinds of actions it facilitates/inhibits.

Third, and closely connected to the previous point, we have to scrutinise how we imagine and act upon our various allegiances. We have to, as feminist scholar Visweswaran (1994, p. 132) has argued, "exhibit and examine our alliances in the same moment." As community psychologists, we tend to position our field as being aligned with marginalised communities, with social justice. At the same time, we are continually trying to justify our relevance as a distinctive sub-field of psychology. We have to pause and remind ourselves that disciplinary boundaries are incidental to and often prohibitive in the realm of transformative social change. Along these lines, the present chapter should not be read as an exhaustive or definitive statement of decolonial ethics. Decoloniality has emerged from and must remain open to being shaped by grassroots struggles for justice. We have to transcend our disciplinary allegiances in order to belong to, accompany, and be accountable to our relationships to struggles for justice.

Finally, decolonial endeavors have to be inflected with radical hope. At times, this demands that we refuse, suspend or transcend the parameters of what is considered practical—increasingly determined by a neoliberal logic. As Audre Lorde (2003, p. 27) reminded us: "*For the master's tools will never dismantle the master's house.* They may allow us to temporarily beat him at his own game, but they will never enable us to bring about genuine change." Therefore, we have to engage in generative, collective utopian imaginings in an effort to dream a decolonising community psychology into place.

Author Note
Small portions of this chapter have been drawn from two previous publications: Dutta, U. (2016). Prioritising the local in an era of globalisation: A proposal for decentering community psychology. *American journal of community psychology, 58*(3–4), 329–338; Dutta, U. (2018). Decolonising "community" in community psychology. *American Journal of Community Psychology, 62*(3–4), 272–282.

References

Adams, G., Salter, P. S., Kurtiş, T., Naemi, P., & Estrada-Villalta, S. (2018). Subordinated knowledge as a tool for creative maladjustment and resistance to racial oppression. *Journal of Social Issues, 74*(2), 337–354. https://doi.org/10.1111/josi.12272

Ahmed, S. (2007a). A phenomenology of whiteness. *Feminist Theory, 8*(2), 149–168. https://doi.org/10.1177/1464700107078139

Ahmed, S. (2007b). The language of diversity. *Ethnic and Racial Studies, 30*(2), 235–256.

Ahmed, S. (2012). *On being included: Racism and diversity in institutional life*. Duke University Press.

Alfred, G. R. (2005). *Wasase: Indigenous pathways of action and freedom*. Broadview Press.

Anderson, L. S., Cooper, S., Hassol, L., Klein, D. C., Rossenblum, G., & Bennett, C. C. (1966). *Community psychology: A report of the Boston conference on the education of psychologists for community mental health*. Boston University.

Andreotti, V. D. O. (2011). (Towards) decoloniality and diversality in global citizenship education. *Globalisation, Societies and Education, 9*(3–4), 381–397. https://doi.org/10.1080/14767724.2011.605323

Anzaldúa, G. (1999[1987]). *Borderlands/La Frontera*. Aunt Lute Publishers.

Anzaldúa, G. E. (2002). (Un)natural bridges, (un)safe spaces. In G. E. Anzaldúa & A. Keating (Eds.), *This bridge we call home: Radical visions for transformation* (pp. 1–5). Routledge.

Arvin, M., Tuck, E., & Morrill, A. (2013). Decolonising feminism: Challenging connections between settler colonialism and heteropatriarchy. *Feminist Formations, 25*(1), 8–34.

Battiste, M., Bell, L., & Findlay, L. M. (2002). Decolonising education in Canadian universities: An interdisciplinary, international, indigenous research project. *Canadian Journal of Native Education, 26*(2), 82–95.

Bell, D. (2017). Community psychology's gaze. In M. Seedat, S. Suffla, & D. Christie (Eds.), *Emancipatory and participatory methodologies in peace, critical, and community psychology* (pp. 119–132). Springer.

Bell, D. (2018). A pedagogical response to decoloniality: Decolonial atmospheres and rising subjectivity. *American Journal of Community Psychology, 62*(3–4), 250–260. https://doi.org/10.1002/ajcp.12292

Bell, D., Canham, H., Dutta, U., & Fernández, J. S. (2020). Retrospective autoethnographies: A call for decolonial imaginings for the new university. *Qualitative Inquiry, 26*(7), 849–859. https://doi.org/10.1177/1077800419857743

Brodsky, A. E., O'Campo, P. J., & Aronson, R. E. (1999). PSOC in community context: Multi-level correlates of a measure of psychological sense of community in low-income, urban neighborhoods. *Journal of Community Psychology, 27*(6), 659–679.

Brodsky, A. E. (2009). Multiple psychological senses of community in Afghan context: Exploring commitment and sacrifice in an underground resistance community. *American Journal of Community Psychology, 44*(3–4), 176–187. https://doi.org/10.1007/s10464-009-9274-4

Brodsky, A. E. (2017). Bridging the dialectic: Diversity, psychological sense of community, and inclusion. *American Journal of Community Psychology, 59*(3–4), 269–271. https://doi.org/10.1007/s10464-009-9274-4

Bulhan, H. A. (1985). *Frantz Fanon and the psychology of oppression*. Plenum Press.

Butchart, A., & Seedat, M. (1990). Within and without: Images of community and implications for South African psychology. *Social Science & Medicine, 31*(10), 1093–1102.

Carby, H. V. (2019). *Imperial intimacies: A tale of two islands*. Verso Books.

Carolissen, R., Rohleder, P., Bozalek, V., Swartz, L., & Leibowitz, B. (2010). "Community psychology is for poor, black people:" Pedagogy and teaching of community psychology in South Africa. *Equity & Excellence in Education, 43*(4), 495–510. https://doi.org/10.1080/10665684.2010.516695

Carolissen, R., Canham, H., Fourie, E., Graham, T., Segalo, P., & Bowman, B. (2017). Epistemological resistance towards diversality: Teaching community psychology as a decolonial project. *South Africa Journal of Psychology, 47*(4), 495–505.

Chavis, D. M., & Wandersman, A. (1990). Sense of community in the urban environment: A catalyst for participation and community development. *American Journal of Community Psychology, 18*(1), 55–81.

Chilisa, B. (2005). Educational research within postcolonial Africa: A critique of HIV/AIDS research in Botswana. *International Journal of Qualitative Studies in Education, 18*(6), 659–684. https://doi.org/10.1080/09518390500298170

Devine-Wright, P. (2009). Rethinking NIMBYism: The role of place attachment and place identity in explaining place-protective action. *Journal of Community & Applied Social Psychology, 19*(6), 426–441. https://doi.org/10.1002/casp.1004

Dutta, U. (2016). Prioritising the local in an era of globalisation: A proposal for decentering community psychology. *American Journal of Community Psychology, 58*(3–4), 329–338.

Dutta, U. (2017). The everyday and the exceptional: Rethinking gendered violence and human rights in Garo Hills, India. In S. Grabe (Ed.), *Women's human rights: A social psychological perspective on resistance, liberation, and justice* (pp. 179–204). Oxford University Press.

Dutta, U. (2018). Decolonising "community" in community psychology. *American Journal of Community Psychology, 62*(3–4), 272–282. https://doi.org/10.1002/ajcp.12281

Esteva, G., & Prakash, M. S. (2014). *Grassroots postmodernism: Remaking the soil of cultures*. Zed Books Ltd..

Fanon, F. (2004). *The wretched of the earth* (R. Philcox, Trans). Grove.

Frankenberg, R. (1993). *White women, race matters: The social construction of whiteness*. University of Minnesota Press.

Gegeo, D. W., & Watson-Gegeo, K. A. (2001). "How we know": Kwara'ae rural villagers doing indigenous epistemology. *The Contemporary Pacific, 13*(1), 55–88.

Giacaman, R. (2019, November). Reconceptualising of health in wars and conflicts. Paper presented at conference entitled "*Palestinian mental health without borders: Challenging barriers and taking down walls*," Al-Bireh in West Bank.

Gordon, A. F. (2008). *Ghostly matters: Haunting and the sociological imagination*. University of Minnesota Press.

Gordon, L. R. (2014). Disciplinary decadence and the decolonisation of knowledge. *Africa Development, 39*(1), 81–92.

Hill, J. L. (1996). Psychological sense of community: Suggestions for future research. *Journal of Community Psychology, 24*(4), 431–438.

hooks, b. (1994). *Teaching to transgress*. Routledge.

Iscoe, I. (1974). Community psychology and the competent community. *American Psychologist, 29*(8), 607–613.
Kessi, S., & Boonzaier, F. (2018). Centre/ing decolonial feminist psychology in Africa. *South Africa Journal of Psychology, 48*(3), 299–309.
Khatun, S. (2018). *Australianama: The South Asian odyssey in Australia*. Oxford University Press.
Kim, J. (2018). Korean indigenous epistemologies with notes on the corresponding epistemologies of indigenous scholarship. In J. Markides & L. Forshythe (Eds.), *Looking back and living forward: Indigenous research rising up* (pp. 105–113). Brill Sense.
Krause, M., & Montenegro, C. R. (2017). Community as a multifaceted concept. In M. A. Bond, I. Serrano-García, C. B. Keys, & M. Shinn (Eds.), *APA handbook of community psychology: Theoretical foundations, core concepts, and emerging challenges* (Vol. 1, pp. 275–294). American Psychological Association.
Lavallée, L. F., & Poole, J. M. (2010). Beyond recovery: Colonisation, health and healing for Indigenous people in Canada. *International Journal of Mental Health and Addiction, 8*(2), 271–281. https://doi.org/10.1007/s11469-009-9239-8
Lorde, A. (2003). For the master's tools will never dismantle the master's house. In R. Lewis & S. Mills (Eds.), *Feminist postcolonial theory: A reader* (pp. 25–28). Routledge.
Lugones, M. (2010). Toward a decolonial feminism. *Hypatia, 25*(4), 742–759. https://doi.org/10.1111/j.1527-2001.2010.01137.x
Lugones, M. (2015, April). *Keynote conversation. The feminist architecture of Gloria Anzaldúa: New translations, crossings and pedagogies in Anzaldúan thought*. Conference held at University of California, Santa Cruz, CA.
Lykes, M. B., & Crosby, A. (2015). Creative methodologies as a resource for Mayan women's protagonism. In B. Hamber & E. Gallagher (Eds.), *Psychosocial perspectives on peacebuilding* (pp. 147–186). Springer.
Maldonado-Torres, N. (2011). Thinking through the decolonial turn: Post-continental interventions in theory, philosophy, and critique—An introduction. *Transmodernity: Journal of Peripheral Cultural Production of the Luso-Hispanic World, 1*(2), 1–15.
Maldonado-Torres, N. (2016). *Outline of ten theses on coloniality and decoloniality*. http://frantzfanonfoundation-fondationfrantzfanon.com/article2360.html
Manzo, L. C., & Perkins, D. D. (2006). Finding common ground: The importance of place attachment to community participation and planning. *Journal of Planning Literature, 20*(4), 335–350. https://doi.org/10.1177/0885412205286160
Martín-Baró, I. (1996). *Writings for a liberation psychology*. (A. Aron & S. Corne, Trans.). Harvard University Press.
McMillan, D. W., & Chavis, D. M. (1986). Sense of community: A definition and theory. *Journal of Community Psychology, 14*(1), 6–23.
Mignolo, W. (2000). *Local histories/global designs: Essays on the coloniality of power, subaltern knowledges and border thinking*. Princeton University Press.
Mitchell, N. (2018). Diversity. In E. R. Edwards, R. A. Ferguson, & J. O. G. Ogbar (Eds.), *Keywords for African American studies* (pp. 68–74). New York University Press.
Mohanty, C. T. (2003). *Feminism without borders: Decolonising theory, practicing solidarity*. Zubaan.
Montero, M. (2002). On the construction of reality and truth. Towards an epistemology of community social psychology. *American Journal of Community Psychology, 30*(4), 571–584.
Morales, A. L. (1998). *Medicine stories: History, culture, and the politics of integrity*. South End Press.
Neal, Z. P., & Neal, J. W. (2014). The (in) compatibility of diversity and sense of community. *American Journal of Community Psychology, 53*(1–2), 1–12. https://doi.org/10.1007/s10464-013-9608-0
Neal, Z. P. (2017). Taking stock of the diversity and sense of community debate. *American Journal of Community Psychology, 59*(3–4), 255–260. https://doi.org/10.1002/ajcp.12132
Pérez, E. (1999). *The decolonial imaginary: Writing Chicanas into history*. Indiana University Press.

Pérez, E. (2010). *Forgetting the Alamo, or, blood memory: A novel.* University of Texas Press.

Pratt, M. L. (1992). *Imperial eyes: Travel writing and transculturation.* Routledge.

Prilleltensky, I., & Nelson, G. (2009). Community psychology: Advancing social justice. In D. Fox, I. Prilleltensky, & S. Austin (Eds.), *Critical psychology: An introduction* (pp. 126–143). Sage.

Quijano, A. (2000). Coloniality of power and Eurocentrism in Latin America. *International Sociology, 15*(2), 215–232. https://doi.org/10.1177/0268580900015002005

Rajan, S. R. (1993). *Real and imagined women: Gender, culture and postcolonialism.* Routledge.

Reyes Cruz, M. (2008). What if I just cite Graciela? Working toward decolonising knowledge through a critical ethnography. *Qualitative Inquiry, 14*(4), 651–658. https://doi.org/10.1177/1077800408314346

Rowe, A. C., & Tuck, E. (2017). Settler colonialism and cultural studies: Ongoing settlement, cultural production, and resistance. *Cultural Studies: Critical Methodologies, 17*(1), 3–13. https://doi.org/10.1177/1532708616653693

Sandoval, C. (2013). *Methodology of the oppressed.* University of Minnesota Press.

Segalo, P., & Kiguwa, P. (2015). XIV. Through our own eyes: A conversation between two South African psychology feminist scholars. *Feminism & Psychology, 25*(1), 78–83. https://doi.org/10.1177/0959353514562819

Shalhoub-Kevorkian, N. (2009). *Militarisation and violence against women in conflict zones in the middle East: A Palestinian case-study.* Cambridge University Press.

Smith, C. W. I. T. R. (2000). Straying beyond the boundaries of belief: Maori epistemologies inside the curriculum. *Educational Philosophy and Theory, 32*(1), 43–51.

Smith, L. T. (2013). *Decolonising methodologies: Research and indigenous peoples* (2nd ed.). Zed Books Ltd..

Sonn, C. C. (2016). Swampscott in international context: Expanding our ecology of knowledge. *American Journal of Community Psychology, 58*(3–4), 309–313. https://doi.org/10.1002/ajcp.12038

Sonn, C. C., & Fisher, A. T. (1998). Sense of community: Community resilient responses to oppression and change. *Journal of Community Psychology, 26*(5), 457–472.

Sonn, C. C., Bishop, B. J., & Drew, N. M. (1999). Sense of community: Issues and considerations from a cross-cultural perspective: Controversy and contentions. *Community, Work & Family, 2*(2), 205–218. https://doi.org/10.1080/13668809908413941

Stivala, A., Robins, G., Kashima, Y., & Kirley, M. (2016). Diversity and community can coexist. *American Journal of Community Psychology, 57*(1–2), 243–254. https://doi.org/10.1002/ajcp.12021

Thomas, E., Pate, S., & Ranson, A. (2015). The crosstown initiative: Art, community, and placemaking in Memphis. *American Journal of Community Psychology, 55*(1–2), 74–88. https://doi.org/10.1007/s10464-014-9691-x

Torre, M. E., & Ayala, J. (2009). Envisioning participatory action research entremundos. *Feminism & Psychology, 19*(3), 387–393. https://doi.org/10.1177/0959353509105630

Townley, G., Kloos, B., Green, E. P., & Franco, M. M. (2011). Reconcilable differences? Human diversity, cultural relativity, and sense of community. *Journal of Community Psychology, 47*(1–2), 69–85. https://doi.org/10.1007/s10464-010-9379-9

Tuck, E. (2009a). Suspending damage: A letter to communities. *Harvard Educational Review, 79*(3), 409–428.

Tuck, E. (2009b). Re-visioning action: Participatory action research and Indigenous theories of change. *The Urban Review, 41*(1), 47–65. https://doi.org/10.1007/s11256-008-0094-x

Tuck, E., & Yang, K. W. (2012). Decolonisation is not a metaphor. *Decolonisation: Indigeneity, Education & Society, 1*(1), 1–40.

Tuck, E., & Yang, K. W. (2014). R-words: Refusing research. In D. Paris & M. T. Winn (Eds.), *Humanising research: Decolonising qualitative inquiry with youth and communities* (pp. 223–248). Sage.

Tuck, E., & McKenzie, M. (2015). *Place in research: Theory, methodology, and methods.* Routledge.

Visweswaran, K. (1994). *Fictions of feminist ethnography*. University of Minnesota Press.
Vizenor, G. R. (1994). *Manifest manners: PostIndian warriors of survivance*. Wesleyan University Press.
Vizenor, G., Tuck, E., & Yang, K. W. (2014). Resistance in the blood. In E. Tuck & K. W. Yang (Eds.), *Youth resistance research and theories of change*. Routledge.
Watkins, M. (2015). Psychosocial accompaniment. *Journal of Social and Political Psychology, 3*(1), 324–341. https://doi.org/10.23668/psycharchives.1821
Watts, R. J. (2017). Utopian vision: A grand solution for a scholarly dilemma. *American Journal of Community Psychology, 59*(3–4), 280–283. https://doi.org/10.1002/ajcp.12139
Wiesenfeld, E. (1996). The concept of "we": A community social psychology myth? *Journal of Community Psychology, 24*(4), 337–346.
Wilson, S. (2001). What is an Indigenous research methodology? *Canadian Journal of Native Education, 25*(2), 175–179.
Yuval-Davis, N. (2006). Belonging and the politics of belonging. *Patterns of Prejudice, 40*(3), 197–214. https://doi.org/10.1080/00313220600769331

Part II
Modes of Enactments and Praxes for Community Psychology

Reflections on Radical Love and Rebellion: Towards Decolonial Solidarity in Community Psychology Praxis

Devin G. Atallah

Introduction

In this chapter, I endeavor to share stories, questions, and insights into decolonial solidarity as the building and bridging of communities in resistance against colonialism and coloniality within my clinical community psychology praxis. I build off of decolonial feminist theorists and activists in reflecting how solidarity can involve courageous crossings and creative enactments of community with radical love and rebellion (e.g. Moraga & Anzaldúa, 1983; Bell et al., 2020; Dutta, 2018; Lugones, 2010; Sandoval, 2000; Hooks, 1990; Shalhoub-Kevorkian, 2009; Wynter & McKittrick, 2015).

Coloniality comprises of the colonial patterns of power and social relationality that continue even after formal colonial administrations end, including intergenerational and interlocking systems of domination that make up the spine of modernity: racism and racialisation, capitalist exploitation, patriarchy, and heterosexism (Lugones, 2010; Maldonado-Torres, 2007). According to Lugones (2010), coloniality is "the process of active reduction of people, the dehumanisation that fits them for the classification, the process of subjectification, the attempt to turn the colonised into less than human beings" (Lugones, 2010, p. 745).

I explore how decolonial solidarity involves the creation of community coalitions in attempts at knowing and loving in struggles against active colonialism and persistent coloniality. More specifically, I reflect on aspects of decolonial solidarity through sharing examples from my work on distinct qualitative research projects contributing to building communities of resistance, which I have been privileged to be involved with over the past several years. These include my work on the frontlines of active settler colonial expansion within refugee camp communities in the West Bank, occupied Palestine; *desde 'el renacimiento de la frontera' con*

D. G. Atallah (✉)
Psychology Department, University of Massachusetts, Boston, MA, USA
e-mail: Devin.Atallah@umb.edu

S. Kessi et al. (eds.), *Decolonial Enactments in Community Psychology*, Community Psychology, https://doi.org/10.1007/978-3-030-75201-9_5

comunidades mapuche en Wallmapu Chile (in 'the rebirth of the border' within Mapuche communities in Wallmapu, Chile); and within a Community Action Team (CAT) that I have helped organise in working towards promoting healing and justice in a public middle school community in Boston, USA.

I would like to highlight that these three places (West Bank, Palestine; Wallmapu, Chile; and Boston, USA) are not simply geographic locations or distinct communities in which I am involved with as a researcher, healer, and activist. These three places represent intersecting yet distinct frontlines of decolonial struggle. Elements of these localities also represent unique homespaces for me where I am differently positioned and differently accountable. Embarking in this chapter, I bring rebellious attitude, personal prose and poetic reflections with critical questions applied to myself as a cultural being, and to my work and training as a social science researcher and psychologist.

In doing so, I hope to begin this chapter holding close onto, and with humility, the importance of working to not only study and understand, but to directly contribute to the creation of counter-knowledges and counter-communities, (re)imagining possibilities from a place of decolonial love and rebellion. I do this while standing on the shoulders of so many radical theorists, feminists, theorists, fighters, and healers who came before me, both within and 'against' the psychology discipline. And in this work, beyond the Wall in Palestine, beyond the borders in Wallmapu Chile, and within the CAT community and anti-racist activisms in Boston, my inspiration is always afforded by, and grounded on the wisdom, strength, generosity, and love of my life partner, the inspiration of my children, and the appreciation nurtured within me for my ancestors, and for all the beautiful people in my life and within my communities that sustain me and hold me accountable. As part of my attempt at engaging in more accountable research and writing, I have woven personal reflections and how they may impact my work throughout this paper, rather than having a separate subsection on 'reflexivity'. A central goal of this chapter is to contribute in an honest and humble way to understandings of decolonial solidarity not only in the theorising of my community clinical psychology research and practice, but also in the way this chapter is written.

Emphasising Decolonial Praxis: De-centering Psychology Definitions and Standards when Seeking to Create Revolutionary Relationships and Understandings

When we turn towards decolonial praxis, a central goal is to begin to create radical communities of solidarity where we can rise together and against the West's incorporating expansion, into a courageous place where love and rebellion coalesce (Lugones, 2010). "Praxis" (e.g. Freire, 1970) includes understandings that knowledges and practices are interconnected with broader webs of complex human relationality, and that cycles between reflection, action, and accountability are required to lead to transformative ways of being that can create sustained changes in our social realities. "Decolonial Praxis" includes reflections, actions, and accountabilities that unfold in the in-between spaces of intersections and crossings, marked by the emergence of radical understandings and relationalities capable of shattering colonially-structured realities.

However, before going further in defining terms used in this chapter, I would like to highlight that this chapter will *not* attempt at operationalising "Decolonial Praxis", "Decolonial Solidarity", or "Decolonial Love" in English and within mainstream community, clinical, or social psychology theories. Instead, I will lean on reflexivity and the stories and narratives of frontliners with whom I work, and who engage in collective action for healing and justice in their everyday resistances to colonialism and coloniality. In fact, when working towards a "decolonial" community clinical psychology praxis, I argue that we do not necessarily need definitions. Definitions and analytical descriptions *do* matter, however, it is also problematic to assume that everything can be captured in acts of scientific operationalising. Defining and classifying can actually serve to deaden and foreclose radical possibilities—possibilities for uncovering unjust root causes of oppressive patterns of power that are masked within normativity and mainstream modes of modernity and persisting colonial consciousness. Indeed, we need questions, stories, poetry, paintings, music, film, silences, connections to lands and complex powerful emotions (not in the psychological sense—but 'emotions' in the sense that may not necessarily have colonial, Eurocentric, or English equivalents). And we need rehumanising relationships, rebellious, revolutionary relationships. When loving together against colonialism, we are not necessarily acting within any study of community, rather, we are *in* community and *in* solidarity. We are seeking to embody, together, the 'things' we are fighting for. Where our communities are *in* us, through us, acting within us. Where we belong to people and to land as we love together in our decolonial praxes and solidarities. And our belonging is bridging; rebellious relationships that create new possibilities for colonial oppressions to be visibilized and for our suffering to speak out and be healed. In part, it is precisely our beloved belonging that is so healing, allowing suffering and injustice to transform, and to teach. Opening in-between spaces for justice to be embodied and lived.

From this point of departure, decolonising approaches to knowledge production understand that we do not need to rely on new definitions, theories, standards, or terms uncovered and coined by Western, Global North academics. Rather, we need to decolonise our thinking and center our understandings on voices of frontliners and embodied knowledges of colonised communities most impacted by oppression (e.g. Atallah et al., 2019; Bell et al., 2020; Dutta, 2018). Therefore, in this chapter, I aim to centre understandings on stories of frontliners and scholar activists who share their narratives about love and relationships as collective action against colonialism for healing and justice. In doing so, I provide examples of ways to understand, read, feel, question and organise around radical loving and solidarity. Stories and examples are threaded/woven across this chapter, rather than providing definitions or operationalisations. Furthermore, when we attempt at scaffolding decolonial counter-knowledges and build understandings from Global South perspectives towards shattering colonially-structured reality, we need to work in (re)imaginative ways—ways that require transdisciplinary or even *undisciplinary* thinking, that often position ourselves 'against' our social scientific disciplines/training, and that impels us to turn outside of mainstream or 'Whitestream' psychology (Atallah, 2016; Bell, 2018; de Sousa Santos, 2018; Ginwright, 2016; Maldonado-Torres, 2016).

Grounded on this disobedient approach to writing, I hope to generate questions, suspicions, and insights into passageways towards transformations in our praxes as activists, as healers, and as scholars/psychologists. Decoloniality is not about building new standards or new 'established' knowledges—or what Aime Cesaire called 'decadent discourses' (de Sousa Santos, 2018; Maldonado-Torres, 2016). In fact, decoloniality involves holding a healthy suspicion of standards, and focusing instead on creating counter-narratives and counter-strategies, and the courageous crafting of "revolutionary relationships, in action, not simply in rhetoric" (Brown, 2017 referencing Toni Cade, p. 71).

Being 'Engulfed by Love' as We Remember and Resist: Decolonial Solidarity as Crafting of Revolutionary Relationships in the Face of Perpetual Pain, Separation, and Devastation

Facing loss and perpetual pain are central to decoloniality. Lost places of belonging. Fragmented lands and lives. This can be experienced within our loss of home and identity, and also, can be rooted in our love and our determination to remember—to never let go of our beloveds who have been murdered in acts of brutal colonial violence. Revolutionary relationships uphold that our community is not only comprised of the living. We belong to the uprooted and the dead as well. Through radical grief and remembrance, we can be engulfed by love so powerful and transformative that new understandings and pathways towards freedom are forged as we actively and defiantly remember and resist.

When engulfed by such powerful and perpetual love, we can experience an abundance that may be felt, for example, through a poem, a painful chasm, an uplifting consciousness, or in the ecstasy of welcomed and deeply respectful affection. This profound sense of abundance can also be sourced from enduring through the agonising loss of our loved ones in never ending journeys through pain that shape our ways of knowing and being. An example of this painful, decolonial loving, is evidenced in a story shared with me by Mohammad (pseudonym), who is a participant in my research in Palestine. Mohammad is a male youth worker and activist in his thirties, born and raised in the refugee camp where my research takes place. A previous project he participated in (see Atallah, 2017a) was entitled: the Palestinian Refugee Family Trees of Resilience project (PRFTR).

The PRFTR project explored how Palestinian refugee families respond to and create opportunities for healing and justice within contexts of historical and ongoing settler colonialism (Atallah, 2017a). PRFTR's methodology engaged the critical constructivist qualitative method of situational analysis (Charmaz, 2006; Clarke, 2005) and decolonising strategies of community engagement (Atallah et al., 2018a, b). Participants in PRFTR were invited to engage in the project through a partnering, grassroots youth organisation, which was founded in the camp by Palestinian refugees themselves more than a decade ago. Mohammad was one of the first youth workers in this organisation (which will remain anonymous to protect the privacy of the community). As Mohammad shared with me, evidenced in the quotes below (which are excerpts from my qualitative research transcripts), his growth as a decolonial activist involved engaging in intergenerational discussions with elders and with schoolteachers about the *Intifada*, and also learning from silence, loss, and love. In particular, Mohammad shares how his passageway into developing decolonial ways of knowing, seeing, and being is rooted in the profundity of loving remembrance he wages for his brother:

> MOHAMMAD: I started to understand the meaning of the *ih-tilal* [the Israeli Occupation] when Israeli soldiers arrested me when I was only 12 years old. This was the beginning of my realisation of the meaning of the word *intifada* [uprising or rebellion] and *ih-tilal*. First, I heard the discussions when people would come to our house, or even it was just my family eating together, or watching the TV with the family or the neighbours, or the chitchat on the streets, all of it was about the *intifada*. As children, we were very curious ...I remember when I asked my history teacher, he tried to explain to us with simple words and simple illustrations, because we were only 12 years old at the time, what the *intifada* was and what was going on. He talked in simple terms... and he promised the next class he would talk in more detail about what the *intifada* meant and to talk about the UN resolution regarding Palestine, something that was never taught to us before, but that next class never actually came because Israeli soldiers arrested him before he could teach it. Then we tried asking the same question with another teacher but the teachers at that time starting to feel very afraid of being arrested....The same situation continued until a neighbour was killed by Israeli soldiers and this was the first time I heard the word *shahid* [martyr]. There was suddenly a *shahid* in the neighbourhood and I never felt the pain that family was going through until my own brother died. My brother's death was a real turning point for me.... When my brother died, I felt that that energy which had been building up all those years exploded. Things were inside, inside until one day it was out....I realised that my brother who died lived on... I began organising student strikes... Throwing stones, becoming really active... and every stone we threw was like a building stone for freedom...

Mohammad shares his journey of resisting the settler colonisation system and questing for justice and healing from overlapping and accumulating losses: from the presence of the absence of his teacher, his neighbour, and eventually, the brutal loss of his own beloved brother. Mohammad highlights how his own arrest as a child marks a certain beginning, an entrance into a site of resistance, and of accessing values and new and old languages emerging from decolonial manifestations of cultural imaginaries, such as *intifada*/uprising or rebellion, and *shahid*/martyr. This site of resistance included profound silence, devastating loss and grief, which Mohammed transformed into a language he later applied to new, radical ways of seeing and being. Mohammad explained how he faced these painful experiences with abundance of love for his brother, together within his refugee camp community, and more broadly with the Palestinian community at the national level. This explosion of activism born from pain, loss, and love supported him in processing new understandings to interpret the nature of the settler colonial oppression, translating the complexity of the occupation's system of compartments into a wordless language he knew in his body—into terms he could feel, understand, and apply towards decolonisation.

Lugones (2010) theorises that the starting point of decolonisation is "coalitional because the fractured locus is in common, the histories of resistance at the colonial difference are where we need to *dwell*, learning about each other" (p. 753; italics in original). Taking a stand, and courageously and creatively spending time in this 'dwelling' place marked by loss, grief, crossings, and coalitions, has long been part of decolonial praxes. From my perspective, and my 'dwelling' place(s), I feel the chasm of 'colonial difference' splitting through me always, knowing that there are parts within me that can never be aligned. I am a Palestinian in the diaspora, born in the USA to a Palestinian father from Chile and a mother of European American, white settler background. My commitment to contributing to the generation of understandings for resisting oppression lives in my body—as an assemblage of coloniser and colonised knowledges. To begin, my research in Palestine with youth workers like Mohammad, takes place in a UN refugee camp in the Bethlehem area that is directly adjacent to the Indigenous village and ancestral place of my father's side of the family.

My paternal ancestors lost lands and lives during *Nakba* ('Disaster' in Arabic) of 1948. For Palestinians, Nakba refers to events where approximately 750,000 people were forcibly displaced off their Indigenous lands during the creation of the state of Israel in 1948 (Pappe, 2006). The government of Israel refers to this event as the 'War of Independence' (Israeli Ministry of Foreign Affairs, n.d.). My paternal grandfather fled Palestine for Chile after *Nakba* in the early 1950s. In this light, the insights and stories that I share from my work in this chapter, generate questions for me in ways that interact with my own intergenerational experiences of displacement, and the ways that native communities and relatives of mine who have remained in Palestine are continually being colonised. I understand these issues as not only being intellectual abstractions, but I feel these colonial violences survived by my family and communities in my body and I try to bring these embodied knowledges into my decolonial solidarity praxis.

In so many ways, I am both coloniser and colonised - born into an incommensurability of being a 'Palestinian Settler' writing this chapter in the colonised Americas—'Turtle Island'. There are parts within Palestinian experiences that no matter how hard I try, I will never understand. However, central to decolonial solidarity is respecting these types of incommensurabilities (Tuck & Yang, 2012). In other words, my efforts to support Palestinian healing and liberation in my community clinical psychology work does not mean I have to understand every aspect of the Palestinian struggle—instead—it means that I have to respect elements of the struggle I can never understand, from where I stand. And while doing that, how do I—a Palestinian living in the Diaspora with Indigenous Arab roots, a Latinx and white Boston resident, a settler on Turtle Island, a cisgender man, an assistant professor affiliated with a Western academic institution—enact decolonial solidarity and support healing transformations in communities in struggle, "not through the tactical eyes of war, but through eyes of love" (Brown, 2017, p. 8)? And when engaging in this research in ways that hold community healing and decolonial loving praxis at the center, how do I honor and privilege the knowledges of grassroots decolonial resisters and radical women on the frontlines, with respect to the ancestors and the ancient, towards creating communities of remembrance and resistance?

Addressing these questions requires, not so much that I define decolonial solidarity, or that I *under*stand healing and resistance, but that I consider *how I stand*, in relation to justice. And not only how I stand, but *where I stand*, and *with whom I stand*. In this light, I hope to look closely at my locations and at my coalitions, and offer this practice as a potential decolonial enactment of community clinical psychology.

Where Do We Stand? and With Whom? Decolonial Solidarity as Standing In-Between, and with Love, Against Colonial Ideologies and Multiple Degrees of Incarceration

In this segment of the chapter, I reflect on decolonial solidarity as involving our facing one of the core dimensions of colonialism head on—contending with the transnational, colonial 'system of compartments' (Fanon, 1963), and the binary logics that draw the lines, build the Walls, change the facts on the ground, and incarcerate our bodies and lands alike. In doing so, I draw on perspectives of frontliners in my research in Palestine, and also on my own personal and family narratives, with questions, prose, poetry, and quotes from research participants themselves. These reflections demonstrate examples of ways in which decolonial solidarity includes our coming together to gasp for air and contest what Said (1993) called, 'the amalgam of the arts of subordination'. These enactments of decolonial solidarity seek to cross compartments and leap in-between the chasms of colonial difference towards healing our wounds and building blocks of a radical relationality across our cut up communities, lands, and hearts.

This revolutionary relationality that we build together, and that we build against the colonial 'system of compartments', is grounded on love—decolonial love. Love that releases energy for resistance, as demonstrated in the example of participant Mohammad from my PRFTR study cited above. Love that offers sustenance for our struggles. Love that builds bridges for our crossings, our transgressions, our enduring past limits of colonised time and space. As Sandoval (2000) theorises, citing Roland Barthes, love is a "gentle hemorrhage", a "'breaking' through whatever controls in order to find 'understanding and community'" (p. 139). In this light, decolonial love can be envisioned as a passageway towards knowing and towards finding a community of resistance. It is also a passageway towards becoming a community of understanding. This labor requires an entrance into a unique site of struggle, a decolonial dwelling space—one which is a shared and sacred—a "free, yet marked and wounded space, a site of shifting…" (Sandoval, 2000, p. 142).

Decolonial theorists/activists and radical Global South feminists have long understood that shifting and crossing in-between, can hold keys for liberation, and that borders are fiction (e.g. Anzaldúa, 1987; Lugones, 2010). Building alliances across colonised communities—Global South to Global South—generates powerful potentials for liberation (Fanon, 1963; Malcolm, 1964). Within these 'sites of shifting', solidarity may include a dimension of harnessing privilege, or at the very least, requires an outflow; the breaking out of the prison of the 'compartment' you may be locked within, and having the privilege of mobility or of hearing voices of colonised sisters and brothers from beyond colonial Walls that devastate and separate. But it cannot stop there. Decolonisation doesn't need tourists. Decolonial solidarity requires a passage into direct struggle, linking across movements within a "process of setting aside victimhood and becoming resisting people practicing ways of being and knowing in their struggle against domination" (de Sousa Santos, 2018, p. 28). A question that emerges from this lens is one that challenges understandings in psychology ideologies of colonised peoples as 'an-other' and relegated to being victims, combatants/terrorists, and/or resilient trauma survivors: What is lost, and who loses, when in conditions of colonial violence, we reduce peoples' complex journeys towards decolonisation, healing justice, and human dignity, to trauma/resilience pathways or to security/terrorism frameworks? How can we recognise when we ourselves, well-intentioned psychology scholars wanting to work in solidarity with communities in struggle, writing in English, are still falling into the trap of thinking 'ecologically', 'systemically', or in terms of 'security' in ways that universalise, disembody, and further displace colonised groups?

Lugones warns, "if we are going to make an-other construction of the self in relation, we need to bracket the dichotomous human/non-human, colonial, gender system that is constituted by the hierarchical dichotomy man/woman for European colonials + the non-gendered, non-human colonised" (p. 749). Decolonial solidarity is *not* about advancing one marginalised social group's agenda, or 'giving voice' to a perspective of an oppressed group. Meaningful South-to-South alliances and exchanges of knowledges and passageways towards decolonial being require solidarities that reconnect people to the fullness and depths of humanity. Finding this 'site of shifting' is what I mean by 'Standing In-Between'. It is a space that is marked by love and transformational potentials. This 'site of shifting' involves stepping out of one's disciplinary frameworks, without simply linking struggles of fragmented groups from a place of political passion or choosing to enact a specific group's fervent ideology.

Such nuanced relationalities are profound rehumanising reminders: standing in-between, with love, can be a sign that the destruction of colonialism is not total. As the Palestinian decolonial feminist, legal scholar and social worker Nadera Shalhoub-Kevorkian (2009) states, "the hegemonic powers, as powerful as they can be with all the Empire's support, will always stand weak, fearful, and naked, for they cannot mute the voices of the suffering frontliners" (p. 32). Shalhoub-Kevorkian highlights how Palestinian women position themselves against oppressive powers and imposed states of war and what the author calls, 'betweenness', or the state of being "locked between the West and the non-West, public and private, male and female, reason and emotion, and home and exile…" (p. 213). In the face of interlocking colonial, classed, and patriarchal oppressions, Shalhoub-Kevorkian (2009) emphasises how this 'betweenness'—the space where women frontliners locate and liberate themselves—is constantly shifting and being redefined by the women frontliners themselves.

This 'site of shifting' is marked by betweenness and oppositional consciousness, which embraces multiplicities—what Anzaldúa (1987) calls '*la conciencia de la mestiza*'. According to Sandoval (2000), building off of the work of Anzaldúa and Barthes, this type of loving relationality is radical and relentless in its push towards solidarity, and is always held through embracing multiplicities and dialectical passageways—not part of any binary, and "occurs outside ideology" (p. 141). In fact, this is a place where Eurocentric ideological logics of dichotomies are defiantly subverted (Lugones, 2010). These are sites of decolonial love—dialectical places of deep connection, struggle, and sustenance—ones that are always uncontainable and disorderly, and can be delicious and creative. These sites are passageways to places where colonised selves and communities step into a unique type of power, or as Sandoval calls it—a place with "zero degree of power" (p. 146).

I was privileged to witness a profound example of 'stepping into power' within a decolonial passageway formed by radical love, when an individual with whom I work in Palestine shared her story of surviving torture in an Israeli prison. Naila (pseudonym) is a leader in our ongoing and developing research in Palestine together. She was also a participant in the PRFTR project (Atallah, 2017a). During one of the interviews I completed with Naila, she shared how she was imprisoned already three times in her life. One of the times she was arrested, Naila described how she 'stepped into her power' in a unique way—one that centered on a piercing love that she wielded from the depths of her soul, directing her focus on getting out of prison as soon as possible to return home for her child. This love manifested as radical strength even in the face of torture. Naila explains:

> NAILA: Once, I was arrested as a new mother to my first child, who was only two-years-old at the time. This time the Israeli soldiers arrested me for something I wasn't involved with at all. My sister-in-law was an active fighter and she died in combat. The Israelis put a curfew on our entire refugee camp after she died and they dropped her body at the cemetery in our camp in a big black plastic bag. I went to the funeral to say good-bye to her at the cemetery, ready to bury her, when the Israelis came and arrested me during the middle of the funeral. They arrested other people in my family as well, like my sister. And my husband was also in prison at this time. So my son, only two years old, was left alone without either his father nor his mother. My mother took care of him until I eventually got out of prison and returned to him. But let me tell you, this is what happened to me immediately when I was thrown in the Israeli military jeep, beaten, and taken to prison from the funeral: I got into this mental state where I was so calm and relaxed, even when I was being brutally tortured. Types of torture that I do not want to talk about right now. There was only one thought in my head—that I was not to talk to the Israeli interrogators, no matter what they did to me. I was not ongoing to say anything. I would not say anything and survive all the torture and interrogations so that I could go home as quickly as possible to my child. It was like something blocked my mind from everything around me, there was just one focus—How to get to my son as quickly as I could.

Naila describes how love inside her arose into a power so great, it became a source of strength out of necessity to resist and survive the unspeakable, colonial patriarchal sexual violence of torture that she faced while incarcerated as a radical Palestinian woman detainee in an Israeli prison. These decolonial armaments of love exemplified in Naila's testimony, when integrated, may work towards shattering powers of unspeakable colonial violence, and can lead into what Sandoval (2000) calls a 'no-place' or into a 'zero-degree of power'. These enactments of struggle require courage and endurance, and a wielding together of every aspect of life to wage love against all-encompassing colonial violence, working towards "the cessation of suffering through transforming the powers of domination that cause it" (Sandoval, 2000, p. 145).

However, it is important to highlight that when acting from an oppositional place marked by decolonial love, as Naila describes in her story shared above, she steps into a power that is not simply a response to oppression. Naila describes a decolonial loving that is not the same thing as mere 'empowerment'. It is a return to a way of being that demands wholeness, a fierce reclaiming of *be-ing* fully human. Naila's story reminds me of another great Palestinian freedom fighter, and a past leader within the Palestinian Liberation Organization named Leila Khalid, who famously said:

> I no longer think it's necessary to prove ourselves as women by imitating men. I have learned that a woman can be a fighter, a freedom fighter, a political activist, and that she can fall in love, and be loved, she can be married, have children, be a mother. You see, at the beginning we were only interested in the revolution. We were not mature enough politically. The question of women is a part of our struggle but not the only part. Revolution must mean life also; every aspect of life.

Naila's love for her child sourced a power that returned her to a state of being as a crossing into being more fully human, a power that transformed the incarcerated space that was so violently enclosing, violating, and torturing her, into a site of resistance guided and strengthened by her ungovernable and uncontainable love—a love for her child, and her subverting the need to choose between being either a mother or a resister… Naila was clearly both, and much much more.

As Naila demonstrates, finding, and redefining these sites of oppression involves a climbing out, or a leaping out, of colonial order of consciousness. As Wynter and McKittrick (2015) imagine, insights born from decolonial solidarities involve a leaping out of structures of violence and consciousness that characterise modernity/coloniality. Wynter and McKittrick argue that Fanon, like Dubois before him, was concerned with developing a 'double consciousness,' contributing to sources of knowing and loving capable of re-framing dialectics between *skins* and *masks*. Wynter and McKittrick (2015) call this Fanon's "reverse paradox" (p. 46).

Applying these understandings of decolonial solidarity in my life and work in communities of color in Boston, in Mapuche communities in Chile, and in colonised and occupied Palestine, I notice how so much of my training in psychology—even benevolent universalist human rights ideas, peace and security languages, and social justice discourses—can contribute to my actively oppressing communities that I work with, and at times, actively and internally oppressing myself.

As I shared earlier, my paternal ancestors are from Palestine. We trace our roots back to an Indigenous village near Bethlehem and Jerusalem called Beit Jala, the same town Naila lives in and a village adjacent to the refugee camp Mohammad is from. In fact, some of the participants in my research in Palestine, particularly the elders, who are currently between eighty and ninety years old, actually knew my beloved great-uncle (my grandfather's older brother), who I will refer to as *Amo* (which means 'uncle' in Arabic) in this chapter.

Amo was an Indigenous Healer (*Rohani* in Arabic) from our village. Thus, as I theorise about healing, radical love, and resistance within conditions of structural violence associated with settler-colonialism—whether in Palestine, Chile, or Boston—I welcome in the complexity of a relationality where my own ancestors planted seeds and labored to promote healing and justice in some of the homespaces where my research takes place. And as I write this chapter, I embrace these intersections and the messiness that follows.

In particular, when theorising about the process of 'stepping into power' and challenging colonial degrees of incarceration, I grasp onto any remnants' of knowledges that lived in my *Amo* –keeper of Indigenous Palestinian village healing practices as *Rohani*—that now may live in me. These knowledges have been passed down through my Palestinian grandparents, aunts and uncles, cousins and friends. Knowledges, for example, rooted in my love and respect for mountain herbs and fruit trees, and their *sumoud* (perseverance in Arabic). Love for the ineffable abundance housed in the preparing and sharing of cherished family foods, and in community, as core strengths housed in my ancestral Palestinian village wisdoms.

Furthermore, as a member of the Palestinian community, when engaging in research in the occupied West Bank, I am frequently socially-classified as a possible threat by Israeli state actants. I am racialised as someone who could be 'excessively violent' and therefore deserves disproportionate segregation, detention, and surveillance. Of course, as a Palestinian born in the USA and with both Chilean and United Statesian passports, I have extensively more privileges than most Palestinians. Even writing about my personal and family challenges with Israeli oppression feels quite irrelevant compared to sufferings overcome by frontliners such as Mohammad and Naila whom I quote above. However, I believe that working towards a decolonial praxis involves my welcoming insights and understandings about decolonial solidarity taking place around me in my research, and also inside of me in my personal, family, and community life. Bringing decolonial solidarity consciousness and radical love with me into colonised Palestine, allows me to not only work in ways that advance justice and healing, but also keeps me together, keeps me alive, and helps me challenge state-sponsored hate, in addition to self-hate.

For example, connecting with decolonial love in Palestine allows me to read the violent whitened geographies and discourses that target my own being, my Palestinianness, my body, my movement, and my long journey home. I feel the decolonial movement embracing me—embodiments of radical love caressing me. I feel my *Amo* welcoming me—engulfed by love. Love that provides perspective and shelter against the colonial violence—including against the destructive power of Israel's creation of new facts on the ground with the building of settlements and the

Wall, checkpoints and fences as the realisation of the world's largest open-air prison continues day after day. Love in the face of the Wall requires the provision not only of shelter and protection - but of perspectives and possibilities to speak out against the killing and arresting, uprooting and humiliating, colonial violences which work to perpetuate the Western normative order of consciousness in the here-and-now.

When I return to Beit Jala, the village where my grandfather was born and raised, and where both my paternal grandparents' grandparents are rooted dating back into millennia, I feel decolonial love in my body. I feel my vision changing too—perhaps this is my building of a 'gaze-from-below', as Wynter and McKittrick (2015) call it—a lens to see straight through the Western normative order of consciousness, while *moving out of place*. Yes, when I return to Palestine, I *move out of place*. And within my defiant movements—this is where I begin to engage in community-based qualitative research. Where I organise intergenerational storytelling gatherings. Where I listen, and then I listen some more. I gaze up at the Israeli Wall. I visit elders, some who are my own relatives, some of who are strangers to me, some of them are participants in my research, but I deeply care for them all. And I dance at weddings. I sometimes watch Israeli tanks eat the concrete as they conquer an overcrowded Palestinian ghetto. I walk into the mountains with friends and cook traditional meals of resilience in clay ovens, just like my grandfather used to do. Just like my *Amo* used to do. I visit the dead. I break bread with local psychologists, social workers, community organisers and physicians strategising ways forward to build radical alliances for health and community healing. I visit my Indigenous village. I morn the murder of my great grandfather Khalil, who was shot and killed while working the land—standing amongst our fava beans and olive trees. I walk in the olive groves in the outskirts of our village in what the state of Israel has annexed into their so-called, 'greater Jerusalem'. I feel the presence of the absence of the killed and displaced, and the *sumoud* of the olive trees. I know that many of the lands may not 'legally' belong to us, but that the settler nation-state annexes lands that we belong to. And the settler nation-state does not have the absolute power to completely silence our intergenerational voices. My footsteps speak. They cry and shout when I cross beyond the Israeli Wall. In doing so, I return not only to land, but to a nowhere place, where we drink coffee with cardamom and with attitude. Where my steps home include steps into power. Where I collect mountain sage and other herbs around the village—like *Zaatar*—which is always wild and free. Like my Indigenous family tree. Cut to pieces, and still standing…

A tree without a name
at the edge of the world
like the silence after a massacre
a secret buried in bones
but they are not bones at all
they are the roots of our tree
a wounded, nowhere tree
stubborn and *sumoud*
engulfed by love…
What would it do to the sky
if our roots could speak?

Decolonial Solidarity as 'Communal Flourishing': Restorative Enactments of Interdependence, Abundance, and Transformation

In this section of the chapter, I reflect on decolonial solidarity and radical love as unfolding in our relationships not only when we step into our power, but when we step into a state of collective being that is restorative and yet ungovernable, a place "where political weapons of consciousness are available in a constant tumult of possibility" (Sandoval, 2000, p. 141). These 'weapons of consciousness' challenge Western control over meanings of relationality, individuality, and interdependency. Various Global South Indigenous understandings of relationality, such as *Ubuntu* rooted in southern African consciousness, "affirms not only our inherent interrelatedness, but also the resulting responsibility we bear to one another, flowing precisely from our deep connection… Neither tyranny of community nor extreme individualism, *ubuntu* invokes a dialectical unity of opposites… independence is inseparable from interdependence, and individuality expands from being solitary to being in solidarity" (Davis, 2019, p. 18).

This profound interdependence births a way of knowing and loving that can be sourced, for example, from the affirmations we feel when our humanity is wholly seen, heard, or held by another human, animal, river, spirit, herb, tree, or other being(s). This sense of being engulfed by love felt in 'communal flourishing' (Lugones, 2010) is restorative and shatters Eurocentric binaries, refocusing our attention and our taste on recipes as methodologies of balance. In many ways, decolonial solidarity involves the return to community and to humanity through enactments of restoration and balance rather than of binaries. As I turn to examples of decolonial solidarity within my attempts at embodying interdependence, balance, and humanising praxis in my research and healing justice work with Mapuche communities in Chile, I would like to provide some brief context and background information on the Mapuche.

Approximately one and a half million people (nearly nine per cent of the total Chilean population) self-identify as Mapuche (Instituto Nacional de Estadísticas Resultados Censo, 2012). Many Mapuche call their lands *Wallmapu*. Around thirty percent of all Mapuche live below the poverty line (Agostini et al., 2010), and face drastic health inequities which are visible in life very low life expectancy rates, infant mortality, tuberculosis, women's health, and mental health (Obach, 2016). The Mapuche are currently among the most marginalised groups in all of Chilean society. Mapuche rebellions to ongoing settler colonialism in recent decades have included organising and activisms for the reclaiming of their language (*Mapudungun*) and traditional healing systems, while also confronting invasive forestry companies and government legislations related to neoliberal exploitations, land rights, and fighting for comprehensive self-determination. The Chilean military has responded severely, even deadly, to Mapuche protests as many young Mapuche have been killed by Chilean soldiers, commandos, or police in recent years (for example: 17-year-old Alex Lemun, 23-year-old Matias Catrileo, 24-year-old Jaime Mendoza

Collio, and more recently in November of 2018, 24-year-old Camilo Catrillanca, the son of a Mapuche chief from the autonomous community of Temucuicui was assassinated). There is also a growing problem of mass incarceration where many Mapuche are arrested, prosecuted, and detained under an "anti-terrorism" law created decades ago during Pinochet's dictatorship (Correa & Mella, 2010).

In my work with Mapuche communities, I have focused on engaging in community-based qualitative research through the creation of a collaborative community action group which includes a non-governmental organisation of Indigenous advocates and researchers, and a Mapuche community health center—the Mapuche House of Traditional Healing (these non-governmental organisations will remain anonymous in this chapter to protect the privacy of the community members). My work with this team has been guided in many ways by Sayen (pseudonym), who is a radical Mapuche woman and the leader of this Mapuche House of Traditional Healing.

Sayen and I have worked over several years to create a coalition together grounded on decolonial solidarity praxis (see Atallah et al., 2018a, b for details on the work of our research collective). At the beginning, we began by outreaching to both Mapuche and Palestinian community members and eventually trying to bring these communities together. As a Palestinian-Chilean-United Statesian psychologist and researcher, I aim to move from a place of accountability, knowing that my Palestinian ancestors who settled in Chile benefited from the destruction and decimation of Mapuche lands, lives, and livelihoods. In my work in Chile, I aim to practice what Lugones talks about as being "creatively in coalition, a way of thinking of the possibility of coalition that takes up the logic of de-coloniality, and the logic of coalition of feminists of color" (p. 755). In doing so, a starting point for me in my building with Sayen and the Mapuche house of traditional healing as a grassroots Indigenous women-owned and run organisation, was from a place of honesty and humility. We shared stories, foods, and even later began sharing struggles. In particular, Sayen expressed interest in learning more about Palestinian decolonial struggles and the role of traditional healing in that process. I felt my *Amo*, the Indigenous Healer/*Rohani* from Palestine especially close in this time. Sayen and I connected, even in small acts and exchanges of herbs and wisdoms, from *Rohani* to *Machi* (the name for an Indigenous Healer in Mapudungun).

At one point, Sayen invited me to visit a Machi who was incarcerated. He is labeled a 'terrorist' by the Chilean state. During our visit into the prison, the *Machi* (with pseudonym 'Ankatu') and I formed deep connections through words, and also through our simply holding space together. At some periods, Machi Ankatu and I sat together in complete silence, feeling our connection in spirit, perhaps, more than naming it with words. At the end of our meeting, Machi Ankatu got up and left the visitation room of the prison and was out of sight to Sayen and I. He returned a few minutes later with a *Trutruca* in his hands. A Trutruca is a Mapuche wind instrument made from bamboo and a cow horn. Machi Ankatu then proceeded to engrave a phrase directly into one side of the cow horn. After about twenty minutes, the phrase he succeeded in engraving into the horn reads on one side, "from the Ankatu family" and on the other side, "for the Atallah family and the Palestinian people"

Fig. 1 Trutruca from Machi Ankatu, picture taken by Devin Atallah

(see Fig. 1). Machi Ankatu explained to me that he uses the *Trutruca* in his spiritual ceremonies and healing practices and that he hoped I would carry the instrument into my life, family, communities, and healing work. I was engulfed by love upon receiving such a gift, and I felt an emergent communal flourishing, especially within a gift created/shared with such love from behind the walls of the colonial degrees of incarceration.

The strength and generosity that Machi Ankatu waged in this loving act shattered Walls in my being. I feel this love even now in remembering and writing about that moment when he handed me the *Trutruca*. Of course, I feel this love each time my children blow their wind into the Trutruca filling our house, or the Boston streets as we protest, with restorative energies for resistance and continuation of opportunities for radical futures. When we finally left the prison, I noticed tears cascading down Sayen's cheeks. She was also engulfed by love—an un-arrestable, un-detainable love. We both were collectively strengthened for our community praxis ahead, and felt deep gratitude for the Machi's generosity and the depths of his love and humanity.

I continue to follow Sayen and other members of the Mapuche House of Traditional Healing for direction and insight into ways forward in building our research together. We always hold our families and communities at the center. For example, at one point in a developing project together, Sayen suggested that we share foods and celebrate Palestinian and Mapuche legacies of decolonial rebellion with about one hundred members of the community (see Atallah, 2017b, for the *Mondoweiss* article I wrote on this process for more details). Sayen suggested that this coming together and feast could lead and teach us forward. Through detailed discussions of our different family foods, Sayen and I eventually put together a menu, and we settled on serving *mansaf*, which is a traditional Palestinian dish of lamb, rice, spices, almonds, and yogurt, which seeks to message a deep sense of welcome to travelers and guests. The type of yogurt used for this dish is *laban jameed* (Arabic for 'dried yogurt'). *Laban jameed* is dehydrated, with salt and turmeric and other spices (as one likes) into hardened rock balls that do not require refrigeration. An elder in a refugee camp in Palestine near my auntie's and uncle's house in Beit Jala prepared the *laban jameed* and a dear friend shared them with me out of love. I brought these homemade balls of *laban jameed* from the refugee camp in Palestine to Chile, and used these gifts to cook for the Mapuche community at the Indigenous health center in an act of decolonial solidarity. Alongside my *mansaf*, Sayen and other members of the Mapuche House of Traditional Healing cooked Mapuche foods including *sopaipias* with *merken* (fried dough with smoked hot peppers) and many other dishes including *mijokiñ*, *charkan*, *catutos*, *yiwiñ kofke*, and a variety of salads. In total, we fed nearly one hundred people at the *Ruka* (Mapuche word for a traditional community house) in the Indigenous health center. For this gathering, we brought together members of our families, communities, and both Mapuche and Palestinian activists in Chile and there was poetry, music, and also group dialogues and intergenerational story-telling. In our feasts and our circle discussions, we furthered our coalition at the site of colonial difference (Lugones, 2010), centering communal flourishing and our shared intent on not only knowing each other, but on thriving and living together in re-humanising ways. As Lugones shares, this was our re-organising of our "movement toward coalition that impels us to know each other as selves that are thick, in relation, in alternative socialities, and grounded in tense, creative inhabitations of the colonial difference" (p. 748). These types of solidarities, involving communal flourishing and celebrations of abundance, open to thick and nuanced understandings of our humanity. They are feasts, *and* they are decolonial enactments of psychology praxes. These solidarities are evidence of our stance towards decolonial justice, instilling profound lessons for liberation.

Turning to my work in Boston, decolonial solidarity has coalesced around my participation and leadership in organising a Community Action Team (CAT) within a public middle school community in Boston, USA. As the CAT, we are united as a community initiative working together towards promoting Healing Justice in the middle school. Our CAT members include teachers and parents from the school, in addition to long-time anti-racist activists from Boston, community researchers and mental health clinicians, and radical youth workers from a nearby grassroots youth empowerment organisation.

Working within the CAT, part of my research process has involved reflecting on my whiteness and the ways in which parts of my ancestry and my ways of being and knowing are deeply rooted in white supremacy culture and legacies of perpetrating colonial violence across Turtle Island and beyond. On my mother's side of the family, my maternal grandmother was born in Hawai'i and is white, or *Haole*, in Hawaiian. She is a member of English and Scottish ancestries, and was born and raised in Hawai'i. She traces her family roots back to colonisers who travelled from England and Scotland to Plymouth Colony contributing to the brutal dispossession of Wampanoag communities off their Indigenous lands in Massachusetts in the

1600s, and who surely participated in genocidal patterns of power oppressing and decimating Black American and Native American communities across North America throughout the subsequent generations.

As a descendent of white colonists here in Boston—where Wampanoag and other Indigenous people of Massachusetts and sounding areas are the rightful keepers of the lands—examples of questions I strive to live and bring into my work on the CAT have been:

- How do I build revolutionary relationships with Black and Brown communities in Boston not only from my positionalities as a Palestinian and Latinx person, but also as a descendent of white settlers?
- What does relational accountability look like for me within the CAT?
- How do I practice decolonial love within such a complex multiplicity of identities, solidarities, and rebellions that we forge together and apart, within and across our shifting practices and the transformative schoolspaces that we seek to cultivate in our CAT?

These are some of the questions that guide how I try to be in the CAT process. For example, I aim to work in ways that decenter research, or any goals of my own and the institutions I work for, within the CAT. Instead, at the core of our CAT process, we aim to build conditions of community, which can challenge relational patterns that create and perpetuate harm and oppression. We question the institutions that we rely on, and we work to build a culture and a container within the CAT process to hold one another, and to thrive despite the precarity, the homelessness, the poverty, the racism, the sexism, the homophobia—the structural violence all around and within our CAT team members' everyday lives. In doing so, we challenge the parameters of what is expertise and who leads and focuses the values and visions of the CAT. These are examples of practices that emerge from our aim to decenter not only 'research', but also whiteness. We are vigilant to institutionalisations of white supremacy in our relationships and practices.

One of the leaders in the CAT, a long-time anti-racism community activist, is Ms. Moore (pseudonym). Ms. Moore is an African American woman in her sixties, who has worked for many decades to unify Boston's minority neighbourhoods. She has been an important leader in the Boston busing and desegregation process within and across the public schools. When beginning our CAT meetings, Ms. Moore often guides our emerging coalition through a process of returning to our shared CAT's touchstones. Through these touchstones, Ms. Moore reminds all of CAT members of the abundance that is always available through decolonial solidarity and radical love, including revelations that *everyone* is needed for the work ahead, and that *everything* we need for decolonial transformations in our community is already here—in the power and expertise of parents and community activists, in teachers and students, and in our circles and in our insistence to co-create a schoolspace that aspires towards becoming a beloved community and homeplace—a site of resistance (hooks, 1990).

An example of transforming spaces within a colonial institution into a site of resistance with opportunities for restorative enactments of interdependence, abundance, and transformation—waging radical love from behind prison Walls—can be seen in a story shared with me by a participant in the PRFTR project in Palestine named 'Fatima' (pseudonym). Fatima is a young women in her early twenties from a Palestinian refugee camp in the West Bank. She describes, in the quote below, how she harnesses love in ways that resist the colonial, gendered violence and many degrees of incarceration that walls her shifting homespaces. Fatima reports being arrested and serving nearly 2 years behind bars on administrative detention (i.e. detention without trial) in an Israeli prison. Fatima was released without ever having faced any charges.

> FATIMA: A few years ago, when I was 20 years old, Israeli soldiers invaded our house in the middle of the night, in the heart of winter.... The soldiers busted through our door and entered the house shouting and asked for me. I was really surprised that I wasn't afraid. I think the soldiers were too. My mother and father were shouting at the soldiers. They took my mother outside because she was shouting so much. A soldier asked for ...he said "Oh you're Fatima" and I said "Yes", and he said, "So you work in an international TV station?" I told him "No, I work with a TV station but not in an international one, just a local one only as writing a few materials for stories about prisoners." I used to go interview families of prisoners and write stories about them for a class I was taking at university.
>
> DEVIN ATALLAH: What drew you to that work?
>
> FATIMA: At the time my father was unemployed because he was sick so I had take to initiative and responsibility to help and earn some money for the family. I was 18-years-old at the time. It wasn't a large amount of money, but I helped write this news and I got some salary. My uncle also used to get arrested a lot, sometimes for six months under administrative detention, and then he would be arrested again so I wanted to know more about prisoners and their lives as well. Basically, I wanted to learn more, and to help the family...

Fatima explains how she became involved with writing stories about prisoners, which she did within an enactment of decolonial solidarity, and as a necessity. Her father's illness, the family's lack of financially resources, and her community's experiences with incarceration opened up a path for her to begin studying and working to help her family, which simultaneously inspired the Israeli occupation authorities to respond and to discipline her, precipitating her arrest. Part of decolonial solidarity includes not only building community across radical relations, ideologies, and re-envisionings, but also includes making choices about how to articulate struggles and how and when to foster alliances "in terms of both their possibilities and their limits" (de Sousa Santos, 2018, p. 32). Fatima made brave choices about how to articulate the struggle of prisoners and expose the colonial violence to show solidarity, and in the end, it resulted in her freedoms being limited and her becoming a target herself. As she continued her interview, Fatima explained numerous ways where she entered a platform of brutal colonial and gendered violence when she was arrested, inspiring her own strength and enactments of profound rebellion against the soldiers, even when her life was at the whim of others; literally, tied up, blindfolded, and defenseless in a jeep resting against the barrel of the gun, yet strong and dignified in her own right. Fatima explained:

> FATIMA: Anyway, back to my arrest story.... So the soldiers told me, "Go get dressed, we'll take you."....my mom made me wear three trousers and lots of jackets. It was snowing, so she wanted me to be warm and prepared. Three socks, and a heavy jacket and my

outside dress and my veil. And I covered my face. But when I went outside the soldiers took my veil off.... I told them that I wanted to say good-bye to my mother and they refused at first but I insisted and so I said good-bye and kissed my little sister who was very attached to me. When I went into the jeep...my mother and others were fighting with the soldiers, a soldier was trying to handcuff me and when I saw that, I started screaming and telling them to leave my family alone... As we were driving off, the door of the jeep was still open. When we drove away, there were seven male soldiers and one female soldier, I starting praying and mom and dad told me later they were praying for me too. The soldier blindfolded me but I could wink and the wrap around my eyes moved a bit so I could see the street without the soldiers noticing. As we left the camp and travelled to the prison I could see the mountain, the snow, some really beautiful scenes outside the blindfold, and I moved a little to get a better view. The soldier was afraid that I was moving because maybe I wanted to try to escape and the soldier stuck his gun into my side, and asked if I was trying to escape, I said, "No, I'm only moving because when your gun touches my side it tickles." And then one of the soldiers asked me, in a mocking way, "Aren't you afraid of the gun?" And I said, "No. I'll die but only once."

Once at the Israeli detention center, Fatima explains how soldiers used threats of sexual abuse to make Fatima comply with dehumanising demands and violations to her rights. She describes how there were children (a newborn and a one-and-a-half-year-old) living in the same prison cell block as her, as well as women spanning multiple generations. They worked to create a homespace together marked by a love so radical, that the children being raised behind bars could feel the wings of freedom from the nurturance that the older prisoners practiced.

Later on in the interview, Fatima shares a love story, so unruly and powerful, she breaks rules and bars, even shattering the Wall—loving like the Wall isn't even there. By stepping into her power with decolonial love, Fatima strips the colonial degrees of incarceration of much of their power. This is a love story about how Fatima found her fiancé in prison. The healing power of creating and sharing love, connecting with rituals and cultural values of romance, weddings, and community celebrations of the joy and promise of loving into a future knowingly marked by oppression and precarity, she pushes limits, engages in a *moving out of place*, and crosses into the ungovernable 'no-place' of decolonial being. Fatima shares:

FATIMA: Once, waiting outside of the court in line with other prisoners, we saw a 13 or 14-year-old boy who soldiers were swearing at, and when the boy talked back to the soldiers to defend himself, about 20 soldiers jumped on him and started beating him with sticks. I saw this before I was about to enter in the courtroom! Just waiting in line! They keep you in this place before going into the courtroom so I was watching and I got really angry because this youth was full of blood after they beat them. I shouted at the soldiers, shaming them for beating up a child. When I was in the court, it was forbidden to talk to the other prisoners but some of them had sisters who were prisoners as well so I wanted to tell them how their sisters were doing. But this wasn't allowed. But I would still shout at the soldiers, and talk to the other prisoners and give them updates about their loved ones. My fiancé was actually one of these male prisoners who I spoke to and gave him updates about his sister who was detained with me, and he fell in love with me because he saw that I was strong. And so we were engaged while in prison. My fiancé got the contact information of my family from my lawyer. He talked to his family about me, and said "I like this girl and I want to ask for her hand in marriage." So his family came to visit my family while we were both in prison. I didn't even remember his face but it was something I wanted to do because he and the other prisoners would also be happy for somebody there, and that would bring

> joy into the prison, so I convinced myself that I wanted to do something that would be a gift to prisoners. So I said "Yes," and there was a party here at my house when I was released, and his friends had a party in prison for him. Then, he was released for 12 days and we got to be together, but then he was re-arrested. We will marry one day, and god willing, I'll invite you to the wedding!

Fatima's engagement story is a powerful example of how, from camp to prison, prison to camp and beyond, she fosters homespaces creatively across differences, re-constructed collectively and persistently in enactments of love so deeply rooted in her humanity—where not even the force of a brutal military occupation, with all its degrees of incarceration, can silence or stop her creation of a caring community and an interdependent and defiant relationality. Fatima shares how the beauty in what she loves is shared, passed down, passed across, passed beyond, and practiced in solidarity. This love almost appears to transcend the totality that tries so hard to control it. Fatima organises in and outside of the prison walls in her enactments of 'betweenness', which involves crossings, and coming together—waging love in ways that transform counter-spaces into radical homespaces. Fatima's story, to me, is evidence of a decolonial solidarity that is the courageous creation of a community-in-struggle-and-in-love. When in community and in love and in struggle in this way, passageways into new moralities and relationalities are forged and interlocking oppressions are repelled and transformed. In doing so, the powers of domination, however, are not the only 'system of compartments' that we transform—our fragmented bodies, selves, families, and communities are also being forever changed.

Ending Reflections and Questions

In all three of the research coalitions with whom I am working, from Palestine to Boston to Chile, the ways in which colonial/structural violences work to institutionalise/normalise states of crisis and precarity demonstrate the importance of our working/researching together in ways that rehumanise, refocus on relational ethics, and actively build communities of care and struggle. I have provided examples of how decolonial solidarity involves working towards knowing and loving one another in ways that (re)imagine colonially-structured realities and contend with critical questions, born from knowledges of people in struggle on the frontlines against racism, patriarchy, classism and the intersecting consequences of accumulating legacies of colonialism. Attempting at a way of knowing, writing, loving, and being in community that honors and uplifts these alternative knowledges and embodied understandings requires our collaboration across oppositional responses to the intersecting and transnational systems of domination that are endemic to settler-colonial societies.

In this chapter, I have shared insights, stories, and questions about decolonial solidarity as intrapersonal and interpersonal counterattacks—the embracing of changing notions of love from decolonial feminist wisdoms and leaderships—requiring that "differences are never seen in dichotomous terms…multiplicity is never reduced…questions proliferate at this time and the answers are difficult" (Lugones, 2010, p. 755). I have followed decolonial feminists and resisting frontliners like my Amo, Ms. Moore, Sayen, Machi Ankatu, Mohammad, Naila, and Fatima, into courageous questions such as:

- How may we know and love each other in ways that shatter dichotomies and dismantle the colonial systems of domination that are embedded in our visions and relations? What does this look like in our practices and in practical, everyday lives and decisions?
- How are we building and being 'creatively in coalition'? What possibilities are born from our coalitions? What are the limits, and limitations, of our coalitions? How are our psychology disciplines, institutional affiliations, and research norms and practices shaping these limits and limitations?
- What are emergences of new relationalities in our communities and in our psychology praxes forged by our radical love and decolonial solidarity?

These questions are connected to a constant questing and re-imagining, about how we ourselves are changing in our homeplaces, in our workplaces, in our relationships, in our questions and concerns, in our strategies and failures, and in our bodies.

As clinical and community psychologists and activists concerned with decolonial healing and justice, I wonder: *How can we follow in the footsteps and leaderships of frontliners contesting colonial oppressions and build our communities of resistance with new moralities and relationalities—not only as agents who respond and act, who theorise and strategise, who reflect and write, who work and network—but also as beautiful living creatures and intergenerational trees for whom cultivating community and cultivating change is a birthright.* As Ms. Moore reminds us in our CAT meetings at the middle school here in Boston where we gather and feast together on freedom foods at least once a month as part of our CAT process—the real struggle is to let the beauty in what we love guide our digestions and our negotiations, to guide how we grow and cultivate our questions, but most importantly, how we live, in community, and into the answers.

References

Agostini, C. A., Brown, P. H., & Roman, A. C. (2010). Poverty and inequality among ethnic groups in Chile. *World Development, 38*(7), 1036–1046. https://doi.org/10.1016/j.worlddev.2009.12.008

Anzaldúa, G. (1987). *Borderlands/la frontera: The new mestiza* (2nd ed.). Aunt Lute Books.

Atallah, D. G. (2016). Toward a decolonial turn in resilience thinking in multifaceted disasters: Example of the Mapuche from southern Chile on the frontlines and faultlines. *International Journal of Disaster Risk Reduction, 19*, 92–100. https://doi.org/10.1016/j.ijdrr.2016.08.027

Atallah, D. G. (2017a). A community-based qualitative study of intergenerational resilience with Palestinian refugee families facing structural violence and historical trauma. *Transcultural Psychiatry, 54*(3), 357–383. https://doi.org/10.1177/1363461517706287

Atallah, D. G. (2017b). Transnational meetings & meals of resistance: Palestinian-Chileans show solidarity with the Mapuche, an indigenous nation of South America in active struggle for self-determination. *Mondoweiss*. Retrieved from https://mondoweiss.net/2017/02/transnational-palestinian-determination/

Atallah, D. G., Bacigalupe, G., & Repetto, P. (2019). Centering at the margins: Critical community resilience praxis. *Journal of Humanistic Psychology*. https://doi.org/10.1177/0022167818825305

Atallah, D. G., Contreras Painemal, C., Albornoz, L., Salgado, F., & Pilquil Lizama, E. (2018a). Engaging critical community resilience praxis: A qualitative study with Mapuche communities in Chile facing structural racism and disasters. *Journal of Community Psychology, 46*, 575–597. https://doi.org/10.1002/jcop.21960

Atallah, D. G., Shapiro, E. R., Al-Azraq, N., Qaisi, Y., & Suyemoto, K. L. (2018b). Decolonising qualitative research through transformative community engagement: Critical investigation of resilience with Palestinian refugees in the West Bank. *Qualitative Research in Psychology, 15*(4), 489–519. https://doi.org/10.1080/14780887.2017.1416805

Bell, D. (2018). A pedagogical response to decoloniality: Decolonial atmospheres and rising subjectivity. *American Journal of Community Psychology, 62*, 250–260. https://doi.org/10.1002/ajcp.12292

Bell, D., Canham, H., Dutta, U., & Fernandez, J. S. (2020). Retrospective autoethnographies: A call for decolonial imaginings for the new university. *Qualitative Inquiry, 26*(7), 849–859. https://doi.org/10.1177/1077800419857743

Brown, A. M. (2017). *Emergent strategy: Shaping change, changing worlds*. AK Press.

Charmaz, K. (2006). *Constructing grounded theory: A practical guide through qualitative analysis*. Sage.

Clarke, A. E. (2005). *Situational analysis: Grounded theory after the postmodern turn*. Sage.

Correa, M., & Mella, E. (2010). *Las razones del illkun/enojo: Memoria, despojo y criminalizacion en el territorio mapuche de Malleco*. LOM Ediciones.

Davis, F. (2019). *Race and restorative justice: Black lives, healing, and US social transformation*. Good Books.

de Sousa Santos, B. (2018). *The end of the cognitive empire: The coming of age of epistemologies of the south*. Duke University Press.

Dutta, U. (2018). Decolonising "community" in community psychology. *American Journal of Community Psychology, 62*, 272–282. https://doi.org/10.1002/ajcp.12282

Fanon, F. (1963). *The wretched of the earth*. Grove Press.

Freire, P. (1970). *Pedagogy of the oppressed*. Continuum.

Ginwright, S. (2016). *Hope and healing in urban education: How urban activists and teachers are reclaiming matters of the heart*. Jessica Kingsley Publishers.

hooks, B. (1990). *Yearning: Race, gender, and cultural politics*. South End Press.

Instituto Nacional de Estadísticas. (2012). Resultados censo 2012. Retrieved from https://www.censo.cl/

Israeli Ministry of Foreign Affairs (n.d.). *The war of independence—Introduction*. Retrieved from https://www.mfa.gov.il

Lugones, M. (2010). Toward a decolonial feminism. *Hypatia, 25*(4), 742–759.

Malcolm, X. (1964). *The Malcolm X collection: Papers 1948-1965*. Schomburg Center for Research in Black Culture, Manuscripts, Archives and Rare Books Division, the New York Public Library.

Maldonado-Torres, N. (2007). On the coloniality of being. *Cultural Studies, 21*(2), 240–270.

Maldonado-Torres, N. (2016). *Outline of ten theses on coloniality and decoloniality*. Retrieved from http://frantzfanonfoundation-fondationfrantzfanon.com/article2360.html

Moraga, C., & Anzaldúa, G. (1983). *This bridge called my back: Writings by radical women of color*. Women of Color Press.

Obach, A. (2016). Vulnerabilidad social y salud en pueblos indígenas [Social vulnerability and health of indigenous peoples]. In B. Cabieses, M. Bernales, A. Obach, & V. Pedrero (Eds.), *Vulnerabilidad social y su efecto en salud en Chile: Desde la comprensión del fenómeno hacia la implementación de soluciones* (pp. 349–366). Universidad del Desarrollo.

Pappe, I. (2006). *The ethnic cleansing of Palestine*. Oneworld Publications.
Said, E. (1993). *Culture and imperialism*. Vintage Books.
Sandoval, C. (2000). *Methodology of the oppressed*. University of Minnesota Press.
Shalhoub-Kevorkian, N. (2009). *Militarisation and violence against women in conflict zones in the Middle East: A Palestinian case-study*. Cambridge University Press.
Tuck, E., & Yang, W. K. (2012). Decolonisation is not a metaphor. *Decolonization: Indigeneity, Education & Society, 1*(1), 1–40.
Wynter, S., & McKittrick, K. (2015). Unparalleled catastrophe for our species. In K. McKittrick (Ed.), *Sylvia Wynter: On being human as praxis* (pp. 9–89). Duke University Press.

Accompanying Aboriginal Communities Through Arts and Cultural Practice: Decolonial Enactments of Place-Based Community Research

Christopher C. Sonn, Amy F. Quayle, and Pilar Kasat

For community psychology to contribute to the decolonial agenda the discipline needs to expand its ways of knowing and doing. Many critical authors have called for epistemic justice, ethical engagement with differently positioned communities, and the development of critical and reflexive methodologies that can contribute to empowerment and social justice (e.g, Bretherton & Law, 2015; Fine, 2018; Seedat et al., 2017). Along with many others, we have heeded these calls, critically reflecting on the discipline of community psychology, and seeking to create ways of working in and across our own gendered, raced, classed communities that can elucidate and disrupt mechanisms of oppression, and promote empowerment and social justice through community-based research and practice.

This chapter outlines critical writing that has framed our research and praxis with Indigenous people in Australia. The chapter then describes a body of work with the Community Arts Network (CAN) in Western Australia, which spans over a decade, and that has focussed on supporting the self-determination and empowerment of Indigenous Australians. Australia's Indigenous people are Aboriginal and Torres Strait Islander peoples and make up 3.3 percent of the Australian population (Australian Bureau of Statistics, 2018). Arts and cultural practices are central to the process of participation and the collaborative praxis of CAN. Its program of work consists of various long-term projects that utilise diverse modalities of arts to promote individual and community empowerment. Through our ongoing involvement

C. C. Sonn (✉)
Victoria University, Melbourne, VIC, Australia

University of the Witwatersrand, Johannesburg, South Africa
e-mail: Christopher.Sonn@vu.edu.au

A. F. Quayle
Victoria University, Melbourne, VIC, Australia

P. Kasat
Curtin University, Perth, WA, Australia

S. Kessi et al. (eds.), *Decolonial Enactments in Community Psychology*, Community Psychology, https://doi.org/10.1007/978-3-030-75201-9_6

over the years, we have developed our understanding of community arts and cultural development (CACD), as this practice is known in Australia, as well as the roles that we can play by engaging with organisations and communities working towards empowerment and social justice for minoritised and marginalised groups.

Two of the authors, Christopher Sonn, and Pilar Kasat, both immigrants to Australia, identify as people of colour, and our involvement as researchers and practitioners with CAN started in the early 2000s and was ongoing until 2016. The third author, Amy Quayle, has been involved with the work of CAN since 2010 as part of her research training in community psychology and her doctoral studies. Amy is a white Australian, who through her personal and theoretical engagement in this area has become deeply aware of her position of power and privilege. Each of us is non-Indigenous and outsiders to the Noongar community whom the projects described in this chapter involved. Since our initial engagement with the agency, we have worked in various roles, including board membership, CAN's managing director, project evaluator, critical partner, and as "participant conceptualisers" (Elias, 1994) in a collaborative inquiry as a form of psychosocial accompaniment (Watkins, 2015). As noted by Watkins (2015), psychosocial accompaniment involves practices that aim "through dialogue to build mutual respect and understanding, promote effective solidarity, and contribute to the empowerment of those marginalized" (p. 324).

Critical Community Psychology and Community Arts and Cultural Development: Theoretical, Conceptual and Methodological Resources

Decolonial and Liberation Psychology

Various interdisciplinary areas of scholarship have been important in our work for generating approaches to community research and action that can contribute to the liberation and promotion of wellbeing of different sectors of society that carry the burden of histories of oppression and forms of structural violence (Burton & Kagan, 2009; Dutta et al., 2016; Maldonado-Torres, 2017; Montero, 2009; Montero et al., 2017; Teo, 2015a). Although not a new critique, many scholars have written about some of the problematic assumptions, such as the mechanistic worldview, assumptions of universality, and individualism, which underpin knowledge production in psychology. Critical scholars have argued that individualistic, objectivist, decontextualised, and ahistorical analysis amounts to epistemic violence (Gergen & Gergen, 2010, 2011; Montero, 2009; Rappaport, 2000; Teo, 2015b; Weis & Fine, 2012) calling for scrutiny of how Western scientific modes of knowing, doing and being have been implicated in colonialism and imperialism (Dutta, 2016; Dudgeon & Walker, 2015; Reyes Cruz & Sonn, 2015; Smith, 1999).

Adams et al. (2015) have called for decolonising psychological science pointing out that through various processes psychology has abnormalised the worldviews of most of the world. Adams and Estrada-Villalta (2017) argued that the writing developed in areas such as liberation psychology and efforts aimed at indigenising psychology are important avenues for recentering psychology's capacity to contribute to decolonisation. Such approaches contribute to the normalisation of ways of knowing, being and doing of groups who have been subject to colonising discourse and practice. In Latin America, Martín-Baró (1994) advocated that psychology should develop a new praxis. From his perspective, a liberatory praxis would recognise peoples' virtues, be based in the lived realities of the oppressed, engage in the recovery of historical memory and seek to de-ideologise taken for granted social realities. He considered such tasks central to the process of reconstructing identities and communities. For Watkins and Shulman (2008), liberation psychologies,

> gather together resources to help people understand possibilities for multiple layers of interpretation through which the world that has been imposed on them can be understood and reorganized. With embodied practices, people are able to evolve ways to resymbolize their worlds through creative conversation and activity in the arts. They are able to commit to transformative efforts through social and political action. (p. 5)

Reyes Cruz and Sonn (2015) advocated a decolonising standpoint in community psychology, which seeks to disrupt:

> essentialist understandings of cultural matters that have served historically to marginalize others. This standpoint brings into clearer view ways in which power/privilege/oppression are reproduced and contested through racialized and ethnicized practices and discourses; that is, how social inequality is maintained and challenged through culture (p. 128).

This standpoint is an ethical, epistemological and political approach and resonates with the liberation paradigm (Montero, 2009), within which liberation psychology sits, as well as other developments described as Southern Theorising (e.g., Connell, 2007; Santos, 2014). Enacting a decolonial standpoint requires the critical engagement with culture to examine how symbolic violence is enacted through discourses of the other and reclaiming ways of knowing, being and doing through various practices including counter- storytelling. As noted by Watkins (2015), an ethics of accompaniment requires a "reorientation of the subjectivity, interpersonal practices, and critical understanding of the accompanier so that (s)he can stand alongside others..." (p. 324).

Culture, Narrative and Stories

People evolve in relationship with local environments of discourse, culture, and custom. As cultural beings, we actively produce culture at the same time as being products of culture. Culture is understood as "the fabric of signs and symbols, language and image, customs and ceremonies, habitations, institutions, and much more that characterize and enable a specific human community to form and sustain itself" (Goldbard, 2013, p. 11). In Boal's critical participatory theatre work, interventions in and through culture are central to social change. For him,

> Culture is not the superfluous, it is not decoration. It is the doing. The manner of doing. And our ways of doing the same thing are not the same, since we all do not dream the same dream, we are not all of the same heart, we are not born in the same land, we do not grow up in the same district, city or state, we do not all eat from the same dish. For this reason, culture is diverse. Culture is everything that exists, truth and lies, everything created by us, in response to our needs, desires, abilities. (Boal, 1998, p. 179)

Narrative is central to cultural approaches to psychology (Bhatia, 2011). Narrative is "a natural vehicle for folk psychology. It deals with the stuff of human action and human intentionality. It mediates between the canonical world of culture and the more idiosyncratic world of beliefs, desires, and hopes" (Bruner, 1990, p. 52). Narratives are conveyed through stories that are central to identity and community-making processes. Stories are individual, social and ideological, and because they are produced in social contexts, they often reflect narratives that are socially and culturally available (Sonn et al., 2013). However, individuals and communities do not always have equal access to resources, such as technology or mass media, for story and narrative production and circulation. The result in some contexts, is that those in less powerful positions may be excluded or marginalised, or their stories silenced, concealed or devalued, while dominant group's stories may work to reinforce racism and existing power arrangements (Bell, 2010). As noted by Watkins & Shulman (2008):

> ...subjectivity is layered with expected scripts and official histories, as well as resistant interpretations; threads of dream and fantasy; sediments of forgotten music, ritual, and story; and bits and pieces of iconic memory outside of conventional narratives. Within buried layers of symbolic meaning, there are resources for lives lived otherwise, a compost where energy is building, where seeds of hope and transformation may take root. (p. 233)

Therefore, decolonial approaches have centered the importance of stories and storytelling as sites for examining oppression but also as sites of resistance, reclamation and transformation. Storytelling is a methodology, which is often located within participatory action research, indigenous and critical approaches in the social science (Kovach, 2009; Zavala, 2016). In these approaches, the researcher-researched binary is disrupted as relationships are based on democratic principles with participants typically repositioned as co-inquirers. In essence, there is a dialogical orientation based in an epistemology of relatedness: "...everything happens within social relationships, ..." and a "dialogue implies another person, an Other, who needs to be accepted not as a subject, but as a social actor, who must be respected, who constructs knowledge, who has a history" (Montero & Sonn, 2009, p. 2). Smith & Sparkes (2006) note that the invitation to tell stories

> ... creates an imperative in which the task is to not only tell one's story, but also to assume responsibility to listen carefully and attempt to grasp what is being expressed and said by others who have contrasting tales. Through this dialogic relation, it is also hoped a fusion, rather than exclusion, of horizons is accomplished. (p. 186)

In critical approaches, there is a concern with elevating silenced, excluded narratives and challenging damaging narratives (see Hooks, 1990)—that is, the goal is counter storytelling for social change.

Community Arts and Cultural Development for Counter Storytelling and Healing

Worldwide, community arts and cultural development (CACD) has been notably inspired by the libertarian ideas of the Brazilian educator Paulo Freire and also influenced by activists, philosophers and writers such as Franz Fanon, and theatre practitioner Augusto Boal (Kasat, 2014; see Montero et al., 2017).

In Australia, CACD is defined by the Australia Council for the Arts (n.d.) as:

> ... a community-based arts practice ... There are many variations of how community arts and cultural development works are made, developed and shared, and as such, there is no one model. What is at the core of this practice, however, is the collaboration between professional artists and communities to create art (para 1, line 1–3)

Community arts is a form of participatory cultural practice in which art is produced and used by local people within their communities as an instrument for social change (Madyaningrum & Sonn, 2011). Goldbard (2006) noted that a range of tools are used in CACD "from aspects of traditional visual- and performing-arts practice to oral-history approaches…to use of high-tech communications media, to elements of activism and community organizing…" (p. 21). Community arts are about claiming ones' right to tell existing stories about self and community and to create new narratives that help with the recovery of historical memory and of:

> … reclaiming a history and filling in a past that helps to make a whole person. Restorations of community histories are important for forming and informing the human subject and developing community solidarity. Not only exercises in nostalgia or feeling good (Thomas & Rappaport, 1996, p. 330).

While CACD is often presented as practice, it also offers a mode of collective inquiry and allows for creative ways of communicating, disseminating, and translating knowledge, which ultimately can help democratise research and knowledge production articulated in the argument for performative social science (Gergen & Gergen, 2010, 2011) and decolonial approaches (Maldonado-Torres, 2017; Seedat et al., 2017).

CACD has been a promising methodology for CAN in its efforts to promote arts-based work and to enact its goals of supporting Aboriginal, Noongar communities in Western Australia. In the next section, we describe our work with CAN to illustrate the different ways in which arts and cultural practice have been woven into inquiry and action over time. Our work has involved evaluation and documenting the practice of CACD with Aboriginal people across the Wheatbelt. Importantly, a decolonising standpoint required that we interrogate and critique the practice of CAN given the possibilities for recolonising space and disempowering communities despite good intentions (see Sonn & Quayle, 2013).

Enacting a Decolonial Agenda: Building Relationships, Making Art, and Reigniting Community

Our focus is on describing the arts practices used, and the insights developed about the ways in which arts and cultural practice can contribute to the decolonial agenda. CAN is a leading CACD organisation that seeks to partner with communities to produce creative projects using arts and deliver training opportunities and mentor artists, cultural workers and government professionals. Further, it manages state and federal government funds aimed at providing opportunities for communities to develop locally owned arts community projects. While each of the projects we outline here can be examined within its own parameters, they can all be understood as part of an overall program of work that has developed in an iterative and generative fashion. Each project was developed in response to community wishes and aspirations to support psychosocial wellbeing of individuals and communities and scaled according to the availability and access to financial resources to undertake the work. It is important to highlight that CAN made a very deliberate decision following extensive feasibility work to invest in and respond to the cultural needs of the Aboriginal, Noongar communities in the Wheatbelt region of Western Australia. Since then, the projects have been developed in response to community needs and anchored in growing partnerships based on relational ethics generated through various cycles of dialogue, reflection, and relationship building inherent in community-based arts practice.

Context: Aboriginal Communities of the Wheatbelt

Australia's Indigenous people are Aboriginal and Torres Strait Islander peoples and make up 3.3% of the Australian population (Australian Bureau of Statistics, 2018). Noongar people are the traditional custodians of the south west corner of Western Australia. They are one of the largest Aboriginal cultural blocks in Australia, "made up of 14 different language groups: *Amangu, Yued/Yuat, Whadjuk/Wajuk, Binjareb/Pinjarup, Wardandi, Balardong/Ballardong, Nyakinyaki, Wilman, Ganeang, Bibulmun/Piblemen, Mineng, Goreng and Wudjari and Njunga* (which may be spelt in different ways)" (South West Aboriginal Land and Sea Council, 2017a, About Noongar, para. 2). Noongar country (i.e., *Noongar boodja*) is made of the entire southwestern part of Western Australia. This area encompasses much of the Wheatbelt region of Western Australia, the state capital Perth, the regional cities of Bunbury and Albany, and hundreds of kilometres of coastline. There are approximately 30,000 Noongar people living in the south west of Western Australia (Government of Western Australia, 2017).

Moort (family), *boodja* (land) and *Kaartdijin* (knowledge) are central to Noongar culture and identity (Collard, 2007; Collard et al., 2004). As noted by the South West Aboriginal Land and Sea Council (SWALSC),

> To be Noongar is to belong; it is to have connection to our boodja-country, our moort-family, and to kaartdijin-knowledge... It is to have pride and to survive. We need to celebrate it-the journey. Ask a Noongar person what our identity means to us and invariably we will talk about the stories. The stories are part of the kaartdijin passed down from the Elders and moort (family). They tell of how to survive in the bush. They are campfire stories of the stars and the seasons (South West Aboriginal Land and Sea Council, 2017b, Identity, para. 2–3).

The entrenched disadvantage experienced by Aboriginal people in Australia more broadly and Indigenous people in other settler colonial societies (Bell, 2014; Czyzewski, 2011), is evident in these Wheatbelt towns as microcosms of broader Australian society. In the southern Wheatbelt town of Narrogin, the Noongar community has experienced high levels of individual and community distress as reflected in the issues of suicide, family feuding, and racism (Commonwealth of Australia. Senate, 2009; McGuire et al., 2013). Such issues have been understood in the context of historical trauma and ongoing structural violence (Kirmayer et al., 2014).

What follows are summaries of projects developed and delivered over time, all enacting CACD practice with and for Noongar people. The projects demonstrate the importance of a dialogical orientation in the process of ethical and respectful engagement; the generative potential of arts-based methods in amplifying voices of marginalised groups to contribute to healing, cultural renewal and reclamation, and the importance of understanding the challenges individuals and communities face in the context of an ongoing history of colonialism and coloniality.

Laying Foundations Through Consultation and Community Workshops

CAN established its first arts and cultural development program in 2006 in the small rural town of Kellerberrin (which has 10% of Indigenous people while the national average is 3%) located approximately two hundred kilometres from the city of Perth in Western Australia. This decision to establish the arts program followed two years of consultation and engagement with the local Noongar community, local authorities, as well as a demonstrable lack of cultural infrastructure in the town. The relationships with the Noongar people were built though face- to-face conversations, ongoing dialogues, and a series of art workshops as well as additional local meetings with the wider community to ascertain the relevance and the need to establish an art and cultural development presence in the town. Through the consultation, it became clear that the town was an important cultural landmark for the Noongar people. Since 2002, the local Aboriginal corporation had been hosting a biennial cultural Indigenous festival with limited support from the local government but very well patronised by Noongar people across the state, some of whom travelled hundreds of kilometres to attend the event. At that time in 2004/05, it was established that the town had one of the very few Noongar Elders recognised as fluent language speakers (of approx. 200) in the Noongar nation. However, almost no resources from the State government have ever been allocated to the development of Noongar arts and cultural activity. Hence the local people and the Elders felt the need to promote culturally relevant activities to foster cultural transmission to the younger generations and as a way to encourage non- Indigenous people to engage with Aboriginal culture, in a rural community where histories of racism and land dispossession has contributed to highly complex and protracted antagonistic race relations.

This background propelled CAN to respond by facilitating resources, both human and material, that could enable a responsive and explicitly empowering agenda by making visible Noongar people, their stories and viewpoints through art making. At the height of the activities on Noongar country, 40% of all CAN employees were Aboriginal and some in very senior positions. This strategy took some time and began by engaging a Noongar artist to run a visual arts workshop. The aims of the workshops were: to get people together, ascertain the desire to participate in arts and cultural development, and to consult with the Noongar community. This workshop was very successful and well received and, because of this response CAN embarked on a series of art workshops aimed at understanding the processes of empowerment that were occurring as a product of these arts and cultural activities.

A series of longer-term photography workshops were delivered with young women after they showed interest in a one-off photography course run by CAN. These workshops were facilitated by two women, a non-Indigenous professional photographer and a CAN cultural worker, a woman of colour. During the workshops the women learnt light, colour and patterns in photography. Older women expressed interest in silk and fabric screen-printing, and thus textiles workshops followed. The attendance of these workshops varied but all the participants were Aboriginal women and children; their ages ranged from very young to elderly.

The workshops were important activity settings, and subsequent research revealed that people discussed their engagement in those workshops as empowering because they *acquired new skills*, *had fun*, and *felt heard* when they told their stories, and they *learned about social issue*s and how they could be addressed (see Green & Sonn, 2008). Enjoyment and pleasure that results from participating in the arts is often not emphasised because the focus is on addressing social ills (Putland, 2008). These projects delivered nuanced outcomes related to the meanings and processes of empowerment for project participants. Inquiry alongside these projects also highlighted important lessons for the organisation in terms of who it employs and what resources they can leverage to invest in CACD activities for Indigenous people. Green and Sonn (2008) reported the challenges for staff that are produced because of uncritical practice and limited reflection on white subject positions in the context of Indigenous and non-indigenous working relationships. They advised that staff needed to be supported in critical reflective practice as a move towards cultural competence and antiracism. This necessarily must involve the critical engagement with the writing of indigenous scholars who have highlighted that essentialist conceptions of culture and the ideology of race had played a role in shaping Indigenous and non-indigenous relationships (e.g., Smith, 1999). Noongar people reported the ways racism and entrenched patterns of racialised interactions continue to shape everyday social relations. The level of participation in the arts and cultural activities demonstrated that the community supported this work. Therefore, armed with the outcomes and the community mandate to do more, CAN secured significant government resources both from the arts and culture portfolio and from Indigenous affairs to scale up these workshops, and include three other neighbouring towns, leading to the Voices of the Wheatbelt project.

Voices of the Wheatbelt: Making Place Through Photography

CAN was able to successfully compete for Federal Government funding to upscale the photography workshops across more towns. The voices of the Wheatbelt project built on the foundations laid with the earlier work. The Voices of the Wheatbelt project was especially designed to privilege Noongar voices and experiences of their local communities and it focussed on the themes of sense of place and belonging. CAN developed a similar program of activities based on learning photography and exploring themes of place and belonging. Photography as medium had been tried and tested in the previous projects, and the young people responded well to it because they expressed interest in digital photography. The program was delivered by a non-Indigenous photographer in schools to over 100 children during 16 weeks within five schools in four Wheatbelt towns. Importantly, two non-Aboriginal and two Aboriginal women artists were recruited to facilitate additional workshops and from this, an interest in song writing was identified in the workshops. The interest in song writing was not surprising given the popularity of Karaoke as an activity amongst Noongar people and across generations. Song-writing and Karaoke are more than just popular activities. Song writing entails active reflection and naming the world and our experiences in it and ways in which we are becoming. It is a means to enhance young people's capacity to be reflective, express their feelings and thoughts, and critically read their social and cultural environment (Sonn et al., 2017).

Working across neighbouring towns was significant at the time because providing opportunities for Noongar people to come together across the Wheatbelt brought family groups together. Movement across the communities is limited because of big distances between places, many people do not have cars, and there is limited public transport in these towns. This practical community linking, and networking opportunities provided by CAN in the form of buses, refreshments and accommodation when necessary, were also happening against a backdrop of the beginnings of the biggest Native Title negotiations in the country.[1] This meant opportunities for people

[1] Native Title negotiations have concluded with a Settlement (almost finalised) with the State government of Western Australia, which is said to resolve Noongar native title claims of the South West region affecting more than 30,000 Noongar people with a package containing a series of benefits including: Recognition through an Act of Parliament, a perpetual trust, access to land, joint management of heritage areas and the developmemt of a Noongar Cultural Centre (South West Aboriginal Land and Sea Council, n.d.).

to be involved in discussing highly political and complex sovereignty issues. CAN played a linking and mediating role, and through its activities supported and activated the networks across the communities in the region.

In a review of the Voices of the Wheatbelt, Sonn et al. (2017) concluded that the arts project entailed the creation of settings for people, who may not typically interact with one another, to come together. Through the various workshops and the subsequent techniques used to explore with participants the meanings of images, the research showed the complexity of life in these towns and the various ways people give meaning to place—through memories and narratives of the past, everyday social relations, and the natural physical environment (Sonn et al., 2017). For Noongar people, this also meant making visible a history of dispossession and exclusion and restoring concealed places of cultural significance.

The workshops across the communities were contact zones (Pratt, 1991) that facilitated people coming together to get to know each other, to share what makes them unique and to discover what they have in common. The photos people took and the process of photo elicitation (Purcell, 2009) and caption writing (see, Fig. 1). were all modalities of meaning making that helped to capture different elements of Wheatbelt life, and invited producers and audiences to explore the community and the landscape through the eyes of those who live, play and work there (Community Arts Network Western Australia, 2014). The Voices of the Wheatbelt project revealed much of the complexity of living in country towns, their histories,

Fig. 1 Writing workshop in Wheatbelt town. Photo taken by Christopher Sonn

relationships with city centres, and the struggle of Indigenous communities on the margins of those places. Significantly, CAN's enactment of CACD resonates with participatory action research in which collaboration, situated engagement, and commitment become central to building trusting relationships (Fine, 2018); in this instance, with Noongar communities across the Wheatbelt in contexts that remain oppressive. It is through this ongoing commitment to genuine collaboration, to being in relation, that the agency was approached to assist with the process of healing following deeply painful events in the Wheatbelt town of Narrogin.

Responding to Social Suffering: Naming Pain and Imagining Hope

A subsequent related project was developed based on an invitation from the Noongar Narrogin community. This project responded to the aftermath of young people committing suicide in one of the towns and issues of family feuding, which became a focus for mainstream media that exacerbated the pain and suffering of communities (Kasat, 2013, 2014). This phenomenon continues to be a vexing issue in Australia and disproportionately impacts upon Indigenous communities. Kasat (2013) described the broader context, the aftermath, and the range of ways in which the non-Aboriginal community (mainstream community) responded to the tragedy.

A key aspiration of the project was to transform stories of pain and suffering into stories of hope (Sonn et al., 2017). In this project, various arts components were developed in consultation with and based in the life stories and experiences of people. These arts components all focused on creating and materialising a new narrative of hope as constituted by the stories of the community. The arts components were carefully considered and facilitated by experienced artists who focused on storytelling techniques to elicit images and metaphors of times when the community 'got along'. This included a set of playing cards, because the community associated good times and togetherness with playing cards. The set of cards captured direct quotations of hope, messages of forgiveness, and affirmation of culture. It also included a *soundscape* composed of people's own voices that depicted the collective community narrative containing the trauma but also their own solutions and answers to it. Finally, it involved the transformation of a community hall decorated with various symbols, including bon fires, which signified 'turning the lights back on' in the town, as per one participant's suggestion, so that Noongar people could come back home.

Acquiring knowledge of local community contexts, developing trust, creating safe spaces for people to tell their own stories and be heard, and using arts as a vehicle to express private pain and hope were central aspects of this project. By establishing personal relationships, following cultural protocols, employing local Noongar people, actively listening and ensuring adequate time was spent in listening to ideas and concerns from Noongar Elders, stories of pain, violence, and loss were heard, as well as allowing the desire for collective hope to emerge (Sonn et al., 2017).

In line with liberation arts, the processes involved in this project included the collection and re-articulation of individual and group stories through creative output, and the production and staging of a collective soundscape to facilitate the creation of a counter-narrative to suicides, family feuding and violence and create a renewed community narrative that fostered hope and healing (Sonn et al., 2017). While this was a significant event for the Noongar community, it is important to point out that there was very little non-indigenous participation in the event, in fact cynicism that such an event would take place without violence was expressed by members of the local police (Kasat, 2013).

Remembering and Counter Storytelling: Memory, Portraits and Legacy

The final project to be discussed in this chapter is the Bush Babies project in Narrogin which extended on this previous work. The Bush Babies project was delivered as part of the Rekindling Stories on Country strategy which aims to create opportunities for intergenerational dialogue and cultural transmission, and celebrate Noongar history, culture, and language. Rekindling Stories on Country emerged in response to the stated needs to Noongar people who said: "Our young people are dying; they need their culture" (Community Arts Network Western Australia, 2014, p. 73). This strategy is the culmination of the work started in 2000 and reflects the key lessons CAN has learnt throughout their engagement with Noongar people about the importance of processes of cultural reclamation and strengthening and creating opportunities and spaces for Noongar voices.

The Bush Babies project used story-telling to record and convey the stories of Noongar Elders from across WA. The Elders are people who were born in the bush on the margins of mainstream society. The project started with a Noongar Elder who wanted to honour the Noongar Bush Babies and midwives who delivered them. Since the original project, there have been several iterations delivered in different towns across the Wheatbelt. The actual process for story making involved different arts practices, but Noongar conceptions of story and country was the foundation. In each of the towns Bush Babies was delivered, it was the Noongar community in these towns who determined the direction of their Bush Babies project. Here, we focus on the Narrogin iteration of the Bush Babies project, which followed on from Narrogin stories. Stemming from a recognition that many Elders in Narrogin had not had the opportunity to have their stories heard, the focus was on documenting the stories of Elders in the community. Importantly, the previous work in Narrogin as part of Narrogin stories as well as another project undertaken by CAN on cultural mapping and community governance had laid the necessary groundwork for the Elders to be ready to share their stories.

Using Storytelling to Record and Create a Community Archive

The Bush Babies project went to Narrogin in 2013, and following Noongar community requests, the focus was on creating opportunities for intergenerational storytelling and cultural transmission. As part of this, Chronicle, a digital storytelling company, facilitated a two-day storytelling workshop at the local High school involving local Elders and Aboriginal and Torres Strait Islander media studies students. Elders shared stories of their lives growing up—often these were stories of removal to a mission as a child or growing up on an Aboriginal reserve on the outskirts of town. Reserves and missions were places that Aboriginal people were forcibly relocated to, with missions typically run by churches and reserves run by government and sometimes churches (Australian Institute for Aboriginal and Torres Strait Islander Studies, 2016). Students were then supported to create short digital stories using photographs and the recorded Bush Baby stories.

CAN recorded additional Bush Baby stories informally often on country. CAN also facilitated workshops with students (Primary school and high school) and family members to explore the Storylines database, which is "an online archive for the State Library's digitised heritage collections relating to Aboriginal history in Western Australia" (State Library of Western Australia, 2016). The facilitation of these workshops came with challenges that needed to be negotiated sensitively and patiently. While the previous project, Narrogin stories had contributed in some ways to a renewed sense of hope, there were ongoing issues of conflict (i.e., feuding) within the community, which were reported as a key concern for Noongar people as part of the National Empowerment Project (McGuire et al., 2013).

A central aspect of the Bush Babies project in Narrogin became the *Honouring the Elders portrait project and exhibitions,* which showcased portraits produced by local artists alongside the recorded Bush Baby stories (see Fig. 2). The portraits project started with a non-Indigenous artist who was interested in turning a photograph of one Noongar Elder into a portrait. This Elder, Nana Purple, was at the time, the oldest known Ballardong woman (see Fig. 3). Sadly, she passed away in 2014 at the age of 100. Local people, the majority being TAFE students, were involved in producing the portraits of several Elders from Narrogin, Kellerberrin and Quairading. Many of the artists were non-Indigenous, while two identified as Aboriginal. One of the local artists involved in the project noted:

> We have a huge age range so we people in their 80s painting people in their 80s…it was a lovely thing to see I really liked that and particularly for people who were parenting in the 50s ah or born in the 1950s in this region or lived here or grew up in that time in this region um the attitudes that, it was hugely racist so for people who came, who have come through that time, who have come to now painting portraits of Aboriginal elders I think that's a major, major achievement. (as cited in Quayle, 2017, p. 104)

The Elders portraits, photographs, hand woven baskets, along with the digital stories produced by students, were exhibited locally in Narrogin and in Perth at the Western Australian Museum. The Honouring our Elders exhibition was also shown across the State as part of Art on the Move (2019), which is "Western Australia's

Fig. 2 Noongar artist Sharyn Egan with her portrait of Noongar Elder Winnie McHenry who inspired CAN's Bush Babies Project. Photo taken by Michelle White, courtesy of CAN

Fig. 3 'Nana Purple' by Graham Smith, inspired by a photograph by Jarrad Seng in The Bush Babies Elders Portrait Exhibition Catalogue 2015: Honoring our Elders born in the bush

only organisation dedicated to touring contemporary visual art exhibitions across the regions, interstate and beyond" (para 1). Rekindling stories on country continues to be part of CAN's work.

Elevating the Elders stories as psychosocial accompaniment. Shulman and Watkins (2001) discussed the task for psychologists in the process of healing in the wake of colonialism as being about learning and practicing "a capacity for deep, respectful, and empathic witnessing in the present to the ancestral narratives and life experience of all those in our communities" (p. 8). For them, this is of importance "in relation to those whom we classify as Other", and for those privileged by a "long history of oppression and discrimination...even if we are not responsible for it" (p. 8). The authors suggested, deep listening, care, an intention to understand and support, and reparative actions are necessary to heal our histories and communities. Importantly this requires a willingness "to break with social codes of silence that have been enforced for generations" (p. 8). Such work has been discussed as psychosocial accompaniment (Watkins, 2015).

Creating opportunities for storytelling through arts and cultural practice provides opportunities for deep, respectful and empathic witnessing. The stories shared through projects like Bush Babies, have been examined to show the various psychosocial strategies individuals and groups mobilise to resist oppression as well as how structural violence becomes embodied (Quayle, 2017). Making use of conversational interviews, the stories collected as part of the project, and archival data, the study sought to expand the platform for Aboriginal storytelling on country. Within this critical narrative inquiry, a number of shared community narratives were identified. The first, 'Life was put on us' showed the various forms of violence and dispossession that have been used historically to maintain relations of domination. The second community narrative, 'It's still going on', highlighted the continuity of structural, cultural, and interpersonal forms of violence that continue to shape and constrain the lifeworlds of Aboriginal people in contemporary Australian society.

The harm caused to individual subjectivities, communities, and transferred and accumulated across generations within families, was also evident in the Elders' stories and was captured within the community narrative, "this is where it all stems from". Three aspects of the psychosocial impact of oppression were identified as part of this narrative. Psychic harm: "It's burnt into our brains, how we was treated, like animals", highlighted the psychological harm caused to individual subjectivities as a result of whiteness/coloniality which has historically denied the humanity of Aboriginal people. Community fragmentation: "No one listens anymore" pointed to the disruptive impact of the dispossession of Aboriginal people of their culture, language, and history, reflected at a community level, and the sense of loss associated with this. Intergenerational impacts: "Our kids are carrying what we been through" showed the intergenerational nature of this psychic harm across and within Aboriginal families.

Within the final shared community narrative ("we survived, we're still here"), Elders' stories pointed to various psychological and behavioral strategies at an individual and interpersonal level used to circumvent experiences of oppression or mitigate its adverse effects. Evident in the Elders' storytelling were symbolic strategies including concealed stories of oppression often involving humour (Duncan, 2014), and resistance stories (Bell, 2010). Importantly, the Elders' stories also highlighted the continuity and survival of people and culture in the face of adversity. Exemplified by stories of country (*boodjar*), kinship (*moort*) and culture (*katitjin*) based in Noongar ontology (Collard et al., 2004), the storytelling showed how Noongar people are actively engaged in processes of cultural maintenance, reclamation, and renewal, or what Vizenor (2008) called survivance. Therefore, the stories shared were not just counter-stories speaking back to whiteness but can be understood as central to constructing identities in the present.

The stories collected and archived so far are resources for Aboriginal communities but are also important counter stories that challenge willful ignorance within the broader Australian community. The process of storytelling about oppressive realities and the various ways in which people resist, subvert, and survive oppression is a central part of the push for epistemic justice (Smith, 1999).

Summary and Conclusion

The authors have, over the years of our partnership with CAN, developed a deeper understanding of the role of arts and cultural practice as central to the tasks of liberation-oriented community psychology (Montero et al., 2017). We have witnessed how CAN has, over time, developed in collaboration with local communities and importantly with Noongar Elders, processes and practices to capture and make visible cultural knowledge that was being dismissed or being lost. The projects have used various creative tools ranging from photography, digital story-telling to portraiture, through a process of psychosocial accompaniment (Watkins, 2015) to create what can be viewed as an archive of materials that responds to the demands for recognition and community concerns about losing culture with its attendant implications for subsequent generations. CACD aligns with participatory research, drawing from a relational ethics and epistemology valuing the lived experiential knowledge of those who are at the sharp end of injustice. Fine (2010) noted that those "marginalized or silenced carry substantial knowledge about the architecture of injustice, in their minds, bodies and souls; in ways that are conscious and floating; individual and collective" (p. 223)

Through CACD and related processes, CAN has taken on this responsibility akin to participatory action research, where each project is carefully crafted in collaboration with the local community, prioritising their needs, and negotiating goals, ways of working and modalities for creativity and representation. Arts and cultural tools mobilised in that process of accompaniment are designed to build capacity, name realities and construct collective memory, and strengthen bonds across generations. This work does not always lend itself to traditional psychological research methodologies, but it has challenged us to place knowledge production with differently positioned groups and across diverse contexts within the broader landscape of knowledge/power dynamics, and to also imagine new ways that we can contest dynamics of exclusion and injustice in pursuit of wellbeing and liberation.

We join with Fine (2010) in her calls for provocative generalisability, which

> …. asserts that researchers and readers, performers and audience, victims, people of privilege and witnesses: we are all positioned in unjust settings. And such an existential truth obligates us to be present, feeling and thinking critically about how unjust distributions of resources and opportunities affect our comfort and discomfort, our dependencies, privileges, joys, our moments of shared pain and potential collective action (p. 220).

CACD can be complicit in perpetuating the status quo, hence it is imperative that CACD practice adopt a decolonial attitude (Maldonado-Torres, 2016) in order contribute to individual and collective liberation in full appreciation of the complex way in which the histories of colonialism and contemporary mechanisms of coloniality continue to shape the lives of people in everyday contexts.

References

Adams, G., Dobles, I., Gómez, L. H., Kurtiş, T., & Molina, L. E. (2015). Decolonizing psychological science: Introduction to the special thematic section. *Journal of Social and Political Psychology, 3*(1), 213–238. https://doi.org/10.5964/jspp.v3i1.564

Adams, G., & Estrada-Villalta, S. (2017). The modernity/coloniality of being: Hegemonic psychology as intercultural relations. *International Journal of Intercultural Relations, 59*, 31–42. https://doi.org/10.1016/j.ijintrel.2017.06.006

Art on the Move. (2019). *Who we are*. Retreived from https://artonthemove.art/who-we-are/

Australia Council for the Arts. (n.d.). *What is community arts and cultural development practice?* Retreived from https://www.australiacouncil.gov.au/artforms/community-arts-and-cultural-development/what-is-community-arts-and-cultural-development-practice/

Australian Bureau of Statistics. (2018). *Estimates of Aboriginal and Torres Strait Islander Australians, June 2016 (cat no. 3238.0.55.001)*. Retreived from http://www.abs.gov.au/ausstats/abs@.nsf/mf/3238.0.55.001

Australian Institute for Aboriginal and Torres Strait Islander Studies. (2016). *Mission and reserve records*. Retreived from https://aiatsis.gov.au/research/finding-your-family/family-history-sources/mission-and-reserve-records

Bell, A. (2014). *Relating indigenous and settler identities: Beyond domination*. Palgrave Macmillan.

Bell, L. A. (2010). *Storytelling for social justice: Connecting narrative and arts in antiracist teaching*. Taylor & Francis.

Bhatia, S. (2011). Narrative inquiry as cultural psychology: Meaning-making in a contested global world. *Narrative Inquiry, 21*(2), 345–352. https://doi.org/10.1075/ni.21.2.13bha

Boal, A. (1998). *Legislative theatre: Using performance to make politics*. Routledge.

Bretherton, D., & Law, S. F. (2015). *Methodologies in peace psychology: Peace research by peaceful means*. Springer.

Bruner, J. (1990). *Acts of meaning*. Harvard University Press.

Burton, M., & Kagan, C. (2009). Towards a really social psychology: Liberation psychology beyond Latin America. In M. Montero & C. Sonn (Eds.), *Psychology of liberation. Theory and applications* (pp. 51–72). Springer.

Collard, L. (2007). Wangkiny Ngulluck Nyungar Nyittiny, Boodjar, Moort and Katitjin: Talking about creation, country, family and knowledge of the Nyungar of south-western Australia. In S. Morgan, T. Mia, & B. Kwaylullina (Eds.), *Speaking from the heart: Stories of life, family and country (pp. 279-297)*. Fremantle Arts Centre Press.

Collard, L., Harben, S., & van den Berg, S. (2004). *Nidja Beeliar Boodjar Noonookurt Nyininy: A Nyunger interpretive history of the use of Boodjar (country) in the vicinity of Murdoch University*. Murdoch University.

Commonwealth of Australia. Senate. (2009). *Select committee on regional and remote Indigenous communities: Narrogin*. Retrieved from http://www.aph.gov.au/Parliamentary_Business/Committees/Senate/Former_Committees/indig/hearings/index

Community Arts Network Western Australia. (2014). *Annual report*. CAN WA. Retrieved from http://www.canwa.com.au/about/annual-reports/

Connell, R. (2007). *Southern theory: The global dynamics of knowledge in social science*. Polity Press.

Czyzewski, K. (2011). Colonialism as a broader social determinant of health. *International Indigenous Policy Journal, 2*(1), 1–14. https://doi.org/10.18584/iipj.2011.2.1.5

Dudgeon, P., & Walker, R. (2015). Decolonising Australian psychology: Discources, strategies, practice. *Journal of Social and Political Psychology, 3*(1), 276–297. https://doi.org/10.5964/jspp.v3i1.126

Duncan, P. (2014). *The role of Aboriginal humour in cultural survival and resistance*. University of Queensland Institutional Repositry.

Dutta, U. (2016). Prioritizing the local in an era of globalization: A proposal for decentering community psychology. *American Journal of Community Psychology, 58*, 329–338. https://doi.org/10.1002/ajcp.12047

Dutta, U., Sonn, C. C., & Lykes, M. B. (2016). Situating and contesting structural violence in community-based research and action. *Community Psychology in Global Perspective, 2*(2), 1–20.

Elias, M. J. (1994). Capturing excellence in applied settings: A participant conceptualiser and praxis explicator role for community psychologists. *American Journal of Community Psychology, 22*(3), 293–318.

Fine, M. (2010). An epilogue of sorts. In J. Cammarota & M. Fine (Eds.), *Revolutionizing education: Youth participatory action research in motion* (pp. 213–234). Routledge.

Fine, M. (2018). *Just research in contentious times: Widening the methodological imagination*. Teachers College.

Gergen, K. J., & Gergen, M. M. (2010). Scanning the landscape of narrative inquiry. *Social and Personality Psychology Compass, 4*(9), 728–735.

Gergen, M. M., & Gergen, K. J. (2011). Performative social science and psychology. *Historical Social Research/Historische Sozialforschung, 36*(4), 291–299. https://doi.org/10.2307/23032295

Goldbard, A. (2006). *New creative community: The art of cultural development*. New Village Press.

Goldbard, A. (2013). *The culture of possibility: Arts, artists & the future*. BookBaby.

Government of Western Australia. (2017). *South West native title settlement*. Retrieved from https://www.dpc.wa.gov.au/lantu/south-west-native-title-settlement/Pages/default.aspx

Green, M. J., & Sonn, C. C. (2008). *Drawing out community empowerment through arts and cultural practice*. Community Arts Network WA.

Hooks, B. (1990). *Yearning: Race, gender, and cultural politics*. South End Press.

Kasat, P. (2013). *Community arts and cultural development: A powerful tool for social transformation*. Murdoch University.

Kasat, P. (2014). Tools for social transformation: Making the personal creative and political. *Griffith Review, 44*, 7–17.

Kirmayer, L. J., Gone, J. P., & Moses, J. (2014). Rethinking historical trauma. *Transcultural Psychiatry, 51*(3), 299–319. https://doi.org/10.1177/1363461514536358

Kovach, M. (2009). *Indigenous methodologies: Characteristics, conversations, and contexts*. University of Toronto Press.

Madyaningrum, M. E., & Sonn, C. C. (2011). Exploring the meaning of participation in a community art project: A case study on the Seeming project. *Journal of Community & Applied Social Psychology, 21*(4), 358–370. https://doi.org/10.1002/casp.1079

Maldonado-Torres, N. (2016). *Outline of ten theses on coloniality and decoloniality*. Foundation Frantz Fanon.

Maldonado-Torres, N. (2017). Frantz Fanon ad the decolonial turn in psychology: From modern/colonial methods to the decolonial attitude. *South African Journal of Psychology, 47*, 432–441.

Martín-Baró, I. (1994). Towards a liberation psychology. In A. Aron & S. Corne (Eds.), *Writings for a liberation psychology: Ignacio Martín-Baró*. Harvard University Press.

McGuire, V., Boladeras, J., Collard, L., Bracknell, C., Butorac. A. (2013). *The national empowerment project: Narrogin*. Retrieved from http://media.wix.com/ugd/396df4_afac6e1f5e174b73bf5f972ff4635c11.pdf

Montero, M. (2009). Methods for liberation: Critical consciousness in action. In M. Montero & C. C. Sonn (Eds.), *Psychology of liberation: Theory and applications*. Springer.

Montero, M., & Sonn, C. C. (2009). About liberation and psychology: An introduction. In M. Montero & C. C. Sonn (Eds.), *Psychology of liberation: Theory and applications*. Springer.

Montero, M., Sonn, C. C., & Burton, M. (2017). Community psychology and liberation psychology: A creative synergy for an ethical and transformative praxis. In M. A. Bond, I. Serrano-García, & C. B. Keys (Eds.), *APA handbook of community psychology: Vol. 1. Theoretical foundations, core concepts, and emerging challenges*. American Psychological Association.

Pratt, M. L. (1991). Arts of the contact zone. *Profession, 91*, 33–40.

Purcell, R. (2009). Images for change: Community development, community arts and photography. *Community Development Journal, 44*(1), 111–122.

Putland, C. (2008). Lost in translation: The question of evidence linking community-based arts and health promotion. *Journal of Health Psychology, 13*, 265–276.

Quayle, A. (2017). *Narrating oppression, psychosocial suffering and survival through the Bush Babies Project*. Victoria University.

Rappaport, J. (2000). Community narratives: Tales of terror and joy. *American Journal of Community Psychology, 28*(1), 1–24.

Reyes Cruz, M., & Sonn, C. C. (2015). (De)colonizing culture in community psychology: Reflections from critical social science. In R. D. Goodman & P. C. Gorski (Eds.), *Decolonizing "multicultural" counseling through social justice*. Springer. https://doi.org/10.1007/978-1-4939-1283-4_10

Santos, B. D. S. (2014). *Epistemologies of the south: Justice against epistemicide*. Routledge.

Seedat, M., Suffla, S., & Christie, D. J. (2017). *Emancipatory and participatory methodologies in peace, critical, and community psychology*. Springer.

Shulman, H. L., & Watkins, M. (2001). Silenced knowings, forgotten springs: Paths to healing in the wake of colonialism. *Radical Psychology, 2*(2), 1–19.

Smith, B., & Sparkes, A. C. (2006). Narrative inquiry in psychology: Exploring the tensions within. *Qualitative Research in Psychology, 3*(3), 169–192.

Smith, L. T. (1999). *Decolonizing methodologies: Research and Indigenous peoples*. Zed Books.

Sonn, C. C., Kasat, P., & Quayle, A. (2017). Creative responses to social suffering: Using community arts and cultural development to foster hope. In M. Seedat, S. Suffla, & D. Christie (Eds.), *Emancipatory and participatory methodologies, peace, critical and community psychology*. Springer.

Sonn, C. C., & Quayle, A. F. (2013). Developing praxis: Mobilising critical race theory in community cultural development. *Journal of Community & Applied Social Psychology, 23*(5), 435–448. https://doi.org/10.1002/casp.2145

Sonn, C. C., Stevens, G., & Duncan, N. (2013). Decolonisation, critical methodologies and why stories matter. In G. Stevens, N. Duncan, & D. Hook (Eds.), *Race, memory and the apartheid archive: Towards a transformative psychosocial praxis*. Palgrave Macmillan.

South West Aboriginal Land and Sea Council. (2017a). *Kaartdijin Noongar: Noongar*. Retrieved from https://www.noongarculture.org.au/noongar/

South West Aboriginal Land and Sea Council. (2017b). *Kaartdijin Noongar: Connection to country*. Retrieved from https://www.noongarculture.org.au/connection-to-country/

South West Aboriginal Land and Sea Council. (n.d.). *Settlement agreement*. Retrieved from http://www.noongar.org.au/settlement-agreement

State Library of Western Australia. (2016). *Storylines*. http://slwa.wa.gov.au/for/indigenous_australians/storylines

Teo, T. (2015a). Critical psychology: A geography of intellectual engagement and resistance. *American Psychologist, 70*(3), 243–254. https://doi.org/10.1037/a0038727

Teo, T. (2015b). Essay on an aesthetics of resistance. In J. Cresswell, A. Haye, A. Larrain, M. Morgan, & G. Sullivan (Eds.), *Dialogue and debate in the making of theoretical psychology*. Captus.
Thomas, R. E., & Rappaport, J. (1996). Art as community narrative: A resource for social change. In M. B. Lykes, A. Banuazizi, R. Liem, & M. Morris (Eds.), *Myths about the powerless: Contesting social inequalities*. Temple University Press.
Vizenor, G. (2008). *Survivance: Narratives of native presence*. University of Nebraska Press.
Watkins, M. (2015). Psychosocial accompaniment. *Journal of Social and Political Psychology, 3*(1), 324–341. https://doi.org/10.5964/jspp.v3i1.103
Watkins, M., & Shulman, H. (2008). *Toward psychologies of liberation*. Palgrave MacMillan.
Weis, L., & Fine, M. (2012). Critical bifocality and circuits of privilege: Expanding critical ethnographic theory and design. *Harvard Educational Review, 82*(2), 173–201.
Zavala, M. (2016). Decolonial methodologies in education. In M. A. Peters (Ed.), *Encyclopedia of educational philosophy and theory*. Springer.

Dialogue and Dialogue Theatre: Processes Toward Decolonial Praxis

Siew Fang Law, Malual Deng, Diego Cifuentes, and Richard Barber

Introduction

Melbourne, Australia, is a city that shines for its youthful, free, cosmopolitan and prosperous public representation. Melbourne is also a major city of a colonized nation, a city with rich Indigenous histories, homes to millions of migrants and refugees, many of whom struggle with public policies and social services. The stories of everyday racism and discrimination are known and owned by those who are exposed to these experiences. In the increasingly sensitive and conservative political climate, these stories do not easily seep through power structures and surface to the mainstream domain.

Dialogue Theatre has been used as an emancipatory process that promotes understandings across power structures and systems. This chapter draws on the South Sudan Voice (SSV) initiative as a case study of Dialogue Theatre to illustrate how the processes contribute to emancipation and decolonisation. The techniques used in Dialogue Theatre are subtle and process-based. To do that, the authors embark on the journey of critical reflections in unpacking the relationships between our *being*

S. F. Law (✉)
The University of Melbourne, Melbourne, VIC, Australia

Institute for Social and Health Sciences, University of South Africa, Johannesburg, South Africa

University of Western Sydney, Penrith, NSW, Australia
e-mail: siewfang.law@unimelb.edu.au

M. Deng
South Sudan Child First Education, Melbourne, VIC, Australia

D. Cifuentes
Victoria University, Melbourne, VIC, Australia

R. Barber
Free Theatre, Melbourne, VIC, Australia

S. Kessi et al. (eds.), *Decolonial Enactments in Community Psychology*, Community Psychology, https://doi.org/10.1007/978-3-030-75201-9_7

(our multiple situated identities) and *doing* (our approaches, practices and assumptions we carry) as part of the core team of SSV.

The chapter begins by contextualizing Dialogue Theatre which originates in Thailand, followed by an overview of what, how and why the SSV project was initiated in Melbourne in 2016. The chapter explores the multiple benefits of dialogues – transformation, data collection – and discusses the ways in which dialogue processes emancipate and decolonise ourselves and our relationships with each other.

Origins of Dialogue Theatre

Dialogue Theatre was developed by Pongjit Saphakhun and Richard Barber with the Makhampom Theatre Group in Thailand over several years from 2011. It represented a synthesis of three decades of practice in "theatre as a community media form... towards transforming Thai society" (Makhampom Foundation's Mission Statement). This process has been based on principles of audience-public engagement in transformative socio-political dialogue. Although Dialogue Theatre represented a response to the multilayers of socio-political polarisation in Thailand, and the need for 'an activist form of dramaturgy which aims to influence and alter the actual world, not just reflect it' (Haedicke, 1998, p. 132), it was inherently designed to disrupt dominant power structures. The choice of the name, Dialogue Theatre, illustrated a commitment to Makhampom's dialogic approach as an explicit shift away from polemic or didactic modes of theatre.

Dialogue Theatre embraces the principle of hybridity, however, this process was less concerned with questions of aesthetics and form, rather emphasizing the process of adapting applied theatre methodology into a stage-based theatre form. A two-part method, made up of a short scripted character-driven play and a live, mediated dialogue between audience members and play characters, effectively represented a combination of performance technique and workshop process. In essence, the method sought to combine a conflict transformation workshop approach, as a collection of applied peace-building games and theatre exercises, with a contemporary performance approach. As an interactive, immersive live theatre form, Dialogue Theatre has resemblances to the Theatre of the Oppressed techniques, aligning with Augusto Boal's advocacy for a "constant search for dialogical forms, forms of theatre through which it is possible to converse" (Boal, 1998, p. 4). It differs from Forum Theatre methods, as audiences in Dialogue Theatre do not stop and change the performance. Instead, the actors-in-characters actively participate in dialogue on political and social justice issues with aims to explore solutions to oppression featured in the performance.

By design, Dialogue Theatre engages community members in the entire process from co-creating the characters, scenario, script, performance, and facilitation of dialogues to debriefing and evaluation. The aim is to use theatre methods to build capacity and create a safe space to engage in important but uncomfortable dialogues. The Dialogue Theatre approach applies the principle of raising critical consciousness of power structures and hierarchy towards promoting group understanding and attitudinal shifts. Through an arts-based approach that seeks to use 'third-person' techniques to encourage open engagement in controversial topics without offence, audiences are opened to face and reflect on their own ingrained biases and assumptions (blind spots). Community-actors are prepared, trained and equipped with improvisational communication techniques and public speaking skills so they can handle difficult questions posed from the audiences themselves and mediate a safe space for critical reflexivity.

Dialogue Theatre promotes a shift from a binary oppressor-oppressed approach to address the nuances, complexities and distortions of contemporary postmodern society. This is derived from a sustained, localized community praxis in northern Thailand that indicated the layers of complexity in social, cultural and political dynamics surrounding the latent structures and systems of discrimination and oppression. A pluralist approach led to the core Dialogue Theatre tenet of multiple character voices being represented on stage. This was complimented by a flexible stage dialogue approach being devised to ensure the diversity of voices represented on stage was also reflected in audience representation and participation. Multiple, adaptable dialogue exercises, such as the 'chat circles', became the vehicle for disrupting social hierarchies within audiences towards a dialogue that allows marginal voices to sit alongside, or even above, privileged voices, in a social dynamic which is both real (in the space) and imagined (in society). This follows Spivak (1988) who challenges postcolonial frameworks and asks "can the subaltern speak?". She argued that the subaltern would not speak by simply being offered a platform or letting people speak.

Dialogue Theatre draws on a decolonising framework, to create mental spaces for individuals to unpack their own coloniality. Coloniality is the subtle, invisible subjectivities, the colonized ways of knowing, behaving, and being that emerged from colonialism and have survived long after its seeming demise (Maldonado-Torres, 2016, 2017). Coloniality rests on epistemic and ontological biases that promote validation of colonial hegemony and superiority. Attending to decolonising practices, Dialogue Theatre draws on techniques such as actors-in-characters to evoke storylines that blur boundaries of existing cognitive structures. A series of subtle boundary blurring techniques can be incorporated in the performance as well as during the post-performance dialogues. For example, blurring the lines between realism and fictions invites audiences to question which part of the performance is for real, and which part is acting; as well as who are the actors or who are our friends. These techniques contribute to cognitive dissonance that helps to shift the audiences' prescriptive cognitive structures (e.g. what are my assumptions in what I see in terms of what is right/wrong, good/bad, appropriate/inappropriate and acceptable/unacceptable). These cognitive shifts open up a new mental space that encourages individuals to want to listen and be informed. When the performance reaches the dialogue sessions, audiences are given space and time to pose questions to the actors-in-characters, as well as respond and interact with each other. Cognitive shifts encourage the audiences to be more prepared to hear unexpected answers and to respond with an open attitude. This process contributes to shifting pre-conditioned fixated prejudiced attitudes, and stimulates curiosity and openness. According to Habermas (1993), a democratic space for egalitarian public interactions is an open, safe and curious space for constructive dialogue. Hence emancipation occurs with audiences of diverse experiences, backgrounds and power positionings.

The following section, South Sudan Voice, offers more details on how Dialogue Theatre (and specifically the dialogue aspects of the theatre) is used to decolonise minds.

South Sudan Voice

South Sudan Voice (SSV) is an initiative of members of the South Sudanese community in Melbourne who came together in 2016 to forge understandings and collaborations with a wide range of individuals, groups and stakeholders about the challenges facing South Sudanese communities in Australia.

Dialogue Theatre was used as a platform for SSV to provide a creative, dynamic forum for critical dialogue on issues affecting South Sudanese Australians in Victoria. Malual Deng prompted the collaboration in 2016, inspired by classes in peace-building and conflict resolution with Siew Fang Law which led to the formation of the collaboration between Malual, Siew Fang, and Richard Barber.

SSV was designed to stimulate positive dialogue with a broad number of communities over critical socio-political issues facing Australia. In the 2018 Global Peace Index, Australia was ranked 13th (after Switzerland and before Sweden), while South Sudan was ranked 136th (after Iraq, before Afghanistan). The ranking process drew substantial international media attention and reinforced the perception of 'freedom is in the West'—in countries that are 'free' from war and conflict. However, mainstream media failed to capture the nuances and subjective experiences of minorities within high ranked countries, impacted by systems and structures that reproduce inequalities. The experiences of SSV actors and their characterizations correspond with the statistics that South Sudanese Australians have higher unemployment rates (Losoncz, 2011), a lower income base (Australian Government Department of Immigration and Citizenship, 2007; Department of Premier and Cabinet, 2018), are subject to interrupted schooling amongst the young (Brown et al., 2006), are racially profiled by the police and on the media (Losoncz, 2011), and often face social marginalization and discrimination (Tipping et al., 2007; Losoncz, 2011).

The SSV project was grounded in the principles of participatory art—an approach to making art following a social change agenda which engages the audience as active participants. Rather than being an approach characterized by prescriptive methods, participatory art is a dynamic method, in which participants have the opportunity to interact through diverse means. For instance, inviting community members outside their conventional spaces (such as churches and schools) and allocating them to non-conventional spaces to create and practice art allows them to explore different ways of building relationship among themselves and with the broader stakeholders. Moreover, Kay (2000) argues that participatory art gives a wide range of possibilities in terms of artistic exercises to practitioners and communities; and adds that participatory art stimulates the social fabric due to it bringing people together and creating an environment in which community members feel supported and make friends.

The benefits of the initiative occurred at multiple levels. It was evident in the case of the SSV community-based project, that one of the main issues to be addressed by the performance was that of intergenerational conflict. This was not only an issue for the community, but also an issue for the cast. By the time the rehearsals started, the older members of the cast were already distanced from the youngest due to the different points of views expressed about education and parenting. However, after several performances and dialogues with different types of audiences, and dialogues between the community members themselves, the relationship became friendlier and both sides started to acknowledge and understand each other's position and perspective.

The story that SSV created for its performances centered on a family's participation in a South Sudanese community fundraising event and portrayed the thematic concerns of the actors. Promoting positive visibility of the South Sudanese Australian community to counter negative profiling by several mainstream media agencies represented one of the primary themes that prompted Malual's initiation of the project. The scriptwriting process involved group-devised techniques. This began with workshopping character identities, combining archetypes, representations of others, and the self, extending into explorations of relationships between characters and scenarios. Richard Barber played the dramaturgical role, curating the workshopped content into script, which was drafted, shared, work-shopped, and improved multiple times in consultation with the core team of community actors. The following excerpts are examples drawn from the project script:

Excerpt 1 from SSV script

The beginning of cultural exchanges and encounters were incorporated and portrayed in scene 1; expositions of everyday racism is written into scene 6.

Scene 1:	*Chol introduces his Anglo-Australian colleague, Angus, to his friend, uncle Moudit*
Moudit:	I can contact my cousin to send you a shirt from juba.
Angus:	No, that's too kind. We just meet and you are already offering me a shirt.
Chol:	It's just the Sudanese way. It's no trouble Angus.
Scene 6:	*Moudit is reading a newspaper at a cafe and Chol joins him*
Moudit:	Did you see there is news about African gangs again.
Chol:	Again? Again? It is like there is something every day now.

Excerpt 2 from SSV script

Youth justice issues were portrayed in regard to the multi-layered and pervasive institutional discrimination against young south Sudanese Australians, particularly by the police.

Scene 2:	*young cousins, Thodee and Adau meet at the train station*
Thodee:	Just come from the youth Club. I need your help with this job I'm trying for.
Adau:	Monday didn't work out?
Thodee:	Nah, you know how hard it is for us. I think they see my name and they put a question mark. Then they see my face and they put a cross.
Thodee:	And I got stopped by the police again after the interview. I told them I was applying for a job as a security guard, not being the security risk.
Adau:	I can't believe you got stopped again. That's crazy Thodee.
Thodee:	Its funny you know. Some police help us form our basketball team but then other police stop us at the station after training.
Adau:	You need a uniform that says "we are a basketball team not a dangerous African youth gang."
Scene 3:	*Thodee's mother, Adut, is on the phone to a friend*
Adut:	(talking on phone) look on channel seven. Yes, you see our boy. You can see the police are beating him. Yes, you can see they hit him…what can we do?...You know if we discipline our kids, they call us a criminal or take them away...The police are like a lion and child protection is the snake.

Excerpt 3 from SSV script

The theme of lateral inter-generational conflict also came to the forefront of the project as the social and cultural challenges facing the actors were represented through the characters.

Scene 3:	*Thodee arrives home and finds his mother, Adut, on the phone*
Adut:	Why are you home so late anyway?
Thodee:	I was just at the youth club. You know I go there on Thursdays.
Adut:	Have you been drinking?
Thodee:	No. but anyway I can have a drink if I want. This isn't South Sudan here mum.
Thodee throws down basketball and walks out.	
Adut:	I hate freedom. I hate freedom.
Scene 7:	*Thodee and Adau arranged a surprise hiphop performance at a fundraising event organised by uncle Moudit*

Adau deejays as Thodee performs hiphop.

Moudit:	(shouts to Adau) can you turn it off? Why did you do that?
Adau:	Because you never give us a chance to perform our stuff.

Adut enters.

Adut:	(angry with Thodee and Adau, shouting) why do you shame me like that? In front of everybody. Why?
Thodee:	You should be proud I perform in front of everyone. The song is about you mum.
Adut:	Why don't you perform Dinka like we say? You should do what your mother tells you.
Thodee:	I'm tired of this. I'm going. (he walks off)
Adau:	(chasing after Thodee) Don't run away from it Thodee. She is your mum.

The 20-min drama ended in a frozen tableau following Adau (a character of a female teenager of South Sudan background) expressing the frustrations of her circumstance, 'caught-in-the-middle'. The stage is transformed into a panel format for the second part of the Dialogue Theatre process—a 90-min moderated dialogue involving actors 'in-character' and audience.[1] This process draws on conflict transformation and Dialogue Theatre approaches (which are explored in the methodology section) to engage a diversity of South Sudanese Australian community groups and other audiences associated with or with interest in engaging with the South Sudanese Australian community in dialogue. Meticulous attention was given to invite audiences that drew a balance between having fair representation of voices (e.g. gender, age, race) and strategic out-reach (e.g. local government and service providers). Instead of reproducing the traditional audiences-actors passive-active dynamism, thorough plans had been built into the process to ensure audiences participated in active listening, dialogue, and ideas formulation with actors-in-characters. The process mediated a discussion that opened to thought-provoking, complex, and solution-oriented engagement, stimulated by the lived experience of community actors.

Dialogue as Conflict Transformation Process

Freire (1970) has contended that dialogue is more than the exchange of ideas—dialogue can be used to enact processes that seek to re-think and re-create reality. In this sense, dialogue is a process, a tool, a form of relational interaction. Furthermore,

[1] Link to SSV videos: https://www.youtube.com/watch?v=Sx8RUoBSCgo&feature=youtu.be.

if dialogue is carried out in a mature, strategic and constructive manner, it can transform the perceived complainers into problem solvers.

> Dialogue is a moment where humans meet to reflect on their reality as they make and remake it. Something else: To the extent that we are communicative beings who communicate to each other as we become more able to transform our reality, we are able to know that we know, which is something more than just knowing. [...] Through dialogue, reflecting together on what we know and don't know, we can then act critically to transform reality (Shor & Freire, 1987, p. 3).

Dialogue has long been used as an effective processes in conflict transformation (e.g. Ramsbotham, 2016; Arai, 2015; Schultz et al., 2018), peacebuilding (Shweta, 2018) and even as a data collection for qualitative research (Sullivan, 2012). By ways of fostering openness and listening to the others, dialogue can be emancipating, liberating, and socially inclusive. An effective dialogical process encourages spontaneity, improvisation, ideas flow, hence promoting fluidity, dynamism and equilibrium. The key is cultivating safe spaces for openness, mutual exchanges, participation and trust building. As enacting a conflict transformation agenda requires diverse epistemological and methodological approaches (Farias et al., 2019), these underpinning principles of dialogues could play a significant role in being part of social change and conflict transformation.

In academia, there has been an increased attention given to critical praxis and discursive research paradigms in social sciences. Through transformative and liberationist perspectives, much work has raised concerns through examples of people's personal experiences of protracted colonialism, neocolonialism, occupation, and oppression. Nevertheless, much academic scholarship and praxis is bound up in dominant paradigms in which western, individualist ontological and epistemological foundations govern the practices and studies of the social sciences (Law & Bretherton, 2017).

Few have engaged in the concept and functions of dialogues (Farias et al., 2019) in decolonial enactment. As dialogue could be used to explore subjectivity and hybridity of concepts and identities, we consider dialogue as consistent with the goals to decolonise, liberate and break down epistemological, structural and systemic barriers.

In SSV performances, the latter two thirds of the sessions involved the actors in role leading an open dialogue with the audience followed by a series of smaller chat circles to generate audience and character dialogue. At the time of writing this chapter, SSV had performed to a range of audiences including the Department of Justice, local Councils and non-profit youth organizations, engaging youth groups, women groups, activist groups, academics, government officials and community service providers.

While a wide range of potential topics might have emerged from these chat circles, to assist a focused conversation, each group was given a small slip of paper which contained one guided question, such as "We all believe in the idea of freedom, but the freedoms in Australia are putting our children and young people at risk. How do we balance this freedom and risk while still maintaining our cultural traditions?". The audiences and actors-in-character were encouraged to ask each other questions and respond to each other when participating in the chat circles. The intention of having the actors remaining in their characters was to ensure psychological safety, so the questions would be treated in terms of the issues facing the characters, thereby enacting a form of third-person mediation to depersonalize the dialogue.

Amongst the topics explored in the chat circles, was the complexity of the relationship between the community and the police force. The dialogue explored: the challenges surrounding the recruitment of culturally diverse members to join the police force; an interest amongst the police members to learn more about South Sudanese Australians' first-hand experiences of freedom (or lack of) in Australia; responses from the South Sudanese members about their socio-cultural concerns of joining the police force alongside the need for police protections when being on the streets; and, complaints by South Sudanese mothers of their children threatening to call the police when parents attempt to discipline them at home. This perspective was challenged by South Sudanese youths who raised their relative lack of freedom as their behaviours were considered to be more restricted as compared with the other 'normal' Australian youths.

Many of the emerging topics did not have clear 'right or wrong' answers. The strength of the dialogue theatre approach is its ability to explore the nuances and subjectivities of matters in an inclusive manner.

Using Dialogue as Data Collection

In keeping with the principles of decolonising praxis when developing this chapter, we drew on the conception of dialogue as a method of data collection and writing processes. To avoid creating an interrogation or one-way interview, we dialogue. This type of data collection approach requires that those involved are seen as egalitarian partners (Freire, 1970). From a critical stance, working towards the ideal of egalitarian dialogue requires that all those involved share responsibility in making sense of their experiences and assumptions and the power relations that exist before and during the study. Thus, dialogue provides space for working towards egalitarian relationships by deliberately recognizing how power plays a role in the researcher-participant relationship (Farias et al., 2019, p. 5).

When writing this chapter, it became clear to us that the timing was mature for the four core members of SSV and the authors of this manuscript to participate in a collective reflection on our individual journeys of being part of SSV in Melbourne. It was agreed that we use dialogue as a method for collective reflection and data collection, that we share the ups and downs, highs and lows of participating in SSV. As such, we drew on our cultural diversity—one white male artist-activist, one South Sudanese man of refugee background, one international student from Colombia and one Asian female researcher – and exchanged perspectives and narratives of liberation, privileges and injustices through our collaborative journey. In doing so, we re-enacted a form of decolonial praxis that gave rise to greater equilibrium, liberation and emancipation in the work we do.

Malual Deng is an Australian citizen of South Sudanese background in his 40s. He is a community leader/elder, father of four children, and guardian of four young people. *Siew Fang Law* is a female, Chinese Malaysian, immigrant in Australia, in her 40s. She is an academic and mother of two. *Richard Barber* is a male Anglo-Australian citizen in his 50s living between Australia and Thailand. He is an activist-artist and father of one. *Diego Cifuentes* is a documentalist and photographer in his 30s of Colombian background. He is currently based in Australia as an international master's student.

We sat in a small circle and 'yarned'. 'Yarning' is an informal conversation that is culturally friendly and recognised by Aboriginal people as meaning to talk about something, someone or provide and receive information. There are shared cultural practices between the Indigenous Australians and migrants of diverse backgrounds, for example, oral traditions, or yarning and storytelling, are very important for *being* and *doing*. Indeed, the Ilbijerri Theatre Company in Australia has adopted the term 'yarning circles' as their equivalent of chat circles during a dialogue theatre workshop process.

We reflected on and talked about our experiences of participating in the SSV project, and particularly focused on unwrapping our notions and experiences of liberation in participating in the project. We devised the dialogue in several smaller parts – beginning with a discussion focusing on the ways in which Dialogue Theatre 'decolonises' and 'emancipates' each of us and collectively. We drew on our experiences of being part of SSV, the different notions of 'freedom' that have emerged in our work, centering our thinking and dialogue.

Our approach resembled Sullivan's (2012) dialogical research method which enabled us to perform as the qualitative analyst to explore a complex and conscious subjectivity. According to Sullivan, "subjectivity, viewed dialogically, is social, relating to self as well as to others. This kind of subjectivity is also complex in so far as it emerges and draws from a network of history, tradition and power" (p. 43, 2012).

Reflecting on South Sudanese Voice as Decolonial Enactment: A Dialogue

Our reflection centers the question of whether South Sudan Voice represents a decolonised praxis. We responded by each unpacking our own respective history of colonization:

> Malual: As Africans, we have been colonised by Europeans, bringing values that view local values as unimportant and limit traditional local systems, resulting in the loss of culture, values, beliefs.

> Diego: The great majority of Latin America was colonized by the Spanish but since independence in Colombia, American culture has influenced the ways we think of and understand 'freedom', as the North American dream.

Malual: Africans around the world are still impacted by colonization today. The way it has shaped the world means many people still can't settle or work together in harmony. Divisions on ethnicity were not defined the same way traditionally.

Richard: I developed my original understanding of colonization in Australia through the lens of the denial and suppression of Indigenous sovereignty. Living in Thailand has given me insights into the importance of post-colonial society and its neglect in Australia.

Siew Fang: As a Chinese Malaysian, I grew up as a second-generation migrant in a three-times-colonized country. Malaya was colonized by the Dutch, Portugese and British for about two-hundred years each.

We reflected on our diverse relationships with colonization and the places we called home, and made sense of our connections with Australia. We discovered shared sense of place and (dis)placement in this colonial land, and the ways in which our past and present experience with colonization impact our worldviews, behaviours and practices.

In making sense of decolonised praxis, we unpacked the legacy of colonization:

Malual:I agree with term of the legacy of colonization. Most post-colonial nations are independent, but still struggle to set up their own systems.

Richard: Colonization reflects the notion that western culture is perceived to be 'proven' as the ideal model, sustained through globalization.

Siew Fang: Decolonization, as a concept, is not well engaged in the Australian context. It remains sensitive in Australia to talk about this topic, in unwrapping systems, structures, cultures of colonization that has a legacy in its effects on migrants, refugees and others.

We reflected on the challenges and complexity of implementing SSV in the context where colonized systems, structures and cultures remain in the backdrop. Hence, our decolonised praxis is not limited to the Dialogue Theatre performance itself, but the entirety of our thought processes (ontology, epistemology, methodology) that shape, unshape and re-shape our interactions and stories. Our different experiences of 'freedom' in Australia shape the ways we make sense of what it meant to be 'free'. For example, the following quotes centering the notions of 'freedom' show the ways in which the concept was colonized, realised and contested.

Diego: The concept of freedom comes from a western idea. I have seen it as dream of North America, the idea that Western is better. Since the Second World War ended, the influence of North America, especially in defining notions such as freedom affected the whole continent, including Colombia.

Malual: As an African, we struggle with the concept of freedom. Africans often consider that there is no freedom in Africa. This is connected to conflict, but they [Westerners] don't consider that "the war is not our war" or understand the influence of foreign interests.

Diego: When here in Australia, everything is affected by capitalism. It's like a power over me. Freedom has specific meanings for the West—freedom to consume, freedom to buy but not to move or work, or get certain jobs, freedom to vote and to choose who is in power (if you have citizenship).

Malual: People think freedom is in the West. But the opposite happens when we arrive in Australia. I think freedom should be in everyday citizenship. We should be able to feel free to apply for jobs like any other with European backgrounds, free to be voted into Parliament as a minister. But discrimination exists in here. If you come from a particular background, people tend to offer you to do second class job.

Diego: I learn that freedom from collective culture is different from individualist culture. Colombia has very conservative culture and rules. Our roots are very collective. Freedom, to Asunta is similar to what for my family and people it was related to decades ago, is related to the community and space. We were free because we were in the community; we could defend our community. Freedom for the individualist Australian culture refers to free speech, that 'I can choose for myself', that 'my life is not your business', and 'you don't have the rights to talk about my feeling because it is my feeling'. I find these clashes often among many migrant communities living in Australia.

Much of the discussion centered the inequality in conceptualizing 'freedom'—the dominant Australian notion refers to the self-righteous sense of freedom, such as freedom of speech and freedom of expressions in public spaces; however, Malual and Diego do not associate this as 'free'. These cognitive tensions, although subjective and fluid, are a profoundly critical point for us to work on as part of our decolonial praxis. We work-shopped together in addressing and transforming our assumptions as a group. For example, we linked our observations of 'freedom' to audience responses during the immersive Dialogue Theatre performance.

"The police are lions; the Child Protection are snakes! I hate freedom!" We recalled the statement made by Asunta, one of the lead female actors who played 'Adut', the character of a struggling single mother of South-Sudanese Australian background in *South Sudan Voice*. Lions and snakes were her most feared animals in Africa. The statement has become one of the most powerful catalyst for dialogue. Asunta discussed her fear of government officials and her concerns about her lack of control over the safety of her sons, and that they could be taken away by either the police or the Child Protection Officer. Her lack of freedom and safety in Australia was an unexpected irony, as she, her family and community have been bound by rules and legislation that are supposed to free them, give rights to their children, but in reality she felt helpless as she witnessed her children getting into trouble on the street, and couldn't engage with her disengaged teenage children. These spoken words illuminate multilayered tensions, complexities and dilemmas.

We recalled significant learnings from the open dialogue held during the post-Dialogue Theatre performance. The open dialogue is designed to emancipate, liberate and promote mutual understandings at deep cognitive levels—the unconscious biases and assumptions. Specifically, the audiences and actors-in-characters dialogue on questions of 'change' in response to actor-character provocations allowing audiences to participate in critical dialogue and active listening. This space is moderated by the trained actors to navigate assumed positive ideas such as 'freedom'. By talking to the characters, rather than talking about the characters, the audiences are engaged in the process of critical conscientization, towards enabling opportunities for attitudinal and behavioural transformation.

To create more spaces for deepening the dialogues, concurrent chat circles were facilitated after the open dialogue with pairs of characters engaging groups of 10–15 audience members in focused conversations on specific themes (i.e. youth concerns on justice or elder struggles with parenting in the Australian context). The final phase of the Dialogue Theatre involves requests for suggestions for improving this situation, with actor-characters challenging audience perspectives, closing the event with personal observations as to what they found useful from the dialogue.

We, the core team, unpacked our own cognitive framing and practices through reflecting on our individual observations. In the process of dialogue, we identified and challenge dominant forms of power that comes with blindspots, a process toward decolonial praxis. For example,

> Richard: The transformation of audiences starts from the moment the characters speak from a position of power in the dialogue panel. The tendency for South Sudanese community audience members to speak first indicates an inversion of dominant power dynamics. And when the characters defend their position or challenge an audience statement, it becomes clear that the status quo has been disrupted.
>
> Diego: The use of photo portraits (comment boards for audience members to post reflections and suggestions to each character), becomes another tool for the dialogue to implement, whether because of language or other factors.
>
> Malual: It is good we think about language. When we speak with Ding, she agrees we should have a translator, who can speak in Arabic. This is important to translate our concerns.

By reflecting on and sharing our practices, the core team gained greater and deepened insights about the ways in which dialogue theatre works with diverse members in complex contexts. Our reflections and dialogue allowed us to step back as third-persons observing the emerging audience-actors conversations, learn from the process, and further improve our inclusive practices.

> Diego: In SSV, the audiences who have seen our performances seem to go through three stages. One stage is 'rejection' – this is your opinion, this is my opinion and that is it. I don't agree with you. Audiences reject the characters. The next stage is 'transformation', where the audiences started to have empathy with the characters. Then in the third stage, there is a more [power] balanced conversation. For example, at one performance, everybody was making assumptions that [Adut] was a bad parent, but then many Australians, especially white people, started to reflect that they also have the same problem with their teenage children. There, people of different cultures started to reflect with Adut and see things they have in common, to try to find balance and solutions together.

We reflected the ways in which the audiences' and actors' responded to Richard's character—Angus—a well-intentional white-man who wants to 'help' the community.

> Richard: I found it interesting that audiences would not confront Angus as a white male, in his role in assuming authority as a sort of mediator. I was waiting for the question as to why Angus even needs to be there.
>
> Siew Fang: I felt Hong May [as alternate character to Angus] is very awkward. It feels similar to Angus and totally agree that Hong May is not needed in that space. This is the ultimate decolonization, when everyone can lead together.

We had the opportunity to unpack our own core group's dynamic and power shifts as Malual reflected on the fact that he gradually felt safe to disrupt and challenge white privilege.

Malual: We can also interrupt [Angus]. Sometimes we feel uncomfortable to do something for ourselves. In terms of colonization, there is the feeling that someone outside will come to fix it. I felt like a leader in the Law Week performance, when I interrupted Angus, to ask to hear audience members to speak.

Richard: This was an important moment in challenging white authority.

Malual: Our group [of community actors] builds confidence with each performance. We can ask how the group feels about who should lead the dialogue process. There is no fear. We respect each other and learn there is no right and wrong. We feel we can interrupt each other, and that its not disrespectful.

Richard's affirmation encourages greater openness and psychological safety for the team. The exchange creates ripple effects on other observing members.

Diego: I found it interesting when I see the hierarchy in the [South Sudanese organisation] community meeting. In the Dialogue Theatre, the decisions must be taken in collaboration, and this is reflected in the dialogue, whereas in the community, they voice of the woman is always silenced or at a specific time. All the women at the meeting sat at the back and spoke at specific times. In the Dialogue Theatre, people speak without hierarchy and Asunta and Achol can speak out on stage. They have an equal voice.

Through the reflective and dialogue process, Malual also learned about his own privilege as a male elder within the South Sudanese community in Melbourne. He discussed the need to further address gender, age and hierarchy inherent in South Sudanese culture in the context of Australia. Such reflection created an opening and a flow-on effect that stimulated the consciousness of others in the core team to openly discuss a broader experience of power and privileges observed from the Dialogue Theatre project.

Malual: Women don't want to be judged under our leadership system. Such as with Adau (a young South Sudanese woman character), and the question whether she should speak first or whether she doesn't want to speak in front of elders. In Australia, you can say anything, but this will still happen often after the meeting.

Richard: We found this was one of the most important things that happened in Thailand. When we performed Dialogue Theatre in communities, the equality of voices on stage distorted the traditional or dominant hierarchies. This not only gave marginal voices a space to be heard but also became a challenge to dominant systems, even an experiment in cultural change, but within a safe, dialogical context. It was imagined, but it was real in that space.

Diego: I can see how the relationship between younger and older actors has changed. Even if there is a conflict in the dialogue, there is a relationship, culture and background of respect. This means they have created something new.

Siew Fang: It was important that we could also invite decision makers, policy makers, polices and journalists as our audiences. Within 2 hours, our conversations are effective and powerful. It shifted between the creations and the realities. I love the way it transforms and crosses boundaries.

Malual reflected that these inequities represent an effective denial of freedom in his stating that the 'opposite happen when we arrive in Australia'. This viewpoint questions mainstream Australian perceptions of freedom. Diego made distinctions that 'freedom' carries different values to the individualist and collectivist. Migrants who come from collectivist cultural backgrounds find themselves having difficulties adapting and adjusting to individualist systems, structures and cultures after arriving in Australia. This has parallels with the experiences of Indigenous Australians, whose collectivist kinship traditions have been undermined by government interventions in the name of child protection.

Young South Sudanese Australians, such as the characters of Adau and Thodee, in 'being caught in between', live at the nexus of these individualist and collectivist values. Freedom to enjoy education opportunities and make lifestyle choices occur alongside frustrations over constraints to freedom due to systemic discrimination. Richard and Siew Fang, despite different backgrounds and identities, also seek to manage individual and collective freedoms. However, their comparative privilege, whether as a white male Australian or as positioned with status in an academic institution, distinguishes their experience from that of the young South Sudanese Australians.

Richard and Siew Fang's consciousness of this privilege, Diego's negotiation of his limited freedoms yet comparative privilege in contrast to the South Sudanese Australian context, and Malual's post-colonial critique of the concept of freedom has fed into a dialogue that seeks to find an equilibrium within the diversity of experiences and values. By its nature, this notion of equilibrium promotes a critical reflexivity. It is also framed by a consensus within the group that an effective response to the issues facing the South Sudanese Australian community requires an interrogation of 'freedom' and a decolonised praxis. Consequently, the project becomes substantially measured by the achievement of critical reflexivity in audiences as a basis of their critical consciousness regarding these themes.

Discussion

The processes of dialogue and Dialogue Theatre can be considered decolonising and emancipatory in the following ways:

1. These processes hold spaces for members of diverse backgrounds and of different power relations to come together through sharing, listening and understanding our inherent structural dilemmas;
2. The processes transform individuals and groups through deep personal and social awareness and conscientization of the influences of institutions, structures, systems and cultures on each of us.

The use of theatre art was also important in safely managing the significant levels of trauma, generational tensions between the young generation and parents, responsibilities as a diaspora in supporting homeland problems, gender-based tensions, the effects of institutionalized discrimination, cultural issues around social adaptation, and youth disenfranchisement and involvement in crime. Many first-generation South Sudanese Australians experienced conflict and violence before fleeting from their war-torn countries. As newly arrived migrants, many have continued to experience trauma and a range of psychosocial challenges and challenges to their physical health and wellbeing (Burnett & Thomson, 2005; Coker, 2004; Goodman, 2004). Moreover, acculturation stress has found to be critically manifested in the form of intergenerational and role conflict and marital difficulties among the South Sudanese families in Australia (Khawaja & Milner, 2012; Khawaja et al., 2008). These themes were prominent in the storyline and audience dialogue, drawn from the collectively devised process and moderated through the process of audience participation. The issue of respect, and the lack of, have been a recurring theme in academic literature exploring the settlement of South Sudanese Australians (Losoncz, 2011) and emerged as a meta-narrative in the Dialogue Theatre process, manifested in the analysis of 'freedom'.

While SSV was designed with the aims to emancipate and transform our understandings among communities and key stakeholders through theatrical performance followed by actors-audiences dialogues, four of the core team members have different experiences and take-aways through our participation in this project. As demonstrated in the above sections, conceptions such as 'freedom' are embedded with intersubjectivity and multiplicity of meanings that intersect with our identities, ingrained within our social positionalities, and shaped by the systems, structures and cultures we are located in.

In addressing personal and socio-political dilemmas that are bounded by plural conceptions and experiences (such as 'freedom'), our practices of dialogue and Dialogue Theatre enabled a deep unpacking and promoted a sense of conscientization. Conscientization is deep seeded in Freire's emancipatory pedagogy approach, and is "the process by which one develops the ability to think critically about issues of power in relationship to privilege and oppression in different spheres of influence (local, national, international etc.)" (Clark, 2011). Conscientization refers to 'the development of understanding and awareness of one's position in social, political, economic, cultural, and historical contexts, with a particular emphasis on a critical understanding of one's position in power, privilege, and oppression (Hardee & McFaden, 2018). Hence, participating in processes that promote dialogue and conscientization could help us to see and identify our own socio-political blind spots.

Our continued researchers-practitioners dialogues alongside the audiences-characters dialogues within SSV performances have illuminated numerous ideological tensions. The ideas of 'freedom' repeatedly emerged. 'Freedom' as a concept presents a stark dilemma in Australia. For example, Australia has often been perceived as a safe, free and peaceful country. Melbourne has been awarded 'the most live-able city in the world' for numerous years. Nevertheless, Adut (the character of a middle-aged single mother of South Sudanese Australian background) declares *'I hate freedom'* dialectically challenging the broader public rhetoric, such as, whether Australia is a 'young and free' nation (as lyric in the Australian national anthem); whether South Sudanese refugees have gained 'freedom', and shared similar 'freedom' as other non-South Sudanese residents in Australia; whether 'freedom' is what makes people feel safe.

Through the processes of dialogue and Dialogue Theatre, we, the core team, have gained greater understandings of each other and our motivations, and were able to engage in nuances and subjectivities when unpacking our *being* and *doing*. For example, as we explored 'freedom' in multi-dimensional paradigms, Richard and Siew Fang challenged their own views on privilege, their freedom in participating in SSV, and their journey of decolonising their own knowings and praxis. In doing so they wrote the character of Angus (a character of a middle-aged Anglo male) which brings the non-Sudanese voice into the script. Angus, as a white male Australian, carries levels of privilege and freedom that resonate with many audience members. His dilemma, as to whether he should be involved at all in the intervention, can prompt non-Sudanese members, particularly white Australians, to undergo a process of cognitive dissonance as they place themselves in his shoes. Having 'Angus' in the script also allow us, the core team, to feel comfortable when debriefing our experiences of, so-called, 'freedom'.

Similarly, South Sudanese and other African Australian audience members faced a similar situation as they negotiated Chol's (a character of a middle-age South Sudanese Australian) compromised approach to working with 'good' police whilst acknowledging 'bad' police and his request to Angus as 'the right person to help' the community. The dilemmas of and between the characters are effectively transferred to the audience, becoming a vehicle for their conscientization.

Although Thodee (a character of a male South Sudanese teenager) reflected similar perspectives regarding 'freedom' in his perceptions of discrimination in finding employment and in his interactions with police, his statement that he can 'drink if he wants' and that 'this is not South Sudan' suggests that there are cultural freedoms he has embraced in Australia that are in tension with the traditional South Sudanese modes advocated by his mother and uncle. As young South Sudanese Australians, Thodee and Adau (a character of a female South Sudanese teenager) negotiate the duality, or even plurality, of identity as South Sudanese, Australian, and children of refugee camps, revealing the multi-dimensional perspectives on freedom within the family and community and their effects on intergenerational tensions. In turn, audience members were challenged to reflect on their own assumptions and perspectives regarding their own freedom as they engaged the actor-characters viewpoints in understanding the socio-political complexities of the South Sudanese Australian community context.

The diversity of characters on stage allows different audience members to identify with specific characters or perspectives while being encouraged to actively listen to all voices and their dilemmas. The nature of the dialogue between characters and audience represents an atypical interaction, as both decolonised and distorting socio-cultural hierarchies. The tendency of audiences to embrace this 'opportunity' to engage in dialogue with 'the Other' appeared to supersede this imagined aspect of the dialogue. In effect, it became apparent that this specific participatory art form facilitated this imagined space as a mechanism to achieve a level of discourse that is typically constrained by embedded social hierarchies. Similarly, the diversity of the core group in the SSV project allowed differences to shine and created a unique safe space for openness.

Conclusion

Through the processes of dialogue and Dialogue Theatre, we became more aware of the relationships between oneself and others, our individual and collective struggles, our different ways of meaning-making, and our *being* in the plural and

Table 1 Main characters in SSV

Name	Role
Chol	(friend) Community leader mediating with authorities
Adut	(mum) single mum struggling to adapt to Australian norms
Mou	(uncle) traditional Dinka community leader homeland
Thodee	(son) youth artist finding refuge from society in music
Adau	(niece) university student caught between two worlds
Angus	(colleague of Chol) experienced community development professional

interconnected socio-political systems and structures. In order for our practices to be truly decolonised, we actively created processes that enabled practices of critical reflexivity, understanding, and reciprocal support. It is these processes that helped us gain self-awareness, socio-political awareness, and know-hows in navigating and negotiating different forms of power, and making small steps toward decolonial praxis (Table 1).

References

Australian Government Department of Immigration and Citizenship. (2007). *Sudanese community profile*. Commonwealth of Australia. https://www.dss.gov.au/sites/default/files/documents/11_2013/community-profile-sudan.pdf

Boal, A. (1998). *Legislative theatre*. Routledge.

Brown, J., Miller, J., & Mitchell, J. (2006). Interrupted schooling and the acquisition of literacy: Experiences of Sudanese refugees in Victorian secondary schools. *Australian Journal of Language and Literacy, 29*(2), 150–162.

Burnett, A., & Thompson, K. (2005). Enhancing the psychosocial wellbeing of asylum seekers and refugees. In K. H. Barrett & W. H. George (Eds.), *Race, culture, psychology and law* (pp. 205–224). Sage.

Clark, C. (2011). Buddhist detachment as a conceptual point of entry into teaching sociopolitically-located multicultural education online. In G. Kurubacak & T. Volkan Yuzer (Eds.), *Handbook of research on transformative online education and liberation: Models for social equality* (pp. 108–125). IGI Publisher.

Coker, E. M. (2004). 'Traveling pains': Embodied metaphors of suffering among southern Sudanese refugees in Cairo. *Culture, Medicine and Psychiatry, 28*(1), 15–39. https://doi.org/10.1023/B:MEDI.0000018096.95218.f4

Department of Premier and Cabinet. (2018). *South Sudan-born. Victoria community profiles: 2016 census*. The Multicultural Affairs and Social Cohesion Division. https://www.multicultural.vic.gov.au/images/2016_Census/CommunityProfiles2016/South-Sudan-Community-Profile-2016-Census.pdf

Farias, L., Rudman, D. L., Pollard, N., Schiller, S., Malfitano, A. P. S., Thomas, K., & van Bruggen, H. (2019). Critical dialogical approach: A methodological direction for occupation-based social transformative work. *Scandinavian Journal of Occupational Therapy, 26*(4), 235–245. https://doi.org/10.1080/11038128.2018.1469666

Freire, P. (1970). *Pedagogy of the oppressed*. The Continuum International Publishing Group.

Goodman, J. H. (2004). Coping with trauma and hardship among unaccompanied refugee youths from Sudan. *Qualitative Health Research, 14*(9), 1177–1196. https://doi.org/10.1177/1049732304265923

Haedicke, S. C. (1998). Dramaturgy in community-based theatre. *Journal of Dramatic Theory and Criticism, 13*(1), 125–132.
Habermas, J. (1993). *Justification and application*. MIT Press.
Hardee, S. C., & McFaden, K. L. (2018). Social foundations of education and service-based field experiences: Critical foundations for socially just educators. In M. Sulentic Dowell & T. D. Meidl (Eds.), *Handbook of research on service-learning initiatives in teacher education programs* (pp. 41–58). IGI Global Publisher.
Law, S. F., & Bretherton, D. (2017). The imbalance between knowledge paradigms of north and south: Implications for peace psychology. In M. Seedat, S. Suffla, & D. Christie (Eds.), *Enlarging the scope of peace psychology: African and world-regional contributions* (pp. 19–36). Springer.
Losoncz, I. (2011). Blocked opportunity and threatened identity: Understanding experiences of disrespect in south Sudanese Australians. *The Australasian Review of African Studies, 32*(2), 118–142.
Kay, A. (2000). Art and community development: The role the arts have in regenerating communities. *Community Development Journal, 35*(4), 414–424. https://doi.org/10.1093/cdj/35.4.414
Khawaja, N. G., & Milner, K. (2012). Acculturation stress in south Sudanese refugees: Impact on marital relationships. *International Journal of Intercultural Relations, 36*(5), 624–636. https://doi.org/10.1016/j.ijintrel.2012.03.007
Khawaja, N., White, K., Schweitzer, R., & Greenslade, J. (2008). The difficulties and coping strategies of Sudanese refugees pre-, during and post-migration: A qualitative approach. *Transcultural Psychiatry, 45*(3), 489–512. https://doi.org/10.1177/1363461508094678
Maldonado-Torres, N. (2016). *Outline of ten theses on coloniality and decoloniality*. Fondation Frantz Fanon. http://frantzfanonfoundation-fondationfrantzfanon.com/IMG/pdf/maldonado-torres_outline_of_ten_theses-10.23.16_.pdf
Maldonado-Torres, N. (2017). The decolonial turn. In J. Poblete (Ed.), *New approaches to Latin American studies* (pp. 111–127). Routledge.
Ramsbotham, O. (2016). *When conflict resolution fails: An alternative to negotiation and dialogue*. Polity Press.
Schultz, M., Hahn, T., Ituarte-Lima, C., & Hällström, N. (2018). Deliberative multi-actor dialogues as opportunities for transformative social learning and conflict resolution in international environmental negotiations. *International Environmental Agreements: Politics, Law and Economics, 18*(5), 671–688. https://doi.org/10.1007/s10784-018-9410-4
Shor, I., & Freire, P. (1987). What is the 'dialogical method' of teaching? *Critical Teaching and Liberatory Education: The Journal of Education, 3*(169), 11–31.
Shweta, S. (2018). Education for peace through transformative dialogue: Perspectives from Kashmir. *International Review of Education, 64*(1), 43–63.
Spivak, G. (1988). Can the subaltern speak? In C. Nelson & L. Grosberg (Eds.), *Marxism and the interpretation of culture* (pp. 271–313). Macmillan Education.
Sullivan, P. (2012). *Qualitative data analysis using a dialogical approach*. Sage.
Arai, T. (2015). Toward a Buddhist theory of structural peace: Lessons from Myanmar. *Transition, Peace and Conflict Studies, 22*(1), 34–59.
Tipping, S., Bretherton, D., & Kaplan, I. (2007). Meaning in life and social connectedness: The experience of Sudanese young people (re)settling in Australia. In D. Lusher & N. Haslam (Eds.), *Yearning to breathe free: Seeking asylum in Australia* (pp. 157–168). The Federation Press.

Constructing Race and Place in South Africa: A Photovoice Study with ‘Coloured’ Men in Bishop Lavis

Simone M. Peters, Shose Kessi, and Floretta Boonzaier

The study described in this chapter adopts a decolonial perspective on community psychology using Photovoice research. Apartheid constructions of ‘coloured’ identities are presented and deconstructed through the experiences of ‘coloured’ men living in Bishop Lavis, a low-income community situated in Cape Town, South Africa. In this chapter, the term ‘coloured’ is problematised to acknowledge the fact that it is an apartheid-constructed term. In the *Population Registration Act* of 1950, a ‘coloured’ person was defined as someone who was not white or native and thus encompassed classifications of the ‘Cape ‘coloured’, Cape Malay, Griqua, Indian, Chinese, “other Asiatic” and Other ‘coloured’, with some individuals from the same family often being registered as originating from different sub-divisions (Erasmus, 2001, p. 75). We begin with a broad discussion of Bishop Lavis and then follow with a discussion on ‘coloured’ men’s position in South Africa. A case is made for considering the marginalisation of these men’s experiences in South Africa and goes on to explore how ‘coloured’ men’s experiences are mediated by gender, sexuality, race, and class. A decolonial perspective is thus argued to be appropriate in unsettling colonial constructed racialised categories and understanding the lived experiences derived from these constructions. Decoloniality is a stance towards undoing the thinking, knowledges and ways of being that characterised apartheid relations of power and centering the histories, perspectives, identities, and places of those who have been marginalised by it. In order to do so, we have opted for Photovoice research, as a form of participatory action research (PAR) that finds its roots in community psychology, liberatory and justice-oriented approaches to knowledge production. Photovoice centres participants’ voices and experiences through the telling of their own stories and foregrounding their own representations. It is a process of consciousness and awareness-raising to the conditions of people historically marginalised by apartheid and colonisation.

S. M. Peters (✉) · S. Kessi · F. Boonzaier
Department of Psychology, University of Cape Town, Cape Town, South Africa
e-mail: ptrsim006@myuct.ac.za

S. Kessi et al. (eds.), *Decolonial Enactments in Community Psychology*, Community Psychology, https://doi.org/10.1007/978-3-030-75201-9_8

Place Identity, and Lived Experience

Place plays a significant role in an individual's life, thus one's experiences and ways of being are directly linked to the quality and quantity of resources and social capital available in residential spaces (Leventhal & Brooks-Gunn, 2003; Pretty et al., 2003). Many scholars have argued that place is significant in the formation of one's identity, which has led environmental psychologists to coin the term place identity (Pretty et al., 2003). Place identity is defined as:

> those dimensions of self that define the individual's personal identity in relation to the physical environment by means of a complex pattern of conscious and unconscious ideas, feelings, values, goals, preferences, skills, and behavioural tendencies relevant to a specific environment (Proshansky, 1978, p. 155).

Proshansky et al. (1983) argue that individuals' experiences and identities are created in relation to the spaces and places in which they are found. Furthermore, Dixon and Durrheim (2000) argue that the place in which someone is located is important for creating and sustaining a sense of self and to reveal ourselves to others. Environmental psychologists and geographers note that questions of "who we are" are always intertwined with questions of "where we are" (Dixon & Durrheim, 2000, p. 27). Korpela (1989, p. 246) expands on the definition of place identity, explaining it as "a psychological structure that arises out of individuals' attempts to regulate their environments". At the core of this psychological structure, Korpela (1989) argues, is the sense to belong, thus "place-belongingness is not only one aspect of place-identity, but a necessary basis for it. Around this the social, cultural and biological definitions and cognitions of place which become part of a person's place-identity is built" (p. 246).

We therefore argue that it is important to analyse place in order to obtain a holistic sense of participants' lived experiences and identities. Place can be described in a multitude of ways and can consist of multiple psychological and physical attributes (Pretty et al., 2003). In this chapter, we consider an analysis of participants' residential space as our analysis of place and identity.

Apartheid forced removals of oppressed groups organised around constructed racialised categories of black, 'coloured', and Indian, provide a pertinent example of how "places are produced as raced, sexualised, classed, nationalised, ethnicised and gendered through mechanisms of oppression" (Gieseking et al., 2014, p. 1). Over time, people come to identify with where they live, and both shape and are shaped by those spaces. Research shows that people create environmental autobiographies, which are the narratives we have of spaces and how they have shaped our being and thinking (Gieseking et al., 2014).

As argued, one cannot examine an individual's experience adequately without a focus on their living environment and the spaces they occupy (Cooper, 2010). In Cooper's (2010) Photovoice study with young people aged eleven to thirteen in Lavender Hill, Cape Town, South Africa, he found that school learners were not only shaped by their race, gender and class identities but also by Lavender Hill itself. Many participants spoke about how their movements were restricted because of ongoing gang violence in the area, which meant that many would restrict their movements by going to school and directly home, and were reluctant to leave their homes. These restrictions of movement and limitations on freedom also had the consequence of limiting young people's education due to missing days at school (Cooper, 2010). Similarly, Sernhede (2011) found that spaces/residential areas have concrete effects on the people who live in them. For example, she found that for people living in stigmatised areas, their employment-seeking opportunities may be compromised by the exclusionary influence of negative images of their residential spaces.

In the popular and sometimes academic imagination, many areas on the Cape Flats,[1] such as Lavender Hill and Manenberg, are often constructed as 'the same' and grouped together as spaces of risk (Boonzaier, 2018). However, these narratives of sameness conceal the complexities and nuances that exist in each particular area. Furthermore, such ideas of sameness, rooted in colonial thinking suggest that these communities lack individuality and agency. Salo's (2003) research in Manenberg for example, showcased how different this area was from other areas on the Cape Flats. She argued that to the outsider, these areas might be constructed as 'the same' but that they are not; to the Manenberg residents, sociospatial boundaries criss-crossed the area and divided it into smaller communities. The boundaries of these small areas were produced and marked (through graffiti) by the eleven main gangs that operate in Manenberg so that each small community, inclusive of its residents and gang, is different from each other. For example, the Hard Livings, a notorious gang in Manenberg, renamed their turf 'East Side' and painted a big mural of Tupac Shakur on a wall to establish its dominance over other turfs and less powerful gangs like the Young Dixie Boys from Rio Street (Salo, 2003). In this work, we make the case for researching Bishop Lavis, a complex community and one that has been largely under researched.

Bishop Lavis: Brief Overview

Like many other historically designated areas for historically classified 'coloured' people, Bishop Lavis, on the Cape Flats, was created to accommodate those historically classified as 'coloured' after they were forcibly moved from their homes in 1954. The apartheid government passed the Group Areas Act No. 41 on the 27 April 1950. This Act forced physical segregation and separation between races by creating different residential areas for each group (Boddy-Evans, 2018; Higgs, 1971). Many people historically classified as 'coloured' were then forced to leave their homes in areas such as District Six, Goodwood, Parow and Vasco and move to new areas designated for them, collectively known as the Cape Flats (Boddy-Evans,

[1] The Cape Flats refers to a flat, sandy stretch of land which is located on the outskirts of the city of Cape Town. It has been described as a "dumping ground", where black people were forced to move during the Group Areas Act (Boddy-Evans, 2018; Higgs, 1971).

2018; Higgs, 1971). Bishop Lavis, was one such designated area. It was named after Bishop Sidney Warren Lavis, a reverend who campaigned for the improvement of poor living conditions throughout the Cape Peninsula in the early 1900s (Ballard, 2011). Bishop Lavis was managed by the Citizens' Housing League, a non-profit welfare organisation, containing religious groups, who were concerned about the housing and welfare of the citizens of greater Cape Town. The area developed in phases and the first 400 houses were built in 1951. The area was amongst the first designated areas created by the apartheid government for historically classified 'coloured' people. It is the third smallest historically designated 'coloured' township in the metropolitan area. The area has been characterised by high unemployment and high crime rates (Statistics South Africa, 2011), not unlike many other areas of the Cape Flats. Bishop Lavis is one of the precincts in the country and province with the highest levels of drug-related crime and malicious damage to property, which is on the increase (Lamb, 2017). Recently, after much public outcry about the high levels of gang violence and the vulnerability of its residents, Bishop Lavis joined the list of 10 other areas on the Cape Flats where Operation Prosper,[2] a strategy to decrease crime by deploying the army, was implemented (Pather, 2019).

It is also important to note that violence is not the only way in which Bishop Lavis may be characterised. There is never really only one story to tell about particular places, despite the fact that mainstream discourse attempts to do this and, in so doing, covers over the complexity that exists in these areas. Attentive to the fact that in doing decolonial research, the knowledge produced must challenge dominant stereotypical narratives and present alternative narratives about marginalised communities and people (Boonzaier & Kessi, 2018; Smith, 1999), this research aimed to produce counter-narratives about Bishop Lavis and places like it. At the same time, we also show how current, deeply held stereotypes held about the residents of Bishop Lavis stem from longstanding colonial discourses. Academic discourses tend to shape media and public opinion, thus obliging us to challenge limited, and negative, stories and representations of particular people and communities.

Why Young 'Coloured' Masculinities?

Chadwick (2010) argues that research undertaken in post-apartheid South Africa with young people has mainly focused on the legacy of apartheid. In addition, very little of this research has been undertaken through a gendered analysis (Chadwick, 2010). Much of the scholarship on 'coloured' masculinities has shown how performances of gender amongst historically classified 'coloured' men are consistent with

[2] Operation Prosper saw the deployment of the army on the Cape Flats. It was initiated by South African Police Minister, Bheki Cele, after pressure was applied on national government from communities on the Cape Flats to effectively combat crime and gangsterism. After much debate and engagement, the army remains deployed on the Cape Flats (Hyman, 2019) although it is still not clear that their presence is having any effect on crime and violence in these areas.

traditional performances of masculinity (Abrahams et al., 2006; Choi & Ting, 2008; Jewkes et al., 2009; Sawyer-Kurian et al., 2009). However, not enough is known about the alternatives to these traditional performances of historically classified 'coloured' masculinity. To reiterate from earlier, we see the term 'coloured' as apartheid constructed terminology, one that has been made meaning of in different ways, including by those who have taken up the term to describe their own identities. Adhikari (2005, p. 5) argues the following:

> The marginality of the 'coloured' community is reflected in South African historiography in that relatively little has been written on the history of this social group and much of what has been written is journalistic, polemical, speculative, poorly researched or heavily biased. In many general histories 'coloured' people have effectively been written out of the narrative...

In a similar way, Switzer (1995), argued that the historically classified 'coloured' community was a marginalised group and has remained as such. The under-representation of historically classified 'coloured' people in research is also reflected in the discipline of psychology. In a review of the research reported in the South African Journal of Psychology, Louskieter (2018) reported that there was a lack of research with and on 'coloured' people, with most psychological research being undertaken with African[3] people. In addition, the research on people classified as 'coloured' has predominantly been conducted and written by white scholars, who have focused on such issues as gangsterism, alcoholism and violent masculinities – focus areas that continue to mark this group as 'other' and 'at risk'.

Moreover, research on 'coloured' masculinities has suggested that the use of violence and aggression is historical (Andrews, 1999; Campbell, 1992; Epstein, 1998). From this perspective, everyday use of violence is a manifestation of colonial power. Violence and aggression have become vital components in the construction of masculinity in South Africa (Andrews, 1999; Campbell, 1992; Epstein, 1998). These features of manhood have become a method of attaining manhood, and have been shaped by the system of apartheid, where government used violence to exert power, maintain order and control black men, thus encouraging its citizens to use violence to maintain control over others (Andrews, 1999). It has been argued that the legacy of apartheid and its violence continues to impact on the lives of 'coloured' men today (Cooper & Foster, 2008; Van Niekerk, 2019). In Cooper and Foster's (2008) research, the 'coloured' male participants argued that "their masculinities carry the scars of a very oppressive history" (p. 21). The authors argue that these men used violence as "a reaction to emasculation they experienced through marginalization" (p. 20). Moreover, Salo (2003) found that 'coloured' men historically were not deemed as providers because the state provided housing, jobs and state grants to 'coloured' women. Not being able to provide for their families has led to

---

[3] South Africa law divided the population into four major racial groups: white, 'coloured', Indian and African people. African refers to South Africans who are from different ethnic tribes such as the isiXhosa, Nguni, Sotho, and Zulu, to name a few.

arguments being made about ‘coloured’ men’s feelings of powerlessness and emasculation, and their inability to attain accepted standards of manhood.

Much research on ‘coloured’ men has focused only on the negative aspects of this identity. By only focusing on gangsterism, violent masculinities and alcoholism, this research perpetuates stereotypes which hold many negative consequences for these ‘coloured’ men, as is argued above. Academic research paints a particular image of ‘coloured’ men as unskilled with minimal schooling, as most likely to perpetuate intimate partner violence (Abrahams et al., 2006), as abusive (Choi & Ting, 2008), and as having the highest perpetration rates of rape and gang rape (Jewkes et al., 2009). As argued, this one-dimensionality covers over the complexity of what it means to be a man, raced as ‘coloured’ in current day South Africa, bearing the weight of history as well as contemporary representations of one’s identity. Salo (2003) argues that one must take a holistic *and historic* approach when researching and trying to understand ‘coloured’ masculinities. If not, we run the risk of reproducing apartheid depictions and simplifying the alternative rationalities for negotiating dominant forms of ‘doing gender’ that these men are often faced with.

Engaging Youth Using Photovoice

Community psychology in South Africa emerged in the 1980s as a critical approach to individualism and positivism of mainstream psychological research (Duncan et al., 2007). It acknowledges the implication of psychology as a discipline in the reproduction and legitimisation of apartheid ideologies, and represents a key shift away from the assumptions of scientific neutrality in research towards a clear agenda of liberation and social justice. Community psychology embraces bottom-up approaches to understanding social change with a focus on empowerment, social capital and critical consciousness (Campbell, 2004). Praxis in community psychology is the understanding that theory and practice are mutually produced and therefore that knowledge emerges through the practical engagement with theory, in a cycle of reflection, research, and social action (Prilleltensky, 2001). In turn, decolonial praxis challenges community psychology to not only reflect on building empowerment, social capital and critical consciousness in local contexts, but also on how these processes are simultaneously influenced by historical and global relations of power shaped by coloniality.

One method with this potential and that has gained popularity in community psychology has been Photovoice. Photovoice is a participatory action research (PAR) methodology, also defined as “a process of research, education, and action” (Brydon-Miller, 1997, p. 658). The aims of Photovoice, which are in line with PAR aims, are to “(1) record and represent everyday realities; (2) promote critical dialogue and knowledge about personal and community strengths and concerns; and (3) reach policymakers” (Wang, 2006, p. 148). In a typical Photovoice project, participants are given cameras or video cameras and are offered a basic training in how to use them. They are then asked to use the cameras to document their experiences and perspectives on a specific issue in their lives. Later in the process, they are required to use the pictures taken to construct a story around a particular theme. These photographs and captions are critically discussed and then exhibited to the general public, including relevant stakeholders, such as representatives from non-governmental organisations (NGOs) and policy makers (Wang, 2006). Participants thus use the photographs to critically reflect on the challenges and assets in their lives and communities, and use these images to raise awareness, influence policy or help with prevention interventions in their various settings or communities. In this way, Photovoice enables participants to take on the role of researchers, to gain technical skills in photography and research, and become active in bringing about change in their communities (Strack et al., 2004). Studies have found that participation in the Photovoice process helps participants to develop critical consciousness, as participants are given the opportunity to critically engage, think and reflect on social problems facing them and mobilise for social action (Carlson et al., 2006).

We employed Photovoice methodology as a way of challenging mainstream research on young people, which is dominated by adult researchers, who decide on the agenda of the research by asking particular questions (often framed around risk and danger) and who have the final say in the analysis and dissemination of the data (Boonzaier & Kessi, 2018). As Boonzaier and Kessi (2018) have argued, research with young people needs to be participatory in nature, and privilege the agendas of the participants themselves. Photovoice projects have been used with young people in South Africa, and have explored a wide variety of issues affecting South African youth. The methodology has been shown to be a powerful tool for engaging with young people (Helman et al., 2018; Kessi, 2011, 2018; Kessi et al., 2019), especially on topics of health (Moletsane et al., 2007), identity (Boonzaier & Kessi, 2018; Helman et al., 2018; Kessi et al., 2019; Langa, 2010), community and racial identity (Bray et al., 2010; Cooper, 2010), sexuality (Boonzaier & Zway, 2015), belonging and space (Boonzaier & Mkhize, 2018; Cornell & Kessi, 2017; Kessi & Cornell, 2015), and safety (Suffla et al., 2012). In their study with high school students, Moletsane et al. (2007) asked learners to document issues of HIV/AIDS stigmatisation in a school and their communities in rural KwaZulu- Natal, a province of South Africa. The project aimed to explore and raise learners' awareness of HIV and AIDS-related stigmatisation. Through the Photovoice process, the learners were able to recognise and identify possibilities for their personal agency for intervention against stigmatisation. Similarly, Langa (2010) found the Photovoice method to be a powerful and accessible method to explore constructions of masculinity. In his project, a group of adolescent boys in Alexandra, a township in the Gauteng province of South Africa, used their photographs to showcase how varied the "life as a boy" is and that different 'types' of boys existed in that space, adding further support to the complexity of masculine identities. In further examples amplifying the utility of the method, Boonzaier and Kessi (2018), Boonzaier and Mkhize (2018), Kessi and Cornell (2015), and Cornell and Kessi (2017) explored the experiences of black students within a particular space, the University of Cape Town. Through these Photovoice projects, black students were able to speak back to dominant and stigmatising representations of blackness and heteronormativity and demonstrate how the untransformed institutional culture and spaces in the university impacted on their identities. In the work by Kessi et al. (2019) with young people, the authors showed how the Photovoice method challenged power and dominant narratives, offered a platform for participants to construct alternative narratives, fostered empowerment by encouraging participants to move away from a victim-blaming approach, and promoted social justice. These studies reveal that Photovoice is a powerful means through which participants can explore their identity, construct alternative knowledges about their lives, and define their own problems and their gendered and sexual identities in a manner that might be constructed as decolonial. Starting from and foregrounding the knowledges of the historically oppressed and stigmatised emphasises the decolonial potential of Photovoice through contributing to the ecology of knowledge production. These studies demonstrate that Photovoice can be used as a decolonial research methodology as it allows participants to not only challenge colonial narratives about racialised and gendered identities but also produce alternative narratives through a process of critical engagement with these colonial constructs. The Photovoice methodology allows a platform for participants to critically reflect and make sense of their experiences through a historical lens, while also instilling a want for social action and social change, which is necessary for a decolonial agenda (Catalani & Minkler, 2010; Kessi, 2018).

Accordingly, we suggest that our Photovoice project represents a form of decolonial praxis rooted in the principles of community psychology whilst retaining a focus on deconstructing and resisting historical and popular representations, in this case of young 'coloured' men and their communities. Here, the use of Photovoice offers a deeper awareness of the complex socio-economic issues surrounding transformation in Cape Town and in South Africa as a whole, and locates these in a global and historical context.

Photovoice projects can follow a variety of steps in the data collection phase. In this research, the first phase involved a focus group that was held with participants on their experiences of living in Bishop Lavis and how they made meaning of their identities. Participants were recruited with the help of Taking Back our Future, a NGO working with young people in Bishop Lavis and other surrounding areas. As part of our recruitment strategy, the first author walked around the neighbourhood and approached young men to tell them about the research, requesting participation. The director of *Taking Back our Future* also referred the first author to potential participants. Eight individuals agreed to participate, but only six ultimately participated in the study. The participants were all young 'coloured' men, aged between 18 and 26, who reside in Bishop Lavis.

Participants were invited to partake in a focus group discussion, where they could share their experiences of living in Bishop Lavis and how they make meaning of their identities. The aim of the focus group was to encourage participants to start reflecting on their community, their location and their identity, in particular how their identity is shaped and influenced by their location. The second phase of the research involved narrative interviewing where participants were asked to write a story about their classed and raced experiences of growing up and living in Bishop Lavis and narrate these experiences in separate interviews with the first author. The purpose of the interviews was to invite young men to continue reflecting on their environment and how it may have shaped their identities, thoughts and ways of being. The third phase involved photography training and production planning. Participants attended a one-day training workshop where a professional photographer provided basic training in the use of cameras, lighting, framing and composition, as well as more theoretical issues such as how to use visual representations to tell a story. The participants then discussed their ideas for their stories, and by the end of the session participants received cameras and were given two weeks to take photographs that represented their experiences of residing in Bishop Lavis.

The data (focus group discussion, narrative interviews and photographs) were analysed using a thematic narrative analysis (TNA). With TNA, the researcher looks for shared stories and themes among participants' narratives, focusing on the content of the narratives and what is being narrated. The thematic analysis also allowed us to theorise across a number of interviews and analyse how participants co-construct and give meaning to the events they narrate (Riessman, 2008). In order to find themes in the narratives, the verbal data were firstly transcribed. Both the verbal and visual data were then read multiple times to identify the common narratives across all the data. The data were then read and analysed using a decolonial and intersectional lens, which recognises the important ways in which identities and experiences intersect with race, gender, class, space and other identities (Kessi & Boonzaier, 2018). A decolonial and intersectional lens also provides us with tools to challenge the essentialising and stigmatising colonial tropes of those already marginalised (Kessi & Boonzaier, 2018).

The final stage of the project was the exhibition (see Fig. 1). The first author met with the participants and discussed their photographs and narratives. In this session, the participants chose the photographs that they wanted to exhibit, which were then printed. The participants chose the date, venue and attendees of the exhibition. The exhibition took place on 22 February 2020 in Bishop Lavis and was attended by 30 people, including four participants, the participants' family members, and other young people from the area.

Constructing Race and Place in Bishop Lavis

The findings presented in this section highlight how participants used the research process as a way of claiming a positive self-image and constructing alternative narratives about themselves and their communities. The participants, who were all given pseudonyms in this research to protect their identity, used their narratives to renegotiate power, by positioning themselves in particular ways that reasserted their versions of self and their community. Three key themes shaped the narrative accounts of participants: (1) Constructions of community and the "lekker" [4] Bishop

[4] Lekker is an Afrikaans term which, translated into English, means nice.

Fig. 1 Photo exhibition

Lavis; (2) Constructions of the "unsafe" Bishop Lavis; and (3) Bishop Lavis: A classed, raced and gendered community.

Constructions of Community and the "Lekker" Bishop Lavis

Here, we explore how, through narratives of community, participants construct and talk about Bishop Lavis, and how this space has impacted on their identity and experiences. The participants constructed Bishop Lavis as "lekker" and used their narratives to construct the area as home, a place where they belonged. Korpela (1989) argues that place-belongingness is at the core of place-identity. When one feels like they belong to a community or place, it has a positive effect on their sense of self and how they relate to their place and others (Korpela, 1989). Bishop Lavis is constructed in the media as a dangerous place but to the young men who reside there, Bishop Lavis is represented as home. Here, home represents a place that has a deep emotional meaning to people. The meanings attached to the word home are complex, multi-dimensional and unique (Bennett, 2018). Home, according to Bennett (2018), can be defined "by memories, contents and people as it is by physical form. It is an evolving idea that continues to change" (para 6). One's meaning of home will look different to another's, as meaning is connected to one's experiences with a place. The participants' representation of Bishop Lavis as home is reflected, for example, in the photograph below (see Fig. 2), taken by Brandon, who shows us movement not just from the angle of the picture but also from the children playing soccer in the photo. This picture challenges views of the township as a place where people are inactive, passive or loitering, and instead denotes a place that is closely linked to a sense of belonging and identity. An excerpt from Brandon's narrative below illustrates this sense of belonging.

Fig. 2 My home—Brandon, 25

Brandon: Bishop Lavis, it's a, it's a lekker place and I know people say it's a dangerous place and that people shoot guns here and that you die here but for me the place is alright. But everywhere you go, nowhere are you safe. I grew up in Lavis so for me it's alright, actually, it's the place where I grew up, where I came to know people. Meet people, good people and the wrong people … so what is lekker about this place, is that I feel free here. I feel at home … The flats, I am not going to say … it's actually a lekker place because everyone is lively, all children, everyone plays with everyone, you understand. There's not that still of your child going missing there or so because everyone knows your child, if your child is running in the streets, there will always be one that will go and get them and bring them home, so for me it's an alright place.

The participants narrated their experiences to reflect a deep sense of belonging in the community of Bishop Lavis. It is the place where they grew up and formed a sense of self, a sense of community, and a sense of right and wrong, as captured in the excerpt from Brandon's narrative. Brandon was born in Bishop Lavis and has lived in the area with his single-parent mother, grandmother and five siblings. He later became a father to three children who now reside in the area, and uses his narratives and pictures to showcase how the area has become an important part of his identity.

The narratives showed that a relation between place and identity is abound in Bishop Lavis, making it important to analyse how the young men produce Bishop Lavis through their speech. The narratives below, extracted from individual interviews with participants, demonstrated how the area has impacted on the kinds of men the participants have become.

Simone: (…) tell me about living in Bishop Lavis. How is like to live in Bishop Lavis?
John: Well for me I would not say I am not regretful that I grew up in Lavis, I got the opportunity to see a lot of things, and that motivated me to work harder, to be better in the future. Well I also had many bad friends, but if it wasn't for them then I wouldn't have made such good decisions later on in my life. And for example, nowadays, to be in a group of friend's people who call it a gang, which makes you look bad and is not good, and you put yourself in a bad light and so people won't get to see who you really are. So, for me I am happy that I grew up in this community so that I can do better and show my children what their father went through and what I don't want as a father for them to go through, so if I can give them a better future forward then I will do it.
Leon: Okay for me like the environment is *lekker* man. Like if it's a nice time, it's a nice time and everybody gets along. That's what I can say about Bishop Lavis—everybody gets along. And like the people man, they … how can I say … they will stand up for you … A *lekker* place. For me, I grew up here man. I loved my childhood here. I learned how to shoot marbles, learned how to fly a kite, and learned how to make a kite, how to build one up and make friends man and roam the area, yes.

Both narrators were born and continue to live in the area. Leon, an unemployed IT specialist, who is struggling to find employment, lives with his mother, while John, an entrepreneur who sells cleaning products, lives with his mother and four siblings. In the narratives above, the narrators all position the area of Bishop Lavis as a "lekker" place while at the same time acknowledging the crime in the area. The picture taken by Leon (see Fig. 3) reflects a brotherhood, which challenges stigmatising assumptions of black men in groups being a gang, an example of how the participants used their narratives to 'speak back' to ideas about young 'coloured' men'.

Fig. 3 Relaxing time with the boys—Leon, 21

Working with these young men through Photovoice methods, revealed how they made sense of their experiences as young 'coloured' men. The spaces created in the project for discussion also revealed the young men's resistance to particular narratives that construct a group of black men as gangsters (Kessi, 2018). In Kessi's (2018) research with young black students from the University of Cape Town, she notes how "decolonisation enabled them to start constructing coherent narratives about themselves" (p. 107). In the narratives of these young 'coloured' men, a decolonial framing gave them the tools to produce alternative narratives, that challenge hegemonic accounts associated with the colonial gaze (Smith, 1999). In Leon's photo story (Fig. 3) above, we are being told that what outsiders might typically see as a 'gang' of young men—which is how they are seen in others' imaginations, is actually "just a group of young men providing a sense of friendship and community to each other." In other words, not all 'coloured' men are gangsters.

In many of the young men's narratives, we see a sense of ambivalence, where participants speak of the high crime rates and the shooting in the area but also express a fondness for the place in which they were born and currently live. In their narratives, the young men help Bishop Lavis gain respectability by discussing the good characteristics of the area, showcasing the area as more than just crime-filled but instead a place of community and family. In examining participants narratives of their everyday experiences, we see the multiple ways that these men negotiate their identities, present their communities and develop a sense of belonging. They use their talk to challenge the perceptions and constructions of their community as only a crime-filled gangland. The discussions happening within the Photovoice space allowed the men to question particular historical constructions of their community, and interrogate where these stereotypes originated from. In these discussions, the men all agreed that dominant knowledges about their community needed to be dismantled. This is what Maldonado-Torres (2007) calls a decolonial turn, which is defined as a multidisciplinary critique of colonialism and coloniality. It is characterised by contesting and challenging the status quo and producing knowledges that seek to undo coloniality (Maldonado-Torres, 2007).

Participants' narratives reflected a sense of community, where people "look out" for each other and learn from each other. These are common narratives found in communities where members feel marginalised (Cooper, 2010; Salo, 2003). The experience of many of the young men in our project, because of their race and class, is one of marginalisation, yet their narratives reflect a sense of belonging in and identification with their particular community. Community members from other areas on the Cape Flats shared ambivalent narratives about their community too, where they spoke of the dangerous nature of the area but at the same time argued that their areas were "lekker" and more than what the media constructs (Cooper, 2010; Jensen, 1999, 2004; Salo, 2003). Keskinen (2018) described how her participants actively resisted the negative stereotypes that existed in their suburb, and found ways to build their lives and neighbourhoods in more respectable ways and perform agency in their societal contexts. This too could be argued for the participants of our project, where they disrupted apartheid and colonial practices and deployed their narratives to resist negative stereotypes about their communities.

The Construction of the "Unsafe" Bishop Lavis

Throughout the Photovoice process, it became apparent that participants wanted to share narratives of how crime in the area has impacted on their everyday lives, and how as young men they constantly had to negotiate the terrain to stay safe and stay out of trouble. Their narratives of everyday violence illuminate the continued legacy of colonial and apartheid violence in democratic South Africa (Fig. 4).

Fig. 4 People in Bishop Lavis do not feel safe anymore—Liam, 20

Gangsterism has taken its curve and has grown tremendously at a large scale in the area and in the country. Children and adults fear even being outside nor do parents send their children alone to school due to the shootings, kidnapping, robbers, etc. It's sad to say that Bishop Lavis isn't the same anymore.

The narrative of an "unsafe" Bishop Lavis was shared by all the participants when recounting the level of crime and gangsterism in the area.

Leon: Just the, the atmosphere man that makes you feel unsafe and the impact is like … many children there don't even want to walk outside now, they're on alert and you never know when you walk man. Like, like now the other day they shot in the next road from me and they never shoot there…

Simone: … What impact does it have on you? That sort of … [Leon interjects]

Leon: …on me?

Simone: Hmm

Leon: Like it brings me down a little man. Like, you never know when to walk, when your life is in danger. Like you have to look over your shoulder every time and it's not safe man. You never know when your life will end or even what time or when to walk. Like it's not as safe as it used to be.

Violence and crime arose as critical issues for the young men of Bishop Lavis. In the narratives provided by Liam and Leon, they both emphasised their lack of freedom and feeling unsafe in the area. Many participants told similar narratives, of a difficult present compared to an idealised past. In these narratives, they compared Bishop Lavis of now to then, with many stating that "it's not what it used to be." Participants remembered how the roads were always filled with children playing in streets, but now empty streets are common, and people feel unsafe, as shown in the image above. The participants who were fathers described feelings of fear for their children's safety and expressed the dream for a safer Bishop Lavis. The negative sequelae of the violence in the area were reported to have impacted their schooling, mental health and physical health. Moreover, here we see an alternative form of masculinity emerging, the vulnerable masculinity, where young men display vulnerability and feeling unsafe, emotions that men are typically encouraged to repress (Mahalik et al., 2003). This is consistent with the narratives that emerged in the research undertaken by Cooper (2010) in Lavender Hill, where participants spoke about how the environment in which they lived impacted on their daily lives, such as having their schooling and playtime interrupted by gang wars in their area. The attention given to crime in South African public discourse, news, young men's talk, community forums and family homes reflects the extent to which it has impacted on the participants' lives and psyches. Moreover, for the young men in this project, if they were not trying to stay safe, they were trying to not get drawn into gangsterism.

> Michael: Well, they [gangsters] give you drugs and take you into, for example to do stuff like being a gangster. They see how you struggle, they give you money and then well, you one of them. So, it's quite an easy process … it's basically part of the struggle but for their own advantage, you see. They are seeing that you are asking people for money, now they see okay, okay, you are a good target and they give you a R100, go buy bread for your mother and come back to the corner and come and sell me drugs. And so, if you sell me drugs, I will give you money and provide for your family. And for somebody who doesn't work, especially the people here. And especially the guys here, they have no other option.

Participants spoke about how poverty was rife in the community and jobs were scarce. Many boys thus become 'targets' for gang bosses who lure them into gangsterism with the promise of money and a family. Gangs became popular in communities on the Cape Flats because they served a function of providing resources and a sense of family to young men who were seeking to belong and wanting to provide for their families (Pinnock, 1984, 1995, 1997). However, all the participants in this project used their talk to actively distance themselves from gangsterism and denounce it. Some of the participants revealed that they had fallen prey to gangsterism because the gang provided them with money, a sense of belonging and protection. After near death experiences, going to prison, being given ultimatums by their families either to leave the gang or leave the house, they spoke about their efforts to desist from gangsterism.

In their expressions of resistance, participants' photo-narratives depicted how they use strategies, such as starting businesses or engaging in sports, to counter the lure of gangsterism. This is shown in the two images below (Figs. 5, 6 and 7).

> John: Nothing is impossible, just do better. Like I am now, I am starting my business now and well the plan goes on. I am selling chemicals now and all that biltong things and I see that it brings in a good income and if I have a job then I can also uplift others and help them with a business too. Then I will tell them it's just so and so, and drop it by them and just get my share and he does his department and so I will uplift people you see … what I discovered is when you work for something, you appreciate it more than when you just got it. So that was a good example for me seeing as I also started from scratch, while I also had nothing. I would always go ask my brother for his music system and lekker chilled and listened to music but after a while I motivated myself to be more self-independent and make my own choices. And you can strive for more in life, so there are better things than gangsterism.

All the participants in this study positioned themselves as having larger goals and dreams for their lives. They constructed gangsterism as "not worth it", or the "easy

Fig. 5 Gangsterism is not for me, I have bigger dreams—John, 21

Fig. 6 Gangsterism is not for me, I have bigger dreams—John, 21

way out" and something for unsuccessful people with no dreams. In their construction of gangsterism, they challenged the stereotypes that all young men from historically disadvantaged 'coloured' areas enter gangsterism as a common rite of passage into manhood (Cooper, 2009). As in the in the previous theme, these young men 'speak back' to popular constructions of 'coloured' masculinities. Participants thus deconstructed colonial images of young coloured men as inherently deviant and criminal through the production of counter narratives based on a critical understanding of both the allure of gangsterism as a resource for identity and belonging and its negative and violent effects. These counter-narratives included alternative spaces for belonging, such as young men's narratives about playing sport, starting a

Fig. 7 Soccer has helped keep me out of trouble"—Mark, 19

business, and the importance of gaining respectability in their communities by staying out of trouble and distancing themselves from anything criminal. It is apparent that crime in the area affects their lived experiences; however, they have attempted to create strategies to avoid danger and also to create more positive forms of identification for themselves as men constructed as 'coloured'.

Bishop Lavis: A Classed, Raced and Gendered Community

Here, we unpack how participants' photographs and narratives constructed Bishop Lavis as raced, gendered, and classed. From the photographs and narratives, it was evident that most residents in the area are working class and poor. This we see in the photos and narratives that portray the lack of resources, and poor housing and infrastructure in the area. The photos and narratives also depict the gendered nature of their lives since they show women and men doing particular jobs usually reserved for each gender. Moreover, in their narratives, participants represent Bishop Lavis as a 'coloured' area with no African or white people living in the neighbourhood, which shows us the racialised construction of the area (Cooper, 2010). The photographs and narratives also challenged some of the dominant narratives about the area, conveying the unique experiences of living in Bishop Lavis. This is reflected in the extracts below, taken from an individual interview with John and a focus group with participants respectively:

Simone: So what makes someone a 'coloured'?

John: [laughs], so it's actually the inner person that grew up in this community of Bishop Lavis, so my language, like people say your language, your actions and what you do

Simone: But what actions make one a 'coloured' because peoples' actions are different

John: It's actually a whole lot of things, people would think if you pull out your front teeth then you a 'coloured', if you speak gangster language then you a 'coloured', if you speak fluent Afrikaans then you a 'coloured'. So, people actually have a lot to say about what it means and takes to be a 'coloured' so

Simone: Okay, so for you what makes someone a 'coloured' in your eyes?

John: For me, it's how someone grows up and all the things they go through towards life and all the things that go with that, they give it a meaning. It's like I grew up in Bishop Lavis, it's not a white area and it's not an African area, it's in-between, so I fall in there

Brandon: A 'coloured' you know by their mannerism, how he is

Simone: What manners does a 'coloured' have

Brandon: And he also talks, it's in like good, bad, very bad

Leon: But you get different types 'coloureds' man

Brandon: And *kak*[5] bad

Simone: Okay so tell me about the different types of 'coloured'

Brandon: You know *mos*, its good *nuh*

Simone: So, what is a good 'coloured'

Liam: Good 'coloured' is like

Brandon: Like white man 'coloured'

Simone: What is a white man 'coloured', is that what I am?

Brandon: You speaking English, yes like you yes. Yes, like you

The definitions of 'coloured' people always defined them as lacking either 'whiteness' or 'blackness' (Fig. 8). Erasmus (2001) wrote; "It ['colouredness'] has been negatively defined in terms of 'lack' or taint, or in terms of a 'remainder' or excess … associated … with immorality, sexual promiscuity, illegitimacy, impurity and untrustworthiness" (p. 17). The identity of 'colouredness' has been closely linked to shame, as shown by various researchers (Adhikari, 2005; Erasmus, 2001; Mthembu, 2015; Van Niekerk, 2019; Wicomb, 1998). In discussing *God's Stepchildren* by Millin (1924), Erasmus (2001) notes how the author assumes that 'mixed race' people suffer shame and guilt because of their impurity and 'mixed blood'. Van Niekerk (2019) argued that the 'coloured' men in her work, who perpetrated violence against their women partners, "carry a history of oppression, dispossession, and shame through their embodiment of pathologising stereotypes" and in their talk find strategies to gain respectability (p. 178). In her work on 'coloured' identities, Erasmus (2001) looked at the future possibilities for a young 'coloured' girl and found that young women could either choose a life of shame, linked to sexual promiscuity and teenage pregnancy, or respectability, linked to education and good employment. Another form of shame is having yellow or dark skin and coarse hair in a community where many had caramel skin and straight hair, thus 'othering' those who did not fit the 'Capetonian coloured' mould (Erasmus, 2001). Many theorists have

[5] Kak is a slang word that can roughly be translated into extremely or shit.

Fig. 8 My home—Brandon, 25

argued that the terms used to define 'coloured' people during apartheid as the left-over population still influences the lived experiences and perceptions of people historically classified as 'coloured' (Salo, 2003, 2004; Mthembu, 2015; van Niekerk, 2019). When asked what it means to be a 'coloured', varied answers were given, from it being dependent on one's skin colour, to one's language, mannerisms and way of being. One can see how shame is still attached to the 'coloured' identity where participants stated there are good 'coloureds' and it ends with "kak" or bad 'coloureds'. In extract one taken from the focus group, we see how the young men equate good 'coloured' people with whiteness because they have ambitions, education and speak English in comparison to "kak" 'coloureds' who speak Afrikaans, have bad manners, are gangsters and so forth. This is what decolonial scholars have termed the coloniality of power (Maldonado-Torres, 2007; Ndlovu-Gatsheni,

(2013). Ndlovu-Gatsheni (2013) defines coloniality of power as "continuities of colonial mentalities, psychologies and worldviews into the so-called 'postcolonial era' and highlights the social hierarchical relationships of exploitation and domination" (p. 8). The coloniality of power is thus the reproduction of colonial thinking by the oppressed (Quijano, 2001). In their talk, we see how these 'coloured' men have internalised what it means to be 'coloured' and reproduce definitions of 'colouredness' in ways that reproduce colonial thinking. Many participants still saw 'coloured' people as in-between, not really owning a fixed identity. Thus, in contemporary South Africa, meaning made of being 'coloured' still shows the complexity of this identity. Additionally, in their narratives, the participants represented Bishop Lavis as a 'raced' community, evident by the lack of 'white' and 'African' people in the area. This racial segregation, although apartheid laws no longer exist, remains prevalent in Cape Town.

The pictures also constructed the area as 'gendered':

Simone: Okay, so when you guys date, do your women have to be a certain way? Must she be able to cook, she has to be a certain way or she can be whatever?

Leon: It depends on how she is

John: No, she must do her duties

Simone: What is the women's duties?

John: She must keep the house clean, she must be prepared to, to help me when we together in a situation like a relationship, with the kids. And be self-independent and carry out the plans we made in producing the kids. She must care for the kids

Simone: Producing the kids, I like that

Leon: She must put in a lot of effort

Simone: And she must help you

John: Yes, she must help me, my girl

Simone: Okay and you?

John: She must do her part and I will do my part

The pictures taken by the participants reflected a range of gendered constructions. In the pictures (see Figs. 9 and 10) one finds the women are always photographed in the kitchen or hanging the laundry, and the men are either photographed playing soccer or in relation to their informal businesses, which again reinforces the gender roles ideology of women in the kitchen and men doing *work*. These gender norms found in the broader society can also be found in the young men's talk. In participants' talk, we see how they reproduce colonial historical constructions of the 'good' woman, where they construct a 'good' woman as someone who knows her duties—cleaning a house, producing children and caring for them—whereas a 'real' man must provide and protect for his family. These narratives are consistent with research on constructions of being 'real men' (Boonzaier, 2005; Cooper & Foster, 2008; Elliot, 2003; Hunter, 2004; Malinga & Ratele, 2014).

The participants also constructed the area as 'classed', as shown by Michael (pseudonym):

Michael: Well, I am Michael Timothy Parker, I live in Bishop Lavis. I am 22 years old, I have seen what struggle can do in this place. It takes you beyond places that you can't imagine. Like the people in this place, the gangsterism, drugs, it's bringing people down,

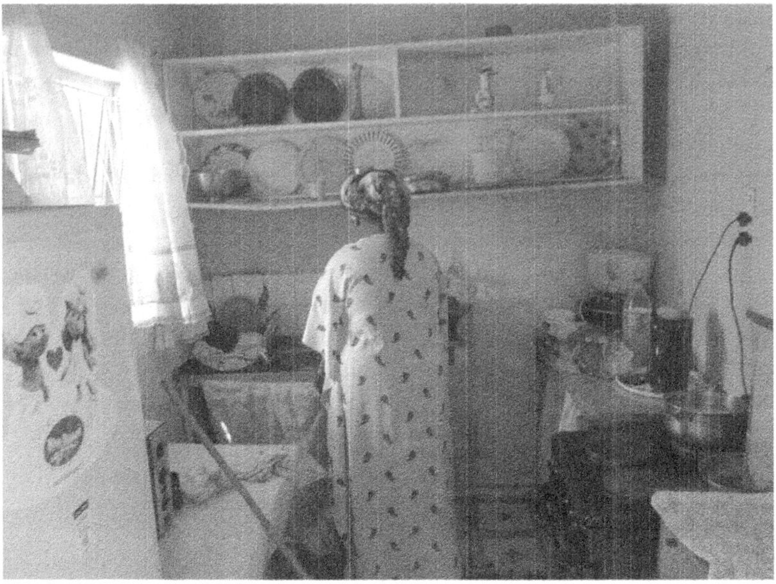

Fig. 9 My mother has always been my role model, defying all odds and can always be found in the kitchen—Michael, 22

> but people still strive to be wonderful people if they come out of this place, because this place is a hole …
> Simone: What do you mean, this is a hole?
> Michael: It just keeps you back, because you know … they want to take the people down who want to achieve something in life

Robins (2000), writing about the Cape Flats, stated that the poor were "trapped in dangerous spaces" and were "unable to shape their environment" (p. 412). When Bishop Lavis was established by the apartheid authorities the area was 'culturally poor'. There was no effective schooling, minimum sport and recreational facilities, and very few churches in the area (MacMaster, 2009) (Fig. 11). In present times, the area houses many unemployed and working-class families, with poverty still being high in the area. This is evident in the majority of the photographs taken, where one sees many informal employment practices, lack of resources, overcrowded flats and poor housing. In their narratives, participants talk of the poverty in the area, the "struggles" that residents have to endure. and construct the area as a "hole" that holds one back from achieving. Despite the hardships and structural violence, which refers to the multiple ways that social, economic and political systems expose particular populations to risks and vulnerabilities (Galtung, 1969) faced by the residents of Bishop Lavis, there is a level of endurance that makes continuing survival possible.

Fig. 10 Where there's a will, there is a way—John, 21

Concluding Thoughts

Photovoice projects have provided politically disempowered young men with opportunities to author individual and collective stories that they consider as best representing how they experience their lives (McIntyre, 2003, p. 64). The Photovoice method also has the potential to be an effective tool to change narratives and reframe stereotypes about marginalised individuals and communities. In our research, the Photovoice process provided the young men with spaces to redefine the stereotypes of their community, which are often over-populated with discourses of violence. The project also provided young people the opportunity to explore alternative ideas about their community, their gender, their racial identity, and other aspects of themselves, such as their aspirations. The project therefore made apparent how Photovoice methodology may be used to challenge and change stereotypes on neighborhoods like Bishop Lavis.

Taking a decolonial approach centred on the narratives young men convey about themselves and their communities, our findings highlight the complexity of 'coloured' men's experiences, and their agency in constructing their community the way they deem best. The decolonising potential of narrative has been noted in previous research (Sonn et al., 2013) and in this chapter we highlight how our participants narrate about their identities and their community in ways that challenge

Fig. 11 Jobs are scarce in Bishop Lavis but we make a plan—Leon, 21

ongoing coloniality. This research may therefore contribute to undoing the stereotyping of 'coloured' young men and the residents of Bishop Lavis, and perhaps contribute towards creating strategies to better support and advocate for young 'coloured' men in this area. At the same time, the findings may be useful for working with men on the issue of men's violence against women, and redefining and creating alternative masculinities.

This work has provided insight into the potential of decolonial, intersectional thinking in community psychology, in particular work with masculinity, identity and violence. It illustrates how using a participatory methodology like Photovoice amplifies the potential of participants (young men in this instance) to counter the ongoing stereotypically racialised ways in which they are popularly constructed in media and academic discourse. These representations of young 'coloured' men illustrate the pervasiveness of coloniality that is countered by the ways in which young men themselves narrate about their own lives. This work also forms part of current and ongoing conversations both within decolonial feminist community psychology (Boonzaier & van Niekerk, 2019), and studies of men and masculinity (Boonzaier et al., 2020) about what it means to counter the hegemony of northern modes of theorising about men in global southern contexts, and how our theorisation has to centre questions that recognise historical trauma and ongoing structural and other forms of violence.

References

Abrahams, N., Jewkes, R., Laubscher, R., & Hoffman, M. (2006). Intimate partner violence: Prevalence and risk factors for men in Cape Town, South Africa. *Violence and Victims, 21*(2), 247–264.

Adhikari, M. (2005). *Not white enough, not black enough. Racial identity in the South African "coloured" community*. Double Storey Books.

Andrews, P. E. (1999). Violence against women in South Africa: The role of culture and the limitations of the law. *Temple Political & Civil Rights Law Review, 8*(2), 425–458.

Ballard, P. (2011). *Sidney Warren Lavis*.

Bennett, L. (2018). *What is the meaning of home?* Retrieved from https://architizer.com/blog/practice/details/what-is-the-meaning-of-home/

Boddy-Evans, A. (2018). *Group Areas Act No. 41 of 1950*. Retrieved from https://www.thoughtco.com/group-areas-act-43476

Boonzaier, F. (2005). Woman abuse in South Africa: A brief contextual analysis. *Feminism & Psychology, 15*(1), 99–103. https://doi.org/10.1177/0959353505049711

Boonzaier, F. (2018). Challenging risk: The production of knowledge on gendered violence in South Africa. In K. Fitz-Gibbon, S. Walklate, J. McCulloch, & J. Maree Maher (Eds.), *Intimate partner violence, risk and security: Securing women's lives in a global world*. Routledge Press.

Boonzaier, F., Huysamen, M., & van Niekerk, T. (2020). Men from the 'south': Feminist, decolonial, intersectional perspectives on men, masculinities and intimate partner violence. In L. Gottzén, M. Börnholt, & F. Boonzaier (Eds.), *Men, masculinities and intimate partner violence*. Routledge.

Boonzaier, F., & Kessi, S. (2018). Challenging representations: participatory research engagements with young people in postcolonial contexts. In T. Shefer, J. Hearn, K. Ratele, & F. Boonzaier (Eds.), *Engaging youth in activism, research and pedagogical praxis*. Routledge.

Boonzaier, F., & Mkhize, L. (2018). Bodies out of place: Black queer students negotiating identity at the University of Cape Town. *South African Journal of Higher Education, 32*(3), 81–100. https://doi.org/10.20853/32-3-2514

Boonzaier, F., & van Niekerk, T. (Eds.). (2019). *Decolonial feminist community psychology*. Springer.

Boonzaier, F., & Zway, M. (2015). Resisting discrimination and negotiating safety: Young black lesbian and bisexual women represent their experiences through photovoice. *African Safety Promotion, 13*(1), 7–29.

Bray, R., Gooskens, I., Kahn, L., Moses, S., & Seekings, J. (2010). *Growing up in the new South Africa*. HSRC Press.

Brydon-Miller, M. (1997). Participatory action research: Psychology and social change. *Journal of Social Issues, 53*(4), 657–666. https://doi.org/10.1111/j.1540-4560.1997.tb02454.x

Campbell, C. (1992). Learning to kill? Masculinity, the family and violence in Natal. *Journal of Southern African Studies, 18*, 614–628. https://doi.org/10.1002/casp.901

Campbell, C. (2004). The role of collective action in the prevention of HIV/AIDS in South Africa. In D. Hook et al. (Eds.), *Critical psychology*. UCT Press.

Catalani, C., & Minkler, M. (2010). Photovoice: A review of the literature in health and public health. *Health, Education and Behavior, 37*(3), 424–451. https://doi.org/10.1177/1090198109342084

Carlson, E. D., Engebretson, J., & Chamberlain, R. M. (2006). Photovoice as a social process of critical consciousness. *Qualitative Health Research, 16*(6), 836–852. https://doi.org/10.1177/1049732306287525

Chadwick, R. (2010). *Gender, sexuality and South African childhoods: A child rights approach*. RAPCAN.

Choi, S. Y., & Ting, K. F. (2008). Wife beating in South Africa an imbalance theory of resources and power. *Journal of Interpersonal Violence, 23*(6), 834–852. https://doi.org/10.1177/0886260507313951

Cooper, A. (2009). "Gevaarlike transition": Negotiating hegemonic masculinity and rites of passage amongst 'coloured' boys awaiting trial on the Cape Flats. *Psychology in Society, 37*, 1–17.
Cooper, A. (2010). "Through a different lens": Snaps of a soccer photovoice project in Lavender Hill Cape Town. *Agenda, 24*(85), 54–61. https://doi.org/10.1080/10130850.2010.9676323
Cooper, A., & Foster, D. (2008). Democracy's children? Masculinities of 'coloured' adolescents awaiting trial in post-apartheid Cape Town, South Africa. *Thymos: Journal of Boyhood Studies, 2*(1), 3–25. https://doi.org/10.3149/thy.0201.3
Cornell, J., & Kessi, S. (2017). Black students' experiences of transformation at a previously "white only" South African university: A photovoice study. *Ethnic and Racial Studies, 4*(11), 1882–1899.
Dixon, J., & Durrheim, K. (2000). Displacing place-identity: A discursive approach to locating self and other. *British Journal of Social Psychology, 39*(1), 27–44. https://doi.org/10.1348/014466600164318
Duncan, N., Bowman, B., Roos, V., Pillay, J., & Naidoo, A. (2007). *Community psychology: Analysis, context, and action*. Juta.
Elliot, W. (2003). *Masculinity: Key South African issues and debates. A Soul city review*. Retrieved from http://www.soulcity.org.za/projects/soul-city-series/previous-series/soul-city-series7/literature-review/masculinity-in-south-africa
Epstein, D. (1998). Marked men: Whiteness and masculinity. *Agenda, 14*(37), 49–59. https://doi.org/10.1080/10130950.1998.9675691
Erasmus, Z. (2001). *'Coloured' by history, shaped by place: New perspectives on 'coloured' identities in Cape Town*. Kwela Books.
Galtung, J. (1969). Violence, peace and peace research. *Journal of Peace Research, 6*(3), 167–191.
Gieseking, J. J., Mangold, W., Katz, C., Low, S., & Saegert, S. (2014). *The people, place, & space reader*. Routledge.
Helman, R., Malherbe, N., & Kaminer, D. (2018). To show like the people that there can be good fathers…here in our place. In T. Shefer, J. Hearn, K. Ratele, & F. Boonzaier (Eds.), *Engaging youth in activism, research, and pedagogical praxis: Transnational and intersectional perspectives on gender, sex and race*. Routledge.
Higgs, B. (1971). *The group areas act and its effects. United Nations Centre against Apartheid*. Retrieved from http://www.aluka.org/action/showMetadata?doi=10.5555/AL.SFF.DOCUMENT.nuun1971_01
Hunter, M. (2004). Masculinities, multiple-sexual-partners, and AIDS: The making and AIDS: the making and unmaking of Isoka in KwaZulu-Natal. *Transformation: Critical Perspectives on Southern Africa, 54*(1), 123–153.
Hyman, A. (2019). *Smiles all round as soldiers finally join the Cape Flats gang war*. Retrieved from https://www.timeslive.co.za/news/south-africa/2019-07-18-soldiers-finally-arrive-on-cape-flats-streets-a-week-after-they-were-sent-in/
Jensen, S. (1999). Discourses of violence: Coping with violence on the cape flats. *Social Dynamics, 25*(2), 75–97. https://doi.org/10.1080/02533959908458676
Jensen, S. (2004). Claiming community: Local politics on the Cape Flats, South Africa. *Critique of Anthropology, 24*(2), 179–207. https://doi.org/10.1177/0308275X04042652
Jewkes, R., Sikweyiya, Y., Morrell, R., & Dunkle, K. (2009). Understanding men's health and use of violence: Interface of rape and HIV in South Africa. *Cell, 82*(442), 3655–3660.
Keskinen, S. P. (2018). Territorial stigmatization, gendered racism and young people's agency in a multi-ethnic school. In T. Shefer, J. Hearn, K. Ratele, & F. Boonzaier (Eds.), *Engaging youth in activism, research, and pedagogical praxis: Transnational and intersectional perspectives on gender, sex and race*. Routledge.
Kessi, S. (2011). Photovoice as a practice of re-presentation and social solidarity: Experiences from a youth empowerment project in Dar es Salaam and Soweto. *Papers on Social Representations, 20*(1), 1–27.
Kessi, S. (2018). Photovoice as a narrative tool for decolonization: Black women and LGBT-student experiences at UCT. *South African Journal of Higher Education, 32*(3), 101–117. https://doi.org/10.20853/32-3-2519

Kessi, S., & Boonzaier, F. (2018). Centre/ing decolonial feminist psychology in Africa. *South African Journal of Psychology, 48*(3), 299–309. https://doi.org/10.1177/0081246318784507
Kessi, S., & Cornell, J. (2015). Coming to UCT: Black students, transformation and discourses of race. *Journal of Student Affairs in Africa, 3*(2), 1–16. https://doi.org/10.14426/jsaa.v3i2.132
Kessi, S., Kaminer, D., Boonzaier, F., & Learmonth, D. (2019). Photovoice methodologies for social justice. In S. Laher, A. Fynn, & S. Kramer (Eds.), *Transforming research methods in the social sciences: Case studies from South Africa*. Wits University Press.
Korpela, K. M. (1989). Place-identity as a product of environmental self-regulation. *Journal of Environmental Psychology, 9*, 241–256. https://doi.org/10.1016/S0272-4944(89)80038-6
Lamb, G. (2017). *Crime stats-the devil is in the detail*. Retrieved from https://www.news24.com/news24/analysis/crime-stats-the-devil-is-in-the-detail-20171102
Langa, M. (2010). Contested multiple voices of young masculinities amongst adolescent boys in Alexandra township, South Africa. *Journal of Child and Adolescent Mental Health, 22*(1), 1–13. https://doi.org/10.2989/17280583.2010.493654
Leventhal, T., & Brooks-Gunn, J. (2003). Moving to opportunity: An experimental study of neighborhood effects on mental health. *American Journal of Public Health, 93*(9), 1576–1582. https://doi.org/10.2105/AJPH.93.9
Louskieter, L. (2018). *PSY3005F: What is critical psychology, week 2notes*. Retrieved from https://vula.uct.ac.za
Mahalik, J. R., Locke, B. D., Ludlow, L. H., Diemer, M. A., Scott, R. P. J., Gottfried, M., & Freitas, G. (2003). Development of the conformity to masculinity norms inventory. *Psychology of Men and Masculinity, 4*(1), 3–25.
MacMaster, L. L. M. (2009). Resilience of faith communities on the cape flats (SA): A pastoral theological perspective 1. *Scriptura, 101*, 288–300.
Maldonado-Torres, N. (2007). On the coloniality of being: Contributions to the development of a concept. *Cultural Studies, 21*(2-3), 240–270.
Malinga, M., & Ratele, K. (2014). 'It has changed me from the person that I was before': Love and the construction of young black masculinities. In S. Petrella (Ed.), *Doing gender, doing love: Interdisciplinary voices*. Inter-Disciplinary Press.
McIntyre, A. (2003). Through the eyes of women: Photovoice and participatory research as tools for reimagining place. *Gender, Place and Culture, 10*(1), 47–66.
Moletsane, R., de Lange, N., Mitchell, C., Stuart, J., Buthelezi, T., & Taylor, M. (2007). Photovoice as a tool for analysis and activism in response to HIV and AIDS stigmatisation in a rural KwaZulu-Natal school. *Journal of Child and Adolescent Mental Health, 19*(1), 19–28. https://doi.org/10.2989/17280580709486632
Millin, S. G. L. (1924). God's Stepchildren. Boni and Liveright.
Mthembu, J. (2015). *Negotiating masculinities: Studying risk behaviours associated with performances of "'coloured's" masculinities*. University of Cape Town.
Ndlovu-Gatsheni, S. J. (2013). *Coloniality of power in postcolonial Africa*. African Books Collective.
Pather, R. (2019). *Mixed emotions as soldiers return to the Cape flats*. Retrieved from https://mg.co.za/article/2019-07-23-mixed-emotions-as-soldiers-return-to-the-cape-flats
Pinnock, D. (1984). *The brotherhoods: Street gangs and state control in Cape Town*. David Philip.
Pinnock, D. (1995). *Gangs, guns & rites of passage: Rethinking youth programmes in South Africa*. University of Cape Town.
Pinnock, D. (1997). *Gangs, rituals and rites of passage*. African Sun Press.
Pretty, G. H., Chipuer, H. M., & Bramston, P. (2003). Sense of place amongst adolescents and adults in two rural Australian towns: The discriminating features of place attachment, sense of community and place dependence in relation to place identity. *Journal of Environmental Psychology, 23*, 273–287. https://doi.org/10.1016/S0272-4944(02)00079-8
Prilleltensky, I. (2001). Value-based praxis in community psychology: Moving towards social justice and social action. *American Journal of Community Psychology, 29*(5), 747–778.
Proshansky, H. M. (1978). The city and self-identity. *Environment and Behavior, 10*(2), 147–169. https://doi.org/10.1177/0013916578102002

Proshansky, H., Fabian, A. K., & Kaminoff, R. (1983). Place, identity, physical world socialization of the self. *Journal of Environmental Psychology, 3*, 57–83.
Quijano, A. (2001). Colonialidad del poder, cultura y conocimiento en América Latina. In W. Mignolo (Ed.), *Capitalismo y geopolítica del conocimiento. El eurocentrismo y la filosofía de la liberación en el debate intelectual contemporáneo*. Signo.
Riessman, C. K. (2008). *Narrative methods for the human sciences*. Sage.
Robins, S. (2000). City sites: Multicultural planning and the post-apartheid city. In S. Nuttall & C. Michael (Eds.), *Senses of culture: South African cultural studies*. Oxford University Press.
Salo, E. (2003). Negotiating gender and personhood in the new South Africa adolescent women and gangsters in Manenberg township on the Cape Flats. *European Journal of Cultural Studies, 6*(3), 345–365. https://doi.org/10.1177/13675494030063005
Salo, E. (2004). *Respectable mothers, tough men and good daughters*. Emory University.
Sawyer-Kurian, K. M., Wechsberg, W. M., & Luseno, W. K. (2009). Exploring the differences and similarities between black/African and "coloured" men regarding violence against women, substance abuse, and HIV risks in Cape Town, South Africa. *Psychology of Men & Masculinity, 10*(1), 13–29. https://doi.org/10.1037/a0013267
Sernhede, O. (2011). School, youth culture and territorial Stigmatization in Swedish metropolitan districts. *YOUNG—Nordic Journal of Youth Studies, 19*(2), 159–180. https://doi.org/10.1177/110330881001900203
Smith, L. T. (1999). *Decolonizing methodologies: Research and indigenous peoples*. Zed Books.
Statistics South Africa. (2011). *Census statistical release*. Statistics South Africa.
Strack, R. W., Magill, C., & McDonagh, K. (2004). Engaging youth through photovoice. *Health Promotion Practice, 5*(1), 49–58. https://doi.org/10.1177/1524839903258015
Sonn, C. C., Stevens, G., & Duncan, N. (2013). Decolonisation, critical methodologies and why stories matter. In G. Stevens, N. Duncan, & D. Hook (Eds.), *Race, memory and the apartheid archives: Towards a psychsocial praxis*. Wits University Press.
Suffla, S., Kaminer, D., & Bawa, U. (2012). Photovoice as community engaged research: The interplay between knowledge creation and agency in a South African study on safety promotion. *Journal of Psychology in Africa, 22*(4), 517–528. https://doi.org/10.1080/14330237.2012.10820563
Switzer, L. (1995). Review of Adhikari, 'Let us live for our children'. *The Journal of African History, 36*, 338.
van Niekerk, T. J. (2019). Silencing racialised shame and normalising respectability in 'coloured' men's discourses of partner violence against women in Cape Town, South Africa. *Feminism and Psychology, 29*(2), 177–194. https://doi.org/10.1177/0959353519841410
Wang, C. C. (2006). Youth participation in photovoice as a strategy for community change. *Journal of Community Practice, 14*, 147–161. https://doi.org/10.1300/J125v14n01_09
Wicomb, Z. (1998). Shame and identity: The case of the coloured in South Africa. In D. Attridge & R. Jolly (Eds.), *Writing South Africa: literature, apartheid, and democracy, 1970-1995*. Cambridge University Press.

Towards Alternative Spatial Imaginaries: The Case of 'Reclaim the City'

Ruth Urson, Shose Kessi, and Shari Daya

Spatial divisions in South Africa's cities are among many markers of the country's patterns of inequality. In cities like Cape Town, spatial apartheid is visible in the continued divisions of land and urban areas along race and class. In 2017, members of a grassroots housing movement called Reclaim the City began to overturn these divisions. Their occupation of the former Helen Bowden Nurses Home, located in the wealthy central suburb of Granger Bay, was followed by occupations of a former hospital in the gentrifying central suburb of Woodstock and a government-owned building in the city centre. These were significant milestones for the young movement, which built on growing coverage in local media of its campaigns for affordable housing near urban centres.

Founded in 2016 under the banner of "Land for people, not for profit", Reclaim the City (RTC) aims to disrupt property power, increase low-income groups' access to centrally-located affordable housing, and end the displacement of evictees to areas outside the city (Reclaim the City, 2018a). RTC is guided by a formal constitution that promotes inclusivity, non-violence, equality, dignity and respect, justice, and transparency, and that supports poor and working-class struggles for affordable housing (Reclaim the City, 2018a). It is supported by Ndifuna Ukwazi, a non-profit research and legal assistance organisation in Cape Town. RTC's activities include weekly legal education and advice plenaries, regular meetings, community events, and creative protest actions. Alongside its legal and educational interventions, RTC's occupations exemplify its activist work in reimagining Cape Town beyond spatial apartheid.

In this chapter, we draw insights from the work of Reclaim the City for decolonial community psychology and introduce the "colonial spatial imaginary" as a tool for framing and challenging post-colonial displacement. We begin by framing space

R. Urson (✉) · S. Kessi · S. Daya
University of Cape Town, Cape Town, South Africa

S. Kessi et al. (eds.), *Decolonial Enactments in Community Psychology*, Community Psychology, https://doi.org/10.1007/978-3-030-75201-9_9

in relation to decolonial community psychology and examining the concept of the colonial spatial imaginary and its relevance to history. We use the case of Reclaim the City to illustrate how their work disrupts the colonial spatial imaginary and advances alternative spatial imaginaries.

Interpretive Framework: Situating Space in Decolonial Community Psychology

Decolonial community psychology is marked by three prominent accents, namely, advancing social justice, decolonising psychological knowledge, and re-examining subjectivity. As an overarching agenda, social justice promotes asking new questions in psychological knowledge production and seeking alternative solutions to social and psychological issues. Accordingly, decolonial community psychology calls for the inclusion of critical and liberatory perspectives in psychological knowledge production (Seedat & Suffla, 2017). By making visible and deconstructing colonial knowledge archives in relation to psychological phenomena, decolonial community psychology builds new ways of knowing (Carolissen et al., 2017; Seedat & Suffla, 2017). Watkins et al. (2018, p. 9) describe accordingly how decolonial projects need to be "attentive to local colonial and neoliberal histories and realities, struggles of resistance and cultural resilience, and the assets and needs of communities and their members."

Decolonial community psychology may therefore re-examine psychological factors like identity, subjectivity, and community in the context of the transgenerational trauma of systems like slavery and colonialism (Kessi & Boonzaier, 2018). "Subjectivity" here refers to personhood and self-awareness in relation to power, others, and desirability (Henriques et al., 1984, p. 3). It is dynamic, both constructed by social, political, and spatial forces, and experienced in certain ways by individuals and communities (Henriques et al., 1984).

We propose that the relationship between subjectivity and space warrants particular attention in decolonial psychology. Throughout history, processes of subjugation have required spatial transformations that achieve exclusion, control, and segregation (Foster, 2004). Spatial legacies of systems like colonialism and apartheid remain in material space and in spatial discourses of borders, central areas, and private space, for example (Foster, 2004). The link to subjectivity is established in the social sciences and geography, where place has been extensively theorised in relation to politics, identity, and subjectivity. In critical and feminist geography, place is argued to be constitutive of identities of class, race, gender, and sexuality, and in political geography, place is argued to shape citizenship processes and power (Paasi, 2003). In environmental psychology, place is often considered for how it becomes part of or influences an individual's identity (Proshansky et al., 1983) through processes of attachment and meaning-making at the individual and collective levels (Scannell & Gifford, 2010). However, social psychological investigations have often ignored our located-ness, or where we are and the meaning this holds for psychological processes (Dixon & Durrheim, 2000).

As Dutta (2016) argues, decentring often United States-based community psychology involves reconsidering both the types of issues we produce knowledge about, and the agents who produce this knowledge. It requires deconstructing the notion of the core and peripheries of knowledge production, to ensure that those on the social, economic, and spatial peripheries are able to produce knowledge that is recognised as legitimate (Dutta, 2016). Accordingly, the decolonial knowledge project is also in itself a geographical project, aimed at reimagining the landscape of knowledge production.

Activist movements such as Reclaim the City, however, have a disputed place in the knowledge production landscape, due to their operation in "invented spaces". While actions in "invited spaces" assist marginalised people in coping with the system that is marginalising them, actions in "invented spaces" aim to disrupt and challenge the status quo (Miraftab, 2004). Formal, legal, or otherwise government-supported invited spaces are considered to produce legitimate knowledge and behaviour, while actions in invented spaces are often considered deviant or illegitimate (Miraftab, 2004). In contrast, Holston (1998, 2009) acknowledges how structural factors may constrain the citizenship of marginalised communities and restrict their realisation of their legal rights. Insurgent citizenship, which operates from invented spaces, allows for the construction of alternative citizenships that can make more just structural and spatial realities possible (Holston, 1998). Accordingly, we examine the implications of RTC's work in both "invited" and "invented" spaces for decolonial community psychology praxis.

Participatory Action Research

Conducted in 2017 and 2018, this project drew on a Participatory Action Research (PAR) framework and photovoice method to understand how RTC members experience and resist evictions in Woodstock. PAR aligns well with a decolonial perspective on knowledge production as outlined in the previous section, in that it is a participant-centred approach geared towards action. PAR destabilises the relationship between researchers and participants, viewing researchers as collaborators with participants, and participants as experts on their own experiences (Watkins & Shulman, 2008). In this way, it allows for the decolonisation of knowledge, by centring the voices of marginalised groups in articulating their own knowledge and experiences and engaging researchers in the participants' context (Watkins & Shulman, 2008). Like decolonial community psychology, PAR seeks to create spaces for the generation of new socially-located understandings and knowledge that can assist in transforming the realities of marginalised communities (Watkins & Shulman, 2008).

We used photovoice, a PAR method, as a primary method of data collection. Photovoice facilitates photography and storytelling in small groups of participants around relevant themes. These tools can operate to produce decolonial psychological knowledge by subverting power dynamics in knowledge production and promoting critical consciousness, an understanding of how structural factors and power dynamics influence personal circumstances and psychological conditions (Montero, 2009; Wang & Redwood-Jones, 2001; Martín-Baró, 1996; Freire, 1970). Producing stories and images, and the critical dialogues that follow, can additionally stimulate action towards social justice (Wang & Burris, 1997).

By allowing participants to represent themselves through the art of photography and storytelling, photovoice creates multi-dimensional representations of experiences imbued with meaning (Wang & Burris, 1997), including those of subjectivity, history, memory, and physical space. In this way, photovoice can create empowering social representations of participants and their communities (Kessi, 2011), that may challenge colonial logic to advance decolonial understandings and identities.

Conceptualising Spatial Imaginaries and Place Identity

In analysing the work of RTC, we draw from work on spatial imaginaries in geography and cultural geography (Watkins, 2015). While imagination is an individual practice, spatial imaginaries are collectively shared internal perceptions, experiences, and beliefs that relate to physical space (Boudreau, 2007). These construct ownership, belonging, and identity in relation to space, and designate the spaces we belong in and the others that belong there too (Gregory, 2004). The literature varies in how it uses spatial imaginaries to refer specifically to representations of the unique features of places, to classify types of places with positive or negative associations in relation to certain ideals, or to represent how places will change, should change, or have changed (Watkins, 2015).

Extending this, spatial imaginaries can be seen to describe sites of struggle over contested meanings, circulated through discourses, practices, representations, and understandings (Davoudi, 2018). From this perspective, spatial imaginaries are embedded with power relations that designate the belonging of different groups in different spaces (Davoudi, 2018). As such, dominant spatial imaginaries may represent the lived experiences, perceptions, and values of some, but influence the spatial realities of others as well (Davoudi, 2018). Spatial imaginaries can similarly be performative, in that they influence how people behave within space, how dominant spatial imaginaries may be normalised, and how people and places are constructed as a result (Gregory, 2004). The consequence may be the exclusion of alternative spatial imaginaries, or the prevention of ideas about what places represent and how they should change from being advanced or internalised by marginalised groups.

For example, Lipsitz (2007) illustrates how, despite the end of formal racial segregation in the United States, race-related spatial inequality is preserved by a dominant "white spatial imaginary". Normalised by housing policies, this idealises homogenous and exclusive spaces. Similarly, Gregory (2004), drawing on Edward Said's (1978) concept of "imaginative geography", describes how spatial imaginaries can affirm or challenge colonial practices. The concept of imaginative geography has similarly been used to understand how spatial imaginaries advanced in the present reinforce colonial imaginaries, by excluding or erasing othered groups from representations of space (see Kothari & Wilkinson, 2010; Gregory, 2004).

This aligns with a critical psychological construction of place identity which conceptualises identity and subjectivity as being located in space and intertwined with place (Dixon & Durrheim, 2000). In this conceptualisation, identities are individually and collectively constructed in relation to space, within a context of power, history, and ideology. Group identities of place, accordingly, can have implications for who is considered in place and out of place in certain areas, serving to include or exclude, and privilege or marginalise, certain groups and individuals. This may affect the construction of subjectivity, such that a sense of self is lost when places change, and that people may be motivated to occupy places that align with their sense of self (Dixon & Durrheim, 2004).

The Colonial Spatial Imaginary

We introduce the "colonial spatial imaginary" as an interpretive tool that brings together a range of understandings related to space and subjectivity. The colonial spatial imaginary encompasses collective perceptions and values that shape place identities, construct belonging in different places, and normalise certain spatial patterns and activities. In the colonial spatial imaginary, place identities operate as colonial representations of places and associated notions of belonging, that may be contested by alternative spatial imaginaries. Accordingly, we emphasise three features of the colonial spatial imaginary: the representation of areas as centres and peripheries, the designation of power and belonging in each of these areas, and the normalisation of certain spatial practices in these areas, particularly patterns of displacement.

Foster (2004) highlights how the spatial discourses associated with subjugation contain metaphors relating to hierarchy, centrality, and sphere (public or private). Discourses of centrality, and the hierarchies they infer, are particularly relevant to the colonial spatial imaginary. The centre-periphery framework classifies desirable and economically productive areas as central, such as productive land, business districts, well-resourced suburbs, and developed urban areas. The periphery, on the other hand, contains the areas which are not central or chosen by their inhabitants, and to which the "other" is relegated when deemed not to belong in the centre. In the colonial era, European towns, forts, and farms were constructed by colonialists as central, and slave accommodation and indigenous settlements as peripheral. Within the colonial spatial imaginary, belonging in the centre is contested but policed to ensure its inhabitants meet certain standards of respectability and economic standing. The association of central areas with respectability, development, and economic value is contrasted with the "wild" peripheries, which are excluded from belonging in the city. In both areas, property ownership is instrumental in distributing power, but this is particularly so in the desirable centres.

The colonial spatial imaginary is performative and influences the enactment of spatial practices and the construction of subjects within certain spaces. It may influence and be influenced by factors like how belonging in central and peripheral urban areas is viewed; how property in each area is valued and distributed; the dominant identities conferred upon different places; and the prioritisation of interests in attempts to transform urban space. Spatial practices, in turn, may increase the value held by the central city areas, while implicitly or explicitly controlling the deficits and threats of the periphery, and moving people from areas in which they are deemed not to belong. In the section that follows, we apply the colonial spatial imaginary lens to trace displacement through Cape Town's colonial, apartheid, and post-apartheid history.

The Colonial Spatial Imaginary in Cape Town's History

From colonialism and apartheid through to the current post-1994 period, the centre-periphery distinction has been maintained by discourses of belonging and respectability and patterns of displacement, such as frontier wars, forced removals, and evictions. The colonial spatial imaginary underpinning these phenomena is not absolute, but dynamic and contested. This is illustrated in the sections that follow.

The Colonial Era: The Cape Colony

Colonialism in the Cape was marked by the establishment of colonial settlements as an ever-expanding centre; and the dislocation of indigenous groups towards the peripheries of colonial settlements. In 1652, settlers from the Dutch East India Company established a port at the Cape (Mountain, 2003). Long before their arrival, the area was inhabited by indigenous San and Khoe (or Khoi) hunter-gatherers and herders. Bantu-speaking groups had also migrated from West African regions and settled towards what is now the Eastern Cape (Laband, 2020).

Colonial farmers viewed the San people as "a nuisance factor" (Mountain, 2003, p. 29) in lands they fabricated to be unoccupied and available for their occupation and appropriation. This was exacerbated by the Company's sponsorship of Dutch farmers to develop farms on viable land further from their initial port and their gradual encroachment on indigenous territories and hunting grounds. Despite fierce resistance as Khoe and San groups fought to defend their land and resources, in many cases, the Dutch dispossessed them of land, water, and food sources (Mountain, 2003). In some cases, this forced Khoe and San groups into servitude (Thompson, 2001).

This colonial spatial imaginary pervaded Dutch settler movement as their expansion of the centre, boundaried by the colonial frontier, led them to Bantu-speaking settlements such as those of the amaXhosa. These interactions escalated into violent wars interspersed with tense habitation of common land. Notably, while colonialists are recorded as regarding land as private property that could be acquired, amaXhosa groups regarded land as communal and allocated by a chief (Laband, 2020). In addition, the colonialists are described as having generally seen themselves as culturally superior to Africans and aimed to exclude Africans from their territories (Laband, 2020). These factors of property ownership and exclusionary segregation, remained salient features of the colonial spatial imaginary over time, to the exclusion of alternative spatial imaginaries.

In 1795, the British arrived as the second wave of colonialists (Mountain, 2003). They were met with resistance from the Dutch, local amaXhosa groups, and remaining Khoe and San people, all organised in complex and diverse allegiances (Laband, 2020; Thompson, 2001). However, by 1809, the British had solidified aspects of the colonial spatial imaginary into policies that racialised spatial planning (Thompson, 2001). In addition to displacing amaXhosa groups from their land towards the peripheries, the British spatially segregated the amaXhosa and Europeans, an early manifestation of the apartheid logic. Slaves, however, were allowed to live in closer proximity to European settlers (Thompson, 2001). When slavery was eventually abolished, South African colonial society had been established as one that tied land ownership to wealth and political power, entrenching former slaves and African groups in poverty and slightly elevating the few that allied themselves with the colonial powers (Thompson, 2001).

Over the following decades, urbanisation of the Cape, expansion of colonial territories, and forced removal of local black groups from their land continued. From the 1850s to 1913, when the Natives Land Act was passed, local European authorities increasingly restricted "native" groups to segregated settlements outside developing urban areas (Mabin, 1992). Cities began to be shaped according to a centre-periphery model, supported by discourses that constructed various "others" in relation to the positioning of cities as a European invention and living space for respectable people (Nahnsen, 2006). A sanitation discourse constructed white Europeans as clean, civilised, and global, and Africans as dangerous, rural, linked to crime and unsanitary conditions, and threatening the civilised nature of the city with their way of life (Miraftab, 2012; Nahnsen, 2006). Such fear-based racial representations partially justified ideological and political race-based segregation (Nahnsen, 2006).

From the 1850s, segregation was also reinforced by industrialisation, which saw central city areas receiving more investment than poor and peripheral areas (Miraftab, 2012). This bolstered a system linking property ownership with political citizenship and power (Miraftab, 2012; Nahnsen, 2006). Those who owned property held power to make political decisions, while those who did not own property could not make such decisions, contributing to the creation of an elite economic and political class. Cape Town was founded within this colonial spatial imaginary that displaced African groups from their land to areas segregated from colonialists, while expanding the property power of colonialists.

The Apartheid Era: Spatial Apartheid in Cape Town

Building on colonial legislation, the Population Registration Act of 1950 cemented the racial classifications of "native", "coloured", "white", and "Indian" that subsequently underpinned the spatial segregation project of the apartheid regime.[1] Most powerfully, the Group Areas Act, also enacted in 1950, demarcated most areas of the city as reserved for particular races. While some groups classified as "coloured" were permitted to live closer to the city, separate settlements were built for groups classified as "native" so as to separate them completely from those classified as "white" when not serving as labour (Miraftab, 2012).

The spatial lines drawn during the colonial era, which reserved colonial centres for European settlers and forced "non-whites" to peripheral areas, were reinforced by apartheid legislation and the discourses underpinning it. Actual outbreaks of disease in peripheral areas, arising from inadequate space and services, were mobilized by apartheid institutions to invoke and naturalise the sanitation discourse, and justify segregation as necessary (Miraftab, 2012). This sanitation discourse was instrumental in the forced removal of about 55,000 residents from District Six in 1966 (Hart, 1988). District Six was a neighbourhood next to Cape Town's city centre inhabited by residents classified under various apartheid racial categories, as well as international immigrants and refugees. It survived over a decade of the Group Areas Act's attempts to remove the majority classified as "native" and "coloured" from urban centres to peripheries. As its population grew and the Group Areas Act increasingly took hold, authorities began to fabricate District Six as a "slum" that had no place in the inner city. They demolished the area, classified it for "white" residence only, and removed residents classified as "native" and "coloured" to a series of group areas on the city's periphery called the Cape Flats (Hart, 1988).

The presence of District Six was not tolerated by the apartheid government because its existence as a low-income, racially-diverse community in the desirable urban centre had contradicted the logic of the colonial spatial imaginary. Its destruction ended this reality (Beyers, 2008). Those displaced lost their livelihoods, property, and proximity to economic centres, to be forcibly removed to peripheral areas with little character, infrastructure, safety, or opportunity. The disruption of communities and collective identity, the psychological trauma caused by the displacement and destruction, and the fear, anger, and bitterness of displaced groups, were long-standing effects of these events (Hart, 1988).

[1] The Population Registration Act legislated the use of three racial classifications, "native", "coloured", and "white", to classify racial groups. Later, the additional classification of "Indian" was added. The principles of racial classification remain in use today, for purposes of recognition and redress, but they are not uncontested. The categories most widely used are "black African", "coloured", "white", and "Indian" with some documents also including "Asian". Since they remain commonly used in both popular and official discourse, we use them in this chapter to discuss both historical and contemporary realities. We do so in full awareness of their contested and often oppressive nature.

Post-Apartheid Cape Town

Cape Town's post-apartheid landscape is remarkably unchanged since apartheid. The latest census data at the time of this project indicates that white groups predominantly inhabit a swathe of higher-income areas stretching from the mountainous southern suburbs across the urban centre to the northern suburbs (Frith, 2011). Woodstock is one of the few central suburbs that continues to defy this pattern. On the broad peripheries of these suburbs are a range of lower-income neighbourhoods and informal settlements with high densities of residents identified as black African and coloured in the 2011 census (Frith, 2011). This is unsurprising, given the economic and political factors that have sustained apartheid-era racial segregation and the continued privileging of the economic elite. Accordingly, the colonial spatial imaginary manifests today through policies that maintain the linkage of property ownership, wealth, and political citizenship; a modern sanitation discourse which seeks to rid the city of "crime and grime"; economic inequality; and patterns of displacement that see the wealthy living in proximity to the city and the poor living on its underinvested periphery.

The Cape Town City Improvement District (CID) or Special Rates Area (SRA) system is one example of the continuing maintenance of colonial links between property ownership and political power. CIDs or SRAs are urban areas in which property owners pay higher levies to the City to fund additional services to be planned and managed for their area only by their own elected board (City Improvement, 2009). Tenants are not included in this decision-making (Miraftab, 2012). This system encourages initiatives that rid CIDs of "crime and grime" and allows CIDs explicitly to propose and orchestrate changes to "public safety measures" and "social services to deal with vagrancy" (City Improvement, 2009).

Miraftab (2012) argues that classifying these policies as innovative undermines South Africa's post-apartheid potential to imagine urban renewal beyond colonial and apartheid lenses. While the CID model can benefit property owners and CID users, it can also mobilise property owners as an elite class in support of unequal and exclusionary spatial practices. Allowing the private sector to arrange superior services in their area and remove people classified as "crime and grime" allows for the enclaving of middle- and upper-class areas. It does little to disrupt the existence of the under-resourced urban periphery. Rather than challenging the colonial spatial imaginary, this perpetuates a colonial urban citizenship contingent on wealth and property ownership (Miraftab, 2012).

More broadly, displacement in the post-apartheid era can physically reinforce the centre-periphery distinctions of colonialism and apartheid. While colonial expansion and apartheid policies related to race and nationality were historical causes of immense displacement, displacement from centres in post-apartheid Cape Town often results from the economic processes of eviction and gentrification. Woodstock is one central suburb of Cape Town that features prominently in eviction reports. Woodstock was a historically working-class, multiracial area that narrowly escaped the scale of forced removals experienced in District Six. Under the Group Areas Act

(GAA), some parts of it were zoned as "white" areas and some as "coloured" areas, but some areas remained "controlled" or "undefined" areas without a formal racial classification (Garside, 1993). People classified as "coloured" were therefore able to rent accommodation or buy property in Woodstock through nominee buyers, and made up a large proportion of its population (Garside, 1993). These factors allowed Woodstock to retain a practical status as a multiracial area, although residents classified as "coloured" still lived with a constant threat of displacement (Elder, 1990).

From the 1980s onwards, gentrification in Woodstock escalated. Under economic pressure to remain competitive, landlords increasingly renovated their properties and raised their rents, or sold their properties to developers, leading to evictions of longstanding working-class residents (Garside, 1993). Businesses and new developments began to target a wealthier clientele. In 1998, the Prevention of Illegal Eviction from and Unlawful Occupation of Land Act ("PIE Act") was passed. This post-apartheid act requires that evictions be classified by a court as "just and equitable" prior to any displacement taking place. Although there is no formally published statistical record of eviction cases at the time of this study, a report by Ndifuna Ukwazi gives some idea of the extent of high-profile evictions in Woodstock in the past 20 years (Pillay et al., 2017). Such evictions have displaced up to 200 residents at a time, whole rows of houses, or families who had lived in their homes for decades (Pillay et al., 2017). Many residents were evicted from poorly maintained buildings by landlords intending to renovate their properties to attract wealthier tenants, some for larger developments, and many also faced unaffordable rent increases (Pillay et al., 2017).

Adding to the plight of evicted residents in Cape Town is the City's solution of Temporary Relocation Areas. Where evictees are to be rendered homeless and are on the waiting list for government-sponsored housing, they are eligible for transitional housing under the Emergency Housing Programme (Cirolia, 2014). However, the only transitional housing at the time of this study existed at Delft Symphony Way Temporary Relocation Area (commonly known as Blikkiesdorp, Afrikaans for "Tin Can Town") and Wolwerivier Temporary Relocation Area. Both are about 30 kilometres outside the city centre, where land is cheaper. These sites have been labelled inhumane and unjust for their living conditions and location. Criticism centres on their distance from residents' communities, schools, and places of work; overcrowding; small, non-secure, leaking and uninsulated shacks; as well as crime and the lack of amenities such as police stations and hospitals (Pillay et al., 2017; Ranslem, 2015; Cirolia, 2014).

Certain parallels of spatial displacement, property-related elitism, and racial exclusion can be drawn between the present-day case of gentrifying Woodstock, and the cases of displacement in District Six in 1966 and the colonial-era Cape. Despite complex differences in cause and intent, Woodstock's patterns of displacement often seemingly comply with the colonial spatial imaginary, by moving low-income, non-property-owning residents, particularly black African, coloured, and Indian residents, to the under-resourced urban periphery. Complicating this is the difficulty for government of sourcing adequate land for public housing in the inner city, due to Cape Town's high inner-city property values and increasing gentrification. Modern-day displacement in Woodstock concentrates wealth, government investment, and opportunity in central areas and, thus far, concentrates low-income housing on the city's under-invested peripheries, which reinforces rather than transforms spatial apartheid (Turok, 2001). This facilitates the economic and spatial exclusion of low-income groups from the inner city, reinforcing the colonial spatial imaginary. This is the context in which Reclaim the City operates.

A Participatory Action Research Project

Our photovoice project began with four members of Reclaim the City, two women and two men. For the purpose of this chapter, they are given the pseudonyms Yonela, Helena, Junaid, and Opopo. A fifth participant, Carol, joined the study for an initial focus group but was unable to participate further. Yonela, Helena, Junaid, and Carol lived in RTC's Woodstock occupation (Cissie Gool House), while Opopo was facing eviction from a nearby apartment. While the others are South African, Opopo is from another African country. During this study, Yonela and Helena both served as House Leaders of Cissie Gool House, responsible predominantly for resolving issues that arose and making sure rules were followed.

During the photovoice project, triangulation was used to capture and contexualise the multifaceted experiences of displacement and resistance within RTC, while centring the experiences of its community members. Key informant interviews were held with the City Councilor and Mayoral Committee Member for Urban Development and Transportation, and the head of political organising at Ndifuna Ukwazi. In addition, given that PAR requires an in-depth and extended involvement, Ruth, the first author, attended RTC's weekly meetings and events for several months before, during, and after the photovoice project.

With permission from the organisation's leadership, Ruth introduced the project during a Woodstock chapter meeting and invited members to join. All members that expressed interest joined the project. The photovoice project was run in 2018 and followed the general stages specified in the literature, namely, an initial focus group discussion took place to establish a topic for photography; participants were trained in photography and had time to take photographs; these photographs were discussed to develop themes over successive meetings; and an exhibition was planned (Wang & Burris, 1997; Castleden et al., 2008). Participants in the photovoice project, interviews, and focus groups gave their written informed consent prior to participating in the project, and exhibitions were planned with photovoice participants and RTC leadership. Consultation with participants was also maintained and invited during the project, to share ownership of the process insofar as possible.

All research was undertaken by Ruth, a white woman, as part of her dissertation for a Master of Arts in Psychological Research. Reflexivity was maintained by Ruth throughout the project, to acknowledge how the research was co-constructed by the researcher and participants, and the role that identity plays (Dwyer & Buckle, 2009). This is particularly salient as being present at meetings, building relationships, and being an "insider" to RTC's goals is a large part of being involved in the movement, rather than being an aloof and distant researcher. While genuinely supporting the movement's goals and participating in its activities, Ruth drew on reflexive processes to maintain an awareness of the power dynamic of being a privileged outsider with a racial identity and economic position that differed from almost all RTC members. Such processes with this awareness, Ruth aimed to avoid overstepping boundaries and to centre participants' voices in the research. This presented its own complexities, as "insider" and "outsider" identities could be dynamic at times in the context of the relationships built over the course of the project. Still, continuous conversations with external academic peers, personal reflection throughout the project, and discussions with participants regarding their preferences for important aspects of the process were emphasised.

Accordingly, two levels of analysis were conducted, each serving a slightly different purpose. First, a preliminary analysis took place in collaboration with photovoice participants. This was intended to facilitate their own development of insights that could be directly applied to their stories and raise their critical consciousness as a group. A secondary analysis conducted by the researchers then combined these insights with themes from participant observation, interviews, focus groups, and the photographs and photo-stories. The aim of the secondary analysis was to develop broader, cross-cutting theoretical insights about the work of RTC as a whole. The section that follows should therefore be read as the authors' interpretations based on the perspectives of the participants gathered in the first level of analysis and the broader range of data collected.

Findings: Challenging the Colonial Spatial Imaginary

The findings of our PAR project illustrate how the colonial spatial imaginary is contested by alternative spatial imaginaries advanced by RTC. To further contextualise RTC's work, we first introduce two particular stories told by RTC members in relation to their experiences of eviction in this section, before describing how RTC works with these stories in the next section. The first story is that of the normalised spatial practice of inordinately harsh and illegal evictions, and second is the displacement of low-income residents from centres to peripheries.

Many RTC members described harsh evictions in which their landlords threw their belongings out of their homes, as Yonela's story and photograph shows (Fig. 1). All photographs and accompanying stories in this chapter were taken and written by the participant whose pseudonym appears with them.

Without owning their homes and, in some cases, being provided with only a verbal rather than written lease agreement, some were particularly vulnerable to this treatment. Similarly, Helena compared the experience of a tenant being evicted with her own experiences in the past, highlighting the lack of a fair legal process for both (Fig. 2).

Fig. 1 A photo-story by Yonela To me this picture reminds me where I'm coming from and what I pass through. This is the house I called home some years ago where I was living with my Aunt, then I was 17 years old. We lived in this house for 9 years and then my Aunt passed away, but before she passed away we were evicted by the landlord and our clothes and furniture were thrown out in the road (Yonela, Photo-story 1)

As an economic and legal process, eviction is associated with the termination of an economic contract or its violation by an unlawful occupant. However, both Yonela and Helena's stories evoke a parallel meaning of eviction, as harsh displacement occurring without reason, respect, or due process. These experiences were echoed by those of other residents who came to RTC for assistance, as Ruth observed in the meetings she attended. This provides insight into the context of displacement in Woodstock, where such stories arise despite the PIE Act mandating that evictions be "just and equitable" and that they only proceed with a court order. In drawing attention to their stories of harsh displacement, Yonela and Helena problematised this particular spatial practice.

Opopo's photo-stories similarly described how his landlord facilitated an illegal constructive eviction, chipping away at the floors and ceiling, ruining his belongings, and littering the flat with building materials, in an attempt to force him to move out. Other RTC members mentioned having had similar experiences with their landlords. However, Opopo also explained how, being African but not South African, he was made to feel out of place both by his white landlord and the legal system. His photo-story below described how court papers are particularly intimidating to

Fig. 2 A photo-story by Helena This photo, it reminds me how I was evicted, because I did not have a place to go, like those people. They did not receive any court order; they just throw their things outside. This is the same situation that happened to me (Helena, Photo-story 1)

"foreigners", who already feel a sense of displacement living in the country as non-South Africans. Notably, he describes how this intimidation leads tenants to leave rather than pursuing the legal process (Fig. 3).

With little recourse to resist displacement, both legal and illegal evictions often displaced those removed from their homes into precarious living situations, or "peripheries" within central areas. These included homelessness; living in shared rooms, *hokkies* (a colloquial term for small makeshift dwellings), or backyard dwellings on other residents' properties in Woodstock; or moving further from the urban centre.

Opopo's story challenges the compliance of tenants with unfair eviction procedures. In contrast, after attending RTC's Advice Assemblies, he chose to pursue the eviction proceedings in court, supported by fellow RTC members and lawyers from Ndifuna Ukwazi. Ultimately, he won his eviction case, but not without pain, as he described in an interview.

> On the stand, he [Opopo's landlord] made a statement that was very painful… he said that he wanna renovate the place and give it to "decent" people- so it was, for me, it was really painful and I say what do you mean, "decent people", explain. He said "It's not a charity organisation," so I said, "I've never said you were a charity organisation, but at least, there's some difficulties that- we've stayed here for this long, and I've explained to you that this is it, but I make sure that I pay the rent- even if yes, the accumulation is there, I'm working

Fig. 3 A photo-story by Opopo Notice of motion (eviction papers): this photo is now the focus of the legal process of eviction. These are court papers and it is not easy as a kwerekwere or as "we" are referred to as "Foreigners". It's very intimidating. Others will just pack their things and leave the house, as they won't have time to go through the courts. "Notice of Motion" Section 4(2) notice. Formal or official start of eviction by getting the above summons. At face value it is traumatising just getting it served to you. To me, the rental issues and arrears, which is explainable, led to the present scenario. We need more stories to be told about the callous treatment of landlord to tenants. Tenants have got to speak out the abuses, but also look at how justly and equitably this can be handled (Opopo, Photo-story 1)

> double hours, just to pay the accumulated…" But he was constantly terrorising me and given the fact that he's a lawyer, he got a firm, he keeps on telling me that he will make my life a living hell. So this was the process of negotiating, or I mean, having a "meaningful dialogue" with him. But he kept on that track. I think he deals with a lot of Africans that are always afraid of him…
>
> (Opopo, Interview)

Implicit in Opopo's story is the narrative of the colonial spatial imaginary, that normalises the upgrading of areas for "decent people" who belong there, and the removal of those who do not belong. While gentrification does exert pressure on landlords to attract higher rents and perhaps decreases their flexibility to late payments, Opopo's landlord did not follow due process or respond to Opopo's efforts at humanistic "meaningful dialogue". Instead, Opopo describes how his landlord resorted to intimidation and shaming tactics, as well as an illegal constructive eviction, to force Opopo to move elsewhere. Rather than internalising this, however,

Opopo challenged this practice by challenging his landlord's characterisation him as being out of place and not a "decent person".

While mistreatment by landlords was often the starting point of eviction stories, the threat or reality of being removed from central Woodstock to the far-flung periphery usually featured too. Opopo, for example, classified the city's periphery as a "dumping ground" for poor residents. These sentiments were echoed by other participants.

> Because like now, in Blikkiesdorp, you see that it's a temporary settlement, but people have been staying there for years, and uh, it means their life doesn't really matter. Uh, the policies legally shapes their life… uh, as soon as those are, they aren't good to stay around here, they should dump them there- they have been dumped, because there are no conditions there for them to come to town, just people who are struggling to live on a daily basis, just calculate their transport…
>
> (Opopo, Focus Group)

Being displaced to the periphery was described as creating immovable barriers to re-entering the city, thereby transforming city residents into waste products of urban development. This characterisation is underpinned by colonial place identities which normalise the central residence of wealthy property owners and peripheral displacement of "poor" black residents.

The Work of Reclaim the City: Building Alternative Spatial Imaginaries

These two stories, of callous treatment from landlords and displacement from centres to peripheries, featured regularly in Reclaim the City's weekly Advice Assemblies. Here, members of RTC gathered in Cissie Gool House to learn about eviction law, provide support to one another, and build solidarity to help one another navigate their evictions. At these meetings, members and facilitators listened and responded empathetically to attendees' stories of eviction, and drew links between their experiences to reframe evictions as systemic rather than individual. Members and facilitators also worked to reframe the place identity of Woodstock, from an exclusive gentrifying neighbourhood to a collective home in which working-class families of many racial identities have existed since before the most recent wave of gentrification. Members shared legal advice and encouraged one another to stand their ground in the face of intimidation by landlords. Accordingly, "No eviction without a court order" was common wisdom passed from longer-standing attendees to new, reinforcing their right to a fair legal process.

These collective practices act as an intervention into the undertones of displacement; the undercurrents of the colonial spatial imaginary in the more visible forces of eviction and gentrification. They are intended to disrupt tenants' feelings of not belonging, their minimisation of their own respectability, and the shame experienced in relation to eviction from the centre, and related complicit patterns of movement. This shame is accompanied by an identity of victimhood and a sense of a lack of control. In its place, RTC builds tenants' sense of deservedness and power to take action against unjust evictions (Reclaim the City, 2018b).

The following focus group extract demonstrates accordingly how Woodstock was defiantly defended by participants, drawing on terms like "our place" and "our home," emphasising ownership and belonging.

> Junaid: *You see, those were the days but, looking now this is 2018, what's gonna happen in 10 years, 5 years' time to Woodstock? So we need to fight for Woodstock.*
> Helena: *Ja, we fight for our Woodstock. This is our place.*
> Junaid: *We need to fight for Woodstock. This is our home.*
> Helena: *This is my hometown.*
> Carol: *I'm not leaving Woodstock-*
> Junaid: *I'm not living in Delft* [Blikkiesdorp]. *They'll have to put me in a body bag and take me out of Woodstock.*
> (Focus Group)

This sense of ownership is reinforced by RTC's adoption of "extra-legal resistance", which regards the existing structures, procedures, and laws as elements flexible to change when they no longer serve the people. Extra-legal resistance is RTC's term for resistance to practices that are not considered fair, equal, moral, or just, even if they are legal. In a public meeting held in Woodstock, one RTC leader captured this perspective by arguing that, "laws evolve and there is nothing that's cast in stone that the city can't change". To cheers and applause, he explained that laws were what allowed apartheid to happen, so people should not be complacent and regard laws as a moral standard, but rather, should evaluate laws and legal practices against the needs of the people. This perspective, which regards rights as a vessel for demanding moral behaviour, rather than the moral standard for behaviour itself, opens up possibilities for change.

While RTC's extra-legal resistance is often controversial for its operation from an "invented" rather than "invited" space, it is often the most impactful in challenging the colonial spatial imaginary, both for its members and the public. The head of organising at Ndifuna Ukwazi, the organisation from which RTC was developed, described how extra-legal resistance opens up the consideration of a more moral spatial reality beyond the bounds of the law.

> We are not addressing the legacy of apartheid. We actually are addressing the fine details of apartheid, that we still have to live with in Cape Town currently. So that's basically our job, whether we build that through the Advice Assembly, where we educate people about their rights, and make them familiar with the courts, assist them in the court, administrative, kind of side. And making sure that the people power that we are talking about is also resisting a sheriff when they come to evict one unlawfully, and even if they follow the law, a lawful eviction that would result in someone being homeless, that's, that for me is morally not acceptable. It cannot be that as a judge you will say this person should be evicted into homelessness, you know, or should be evicted to be sent to Wolwerivier…
> (Key informant interview)

RTC's extra-legal resistance takes different forms. Sometimes, it involved the gathering of members to physically resist evictions as described. However, sometimes it has a distinct sense of creativity and humour. For example, at 5am on a March morning in 2018, RTC activists set up a breakfast table laden with koesisters (Cape Malay doughnuts) outside the home of Councillor Brett Herron. As the City of Cape Town's Mayoral Committee member responsible for housing at the time, RTC invited Herron to sit and discuss dignified housing solutions with them, which he did. Another event, held on Heritage Day, invited the public on a walking tour of Woodstock, which stopped at different sites to hear the stories of those who had faced or were facing evictions.

One of the movement's most prominent forms of extra-legal resistance is the act of occupation. Like other occupations, the occupation of the Woodstock Hospital began as a symbol of people's power, challenging government inertia in providing housing by reclaiming public space to serve its social need. RTC named the occupied Woodstock Hospital Cissie Gool House after Zainunnisa "Cissie" Gool, Cape Town's first black woman City Councillor, who represented her area of District Six until the 1950s. Symbolically, this brought black and coloured low-income residents facing displacement firmly into the visible centre, in contrast to their hidden peripheral status in *hokkies,* homelessness, backyard dwellings, and evictions to the city's periphery. Squatting or occupation can be seen in this way, as a means of advancing otherwise neglected agendas for those who are marginalised and unable to lawfully afford to take up urban space (De Carli & Frediani, 2016). Symbolically, occupations can re-appropriate and politicise public space to challenge increasing privatisation of space (Abellán et al., 2012) and its related process of gentrification (De Carli & Frediani, 2016).

Beyond their symbolism, however, the occupations exemplify how RTC challenges the colonial spatial imaginary with insurgent citizenships that facilitate the taking up of space and influencing of policy in Woodstock, despite not being able to afford gentrified rents or owning property. Like other housing movements before them (Berglund & Peipinen, 2017; De Carli & Frediani, 2016; Abellán et al., 2012; Earle, 2012; Holston, 2009; Miraftab & Wills, 2005), RTC's insurgent citizenship expands upon the marginalising citizenships available to evictees within the limitations of the law.

Cissie Gool House, for example, acts as a housing provision structure for low-income residents, operating both in parallel and as a challenge to more official structures. In describing, accordingly, how RTC provided accommodation to evictees who were rendered homeless, Helena classified the occupation as a housing provider in Woodstock and a necessary resource for evictees. This is in contrast to a current system she characterises as harsh and uncaring (Fig. 4).

Helena's sentiments were echoed by a member of Ndifuna Ukwazi, who addressed Councillor Herron in a public meeting by declaring, "Currently Reclaim the City is the only provider of emergency housing in the inner city". This comment challenged the local government's reliance on emergency housing in areas outside the city. Rather than requiring people to pay rent, occupiers were required to be in need of emergency housing, to contribute to their community by attending RTC meetings and activities, and to follow a number of house rules centred around respect, equality, sobriety, and non-violence. The intake process followed a structured waiting list run by a committee. Yonela described how leaders agreed that anyone could stay in the building.

Fig. 4 A photo-story from Helena This is for one of our evictees who was being evicted and they did not have a place to go anyway. Thanks for our organisation, which provided accommodation for them. It also tells me how the court don't care for poor people (Helena, Photo-story 7)

> ... so we said, in the meetings, that occupation is for everyone. It doesn't matter if you come from Congo, you know, Nigeria, Burundi, wherever, as long as you are evicted or as long as you have a problem.
>
> (Yonela, Interview)

In this way, the occupation attempts to realise RTC's inclusive values in urban space. Within the colonial spatial imaginary, such prime central land comes at a price, excluding those unable to afford it. However, RTC's actions advance the ideal of "Land for people, not for profit", determining such central land to be more valuable if used to resolve the social need for housing. Studies of occupations in Spain and Brazil similarly found that occupations produce an experimental site for the practice of alternative housing, showing that housing for marginalised groups is possible in the inner-city (Abellán et al., 2012; De Carli & Frediani, 2016).

The occupation of Cissie Gool House therefore creates something not tolerated within the colonial spatial imaginary: centrally-located social housing provided regardless of wealth and identity. Helena explained how she believes politicians have something to learn from these ways in which RTC has re-purposed public space. She refers specifically to the former Premier of the Western Cape, who had publicly challenged RTC.

> We gonna tell her, come, have a look, and see what you can do, how many people is evicted here. And how many people is staying here. Come and have a look. Don't say you want to, you don't know, you don't see, because you never come have a look. So let's see *maar, come and have a look for us. And how we took this place and how we did stay here. You see?*
> (Helena, Interview)

This perspective was echoed in the exhibitions of participants' photo-stories that marked the culmination of our photovoice project. The first exhibition was part of a bigger event held by RTC that was unfortunately interdicted. However, its intention was to invite the public to tour the occupation and see how rather than "squatting illegally", RTC had turned an abandoned building into a thriving example of central social housing. The second was a powerful event in which participants showed their photographs and told their stories to fellow RTC members during a meeting in Cissie Gool House. In 2019, the final exhibition formed part of an Open Streets event aiming to highlight streets as public spaces for social interaction and alternative transportation. Participants' photo-stories were exhibited among RTC's collection of portraits of occupiers and titled, "We are not criminals", which drew the exhibitions together to reframe negative representations of occupiers and activists in more humanistic ways. These exhibitions illustrated RTC's attempts to represent themselves to the public and to one another, not as criminals within the current system, but as paving the way for a new system that prioritises the social potential of space over its market value and historical place identities.

RTC's operation from invented spaces often delegitimises their work to those in power. Nevertheless, they have built growing support across local and international civil society and have, in combination with the work of a number of other significant stakeholders, seen some formal successes in the campaign for affordable housing. Notably, in 2017, the City of Cape Town committed to releasing a number of parcels of land for central affordable housing development. The Pine Road informal settlement in Woodstock, for example, was earmarked for development of the first inner city social housing project in Cape Town, set to accommodate 220 families (Development Action Group, n.d.). The 19 families already living in the Pine Road informal settlement were relocated to facilitate this development, not to Blikkiesdorp or Wolwerivier, but rather to the first central transitional housing project established by the City of Cape Town, located in Pickwick Street in Woodstock.

Conclusion: Building Alternative Spatial Imaginaries

Decolonial community psychology is marked by prominent themes of advancing social justice, decolonising psychological knowledge, and re-examining subjectivity. These can be interrelated with space, building on the historical significance of space in colonial systems and a growing body of literature linking place and subjectivity. In this chapter, we introduced the colonial spatial imaginary to conceptualise the link between colonialism and the representation of areas as centres and peripheries, the designation of power and belonging in each of these areas, and the normalisation of certain spatial practices, particularly displacement. While spatial apartheid is no longer formally maintained in Cape Town, it can be bolstered by the colonial spatial imaginary that underpins certain urban policies, spatial patterns, economic phenomena, and the undisrupted actions of private residents. Interwoven with this imaginary are notions of subjectivity, belonging, power, and respectability that influence not only who is considered to belong in a certain space, but whether a person feels that they belong there too.

RTC challenges the colonial spatial imaginary by challenging the designation of belonging in centres and peripheries and the normalisation of certain spatial practices in these areas. While it operates predominantly from the urban centre, it does so to assert new representations of the centre as a space for well-located, dignified social housing. RTC's work challenges the colonial pattern of displacement from the centre, and rejects the creation of peripheries. Instead, it calls for all to be able to participate in urban life from where they live. This advances a notion of belonging based on personhood rather than property power, or "Land for people, not for profit", and encourages transgressive spatial practices that disrupt spatial inequality. RTC thereby challenges how post-apartheid displacement reinforces spatial apartheid in Cape Town, by enacting a deep consideration of alternative spatial imaginaries.

Although RTC activists are not community psychologists, their multi-faceted work can be read as a decolonial enactment of community psychology. RTC's work represents a creative praxis for decolonial knowledge production in relation to space and subjectivity. It challenges the colonial spatial imaginary by conscientising its members to new possibilities, whether legal or "extra-legal", and realising these possibilities in urban space. It brings knowledge from the peripheries into the centre and knowledge from the centre into the peripheries, destabilising the spatial landscape of knowledge production. It does this by conscientising "peripheral" marginalised groups to "central" structural and legal knowledge; by centring the experiences of marginalised groups in its anti-eviction praxis; and by physically centring evictees by establishing occupations and holding demonstrations in central urban neighbourhoods.

RTC's praxis incorporates both legal and extra-legal methods to build alternative spatial imaginaries that empower communities to resist injustice. Despite its primary focus on spatial processes, RTC engages in community-building, identity-building, reframing, and support, recognising that these more psychological processes are interconnected with space. Advice Assemblies, for example, illustrate how collective support, legal and political education, critical conscientisation, and action can empower communities to reframe belonging and stand their ground in the face of displacement. Creative practices like exhibitions, demonstrations, and storytelling aid in advancing representations of belonging, identity, and legitimacy that support alternative spatial imaginaries and circulate new spatial possibilities among the public. RTC's extra-legal activities, such as occupations, operate as creative departures from the bounds and norms of a dominant spatial imaginary, creating communities with shared alternative imaginaries rooted in social justice. These practices facilitate the movement's deconstruction of the colonial logic underpinning post-apartheid urban development, belonging, and subjectivity.

It should be noted that RTC operates in a context of internal difficulties and external constraints, and we have only been able to unpack a few aspects of their complex work. Internally, RTC faces challenges in community-building and the fostering of people's power in the context of divisive and disempowering historical factors. Externally, the widespread scaling and practical implementation of alternative spatial imaginaries needs new supporting systems, new resources, innovation, debate, and political will. In addition, a thorough disruption of the colonial spatial imaginary also requires a focus on the transformation of the under-resourced peripheries. Finally, RTC is not the only organisation working to address spatial apartheid, but it is notable for the ways in which it does this.

The colonial spatial imaginary is a tool for engaging with multiple knowledge systems in relation to urban land conflicts and other spatial issues. For community psychologists engaging with decolonisation, the case of RTC highlights how the experiences of marginalised groups in contexts of historical segregation and displacement can be influenced by marginalising representations, notions of belonging, and spatial patterns and actions. Resisting the colonial spatial imaginary, in turn, does not simply involve either the reallocation of groups across space in more just ways, or the reimagination of identities in space. Rather, it involves both, and a complex process of problematising, critically conscientising, resisting, transgressing, and re-representing marginalised groups in space. RTC represents a powerful attempt to address the spatial and psychological influences of colonial, apartheid, and post-apartheid displacement and to advance new spatial imaginaries.

References

Abellán, J., Sequera, J., & Janoschka, M. (2012). Occupying the #Hotelmadrid: A laboratory for urban resistance. *Social Movement Studies, 11*(3-4), 1–7. https://doi.org/10.1080/14742837.2012.708831

Berglund, E., & Peipinen, V. (2017). Spaces of insurgent citizenship – activism and acceptability in Helsinki. *The Finnish Journal of Urban Studies, 55*(3), 64–70.

Beyers, C. (2008). The cultural politics of "community" and citizenship in the District Six Museum, Cape Town. *Anthropologica, 50*(2), 359–373.

Boudreau, J. (2007). Making new political spaces: mobilizing spatial imaginaries, instrumentalizing spatial practices, and strategically using spatial tools. *Environment and Planning, 39*, 2593–2611. https://doi.org/10.1068/a39228

Carolissen, R., Canham, H., Fourie, E., Graham, T., Segalo, P., & Bowman, B. (2017). Epistemological resistance towards diversality: Teaching community psychology as a decolonial project. *South African Journal of Psychology, 47*(4), 495–505. https://doi.org/10.1177/0081246317739208

Castleden, H., Garvin, T., & First Nation, H. (2008). Modifying photovoice for community-based participatory indigenous research. *Social Science & Medicine, 66*(6), 1393–1405. https://doi.org/10.1016/j.socscimed.2007.11.030

Cirolia, L. R. (2014). South Africa's emergency housing programme: A prism of urban contest. *Development Southern Africa, 31*(3), 397–411. https://doi.org/10.1080/0376835X.2014.887998

City Improvement. (2009). *FAQ*. Retrieved from http://cityimprovement.co.za/wordpress/?page_id=42

Davoudi, S. (2018). Imagination and spatial imaginaries: A conceptual framework. *Town Planning Review, 89*(2), 97–124. https://doi.org/10.3828/tpr.2018.7

De Carli, B. A., & Frediani, A. A. (2016). Insurgent regeneration: Spatial practices of citizenship in the rehabilitation of inner city São Paulo. *GeoHumanities, 2*(2), 331–353. https://doi.org/10.1080/2373566x.2016.1235984

Development Action Group. (n.d.). *Reimagining neighbourhoods through facilitation*. Retrieved from https://www.dag.org.za/re-imagining-neighbourhoods-through-facilitation/#

Dixon, J., & Durrheim, K. (2000). Displacing place-identity: A discursive approach to locating self and other. *British Journal of Social Psychology, 39*(1), 27–44. https://doi.org/10.1348/014466600164318

Dixon, J., & Durrheim, K. (2004). Dislocating identity: Desegregation and the transformation of place. *Journal of Environmental Psychology, 24*(4), 455–473. https://doi.org/10.1016/j.jenvp.2004.09.004

Dutta, U. (2016). Prioritizing the local in an era of globalization: A proposal for decentering community psychology. *American Journal of Community Psychology, 58*, 329–338. https://doi.org/10.1002/ajcp.12047

Dwyer, S. C., & Buckle, J. L. (2009). The space between: On being an insider-outsider in qualitative research. *International Journal of Qualitative Methods, 8*(1), 54–63. https://doi.org/10.1177/160940690900800105

Earle, L. (2012). From insurgent to transgressive citizenship: Housing, social movements and the politics of rights in São Paulo. *Journal of Latin American Studies, 44*(1), 97–126. https://doi.org/10.1017/S0022216X11001118

Elder, G. (1990). The grey dawn of South African racial residential integration. *GeoJournal, 22*(3), 261–266. https://doi.org/10.1007/BF00192824

Foster, D. (2004). Liberation psychology. In D. Hook (Ed.), *Introduction to critical psychology*. UCT Press.

Freire, P. (1970). *Pedagogy of the oppressed*. Continuum.

Frith, A. (2011). *Dot map of South Africa*. Retrieved from https://dotmap.adrianfrith.com/?dots=race&lat=-33.9494&lon=18.6763&zoom=5.09

Garside, J. (1993). Inner city gentrification in South Africa: The case of Woodstock, Cape Town. *GeoJournal, 30*(1), 29–35. https://doi.org/10.1007/BF00807824

Gregory, D. (2004). *The colonial present*. Blackwell Publishing.

Hart, D. (1988). Political manipulation of urban space: The razing of District 6, Cape Town. *Urban Geography, 9*(6), 603–628. https://doi.org/10.2747/0272-3638.9.6.603

Henriques, J., Hollway, W., Urwin, C., Venn, C., & Walkerdine, V. (1984). *Changing the subject: Psychology, social regulation, and subjectivity* (2nd ed.). Taylor & Francis.

Holston, J. (1998). Spaces of insurgent citizenship. In L. Sandercock (Ed.), *Making the invisible visible: A multicultural planning history*. University of California Press.

Holston, J. (2009). *Insurgent citizenship: Disjunctions of democracy and modernity in Brazil*. Princeton University Press.

Kessi, S. (2011). Photovoice as a practice of re-presentation and social solidarity: Experiences from a youth empowerment project in Dar es Salaam and Soweto. *Papers on Social Representations, 20*(1), 7.1–7.27.

Kessi, S., & Boonzaier, F. (2018). Centre/ing decolonial feminist psychology in Africa. *South African Journal of Psychology, 48*(3), 299–309. https://doi.org/10.1177/0081246318784507

Kothari, U., & Wilkinson, R. (2010). Colonial imaginaries and postcolonial transformations: Exiles, bases, beaches. *Third World Quarterly, 31*(8), 1395–1412. https://doi.org/10.1080/01436597.2010.538239

Laband, J. (2020). *The land wars: The dispossession of the Khoisan and AmaXhosa in the Cape Colony*. Penguin Random House.

Lipsitz, G. (2007). The racialization of space and the spatialization of race: Theorizing the hidden architecture of landscape. *Landscape Journal, 26*(1), 10–23. https://doi.org/10.3368/lj.26.1.10

Mabin, A. (1992). Comprehensive segregation: The origins of the Group Areas Act and its planning apparatuses. *Journal of Southern African Studies, 18*(2), 405–429. https://doi.org/10.1080/03057079208708320

Martín-Baró, I. (1996). *Writings for a liberation psychology*. Harvard University Press.

Miraftab, F. (2004). Invited and invented spaces of participation: Neoliberal citizenship and feminists' expanded notion of politics. *Wagadu, 1*, 1–7.

Miraftab, F. (2012). Colonial present: Legacies of the past in contemporary urban practices in Cape Town, South Africa. *Journal of Planning History, 11*(4), 283–307. https://doi.org/10.1177/1538513212447924

Miraftab, F., & Wills, S. (2005). Insurgency and spaces of active citizenship: The story of Western Cape Anti-Eviction Campaign in South Africa. *Journal of Planning Education and Research, 25*(2), 200–217. https://doi.org/10.1177/0739456x05282182

Montero, M. (2009). Methods for liberation: Critical consciousness in action. In C. Sonn & M. Montero (Eds.), *Psychology of liberation*. Springer.

Mountain, A. (2003). *The first people of the Cape: A look at their history and the impact of colonialism on the Cape's indigenous people*. David Philip Publishers.

Nahnsen, A. (2006). *Emergency on planet Cape Town? (Re-)conciliation as a tool for urban planning in a post-apartheid city*. BIS-Verlag.

Paasi, A. (2003). Region and place: Regional identity in question. *Progress in Human Geography, 27*(4), 475–485. https://doi.org/10.1191/0309132503ph439pr

Pillay, S., Russell, S., Sendin, J., Sithole, M., Budlender, N., & Knoetze, D. (2017). *I used to live here: A call for transitional housing for evictees in Cape Town*. Ndifuna Ukwazi.

Proshansky, H. M., Fabian, A. K., & Kaminoff, R. (1983). Place-identity: Physical world socialization of the self. *Journal of Environmental Psychology, 3*(1), 57–83. https://doi.org/10.1016/s0272-4944(83)80021-8

Ranslem, D. (2015). 'Temporary' relocation: Spaces of contradiction in South African law. *International Journal of Law in the Built Environment, 7*(1), 55–71. https://doi.org/10.1108/ijlbe-12-2013-0041

Reclaim the City. (2018a). *Reclaim the city interim constitution*. Retrieved from http://reclaimthecity.org.za/featured-content/

Reclaim the City. (2018b). *Eviction resistance*. Retrieved from http://reclaimthecity.org.za/transitional-housing/

Said, E. (1978). *Orientalism*. Routledge & Kegan Paul.

Scannell, L., & Gifford, R. (2010). Defining place attachment: A tripartite organizing framework. *Journal of Environmental Psychology, 30*(1), 1–10. https://doi.org/10.1016/j.jenvp.2009.09.006

Seedat, M., & Suffla, S. (2017). Community psychology and its (dis)contents, archival legacies, and decolonisation. *South African Journal of Psychology, 47*(4), 421–431. https://doi.org/10.1177/0081246317741423

Thompson, L. (2001). *A history of South Africa* (3rd ed.). Yale University Press.

Turok, I. (2001). Persistent polarisation post-apartheid? Progress towards urban integration in Cape Town. *Urban Studies, 38*(13), 2349–2377. https://doi.org/10.1080/00420980120094551

Wang, C., & Burris, M. A. (1997). Photovoice: Concept, methodology, and use for participatory needs assessment. *Health Education & Behavior, 24*(3), 369–387. https://doi.org/10.1177/109019819702400309

Wang, C., & Redwood-Jones, Y. A. (2001). Photovoice ethics: Perspectives from Flint photovoice. *Health Education & Behavior, 28*(5), 560–572. https://doi.org/10.1177/109019810102800504

Watkins, J. (2015). Spatial imaginaries research in geography: Synergies,tensions, and new directions. *Geography Compass, 9*(9), 508–522. https://doi.org/10.1111/gec3.12228

Watkins, M., James, S., & Ciofalo, N. (2018). Engaging the struggle for decolonial approaches to teaching community psychology. *American Journal of Community Psychology, 10*, 1–11. https://doi.org/10.1002/ajcp.12295

Watkins, M., & Shulman, H. (2008). *Toward psychologies of liberation*. Palgrave Macmillan.

A Decolonising Approach to Health Promotion

Elelwani L. Ramugondo and Isla Emery-Whittington

Introduction

There was much promise in the 1980s when health promotion emerged as a strategy towards better global health indicators. Much of the optimism lay in the expectation that a sociological understanding of health would dismantle biomedicine as the dominant approach to health. A major criticism of biomedicine lies in the fact that it leaves no room for the social, behavioural, psychological, economic or political understandings of illness. A re-emergence of the decolonial discourse globally, however, provides space for re-imagining health promotion. This discourse allows for a move away from the internalized dominant discourse of biomedicine, which is individualistic, apolitical and ahistorical, and a return to indigenous epistemes. Central to the decolonial discourse is appreciating how power, knowledge, being and doing are intertwined, and a recognition of how coloniality often re-inscribes itself in the way people live their lives, remember their pasts, and imagine their futures. For indigenous populations who were colonised at some point in their history, coloniality leads to devastating health outcomes, and defines the very struggle of everyday survival within globalised western modernity.

The current resurgence and a deepening of decolonial scholarship is partly due to student-led protests in South Africa, sparked off by the hashtag #RhodesMustFall campaign at the University of Cape Town. Insisting that this campaign was about more than the removal of the statue of Cecil John Rhodes, a well-recognised symbol of British imperial conquest of Southern Africa and white male supremacy, the

E. L. Ramugondo (✉)
Health Sciences Faculty, University of Cape Town, Cape Town, South Africa
e-mail: elelwani.ramugondo@uct.ac.za

I. Emery-Whittington
SHORE & Whariki Research Centre, Massey University- Te Kunenga ki Pūrehuroa, University of New Zealand, Wellington, New Zealand
e-mail: isla@whittington.net.nz

S. Kessi et al. (eds.), *Decolonial Enactments in Community Psychology*, Community Psychology, https://doi.org/10.1007/978-3-030-75201-9_10

students called for the academy in its entirety to be decolonised. This demand gave impetus to students and academics revisiting work by post-colonial scholars in Africa such as Amina Mama, Ngugi wa Thiong'o, Mahmood Mamdani, Steve Biko, Walter Rodney, Ali Mazrui and Frantz Fanon, as well as engagements with Indian and Latin-American decolonial scholars such as Gayatri Spivak, Chandra Raju, Sylvia Wynter, Enrique Dussel, Nelson Maldonado-Torres and Ramon Grosfoguel.

Taking a longue durée view on global struggles against injustice, Hira dubs decoloniality, the third narrative of liberation (Hira, in press). In so doing, Hira draws a distinction between this third narrative and Liberalism, as well as Marxism, both of which have neglected collective and historical racial oppression. Hira's critique of Marxism echoes Robinson, who pointed out Marx's failure to scrutinize oppression beyond class struggles in Europe, while highlighting the racial character of capitalism (Robinson, 1983). On the third narrative of liberation, Ndlovu-Gatsheni (2015), an African decolonial thinker and scholar, describes decoloniality as "a long-standing political and epistemological movement aimed at liberation of (ex-) colonized peoples from global coloniality" (p. 485). Latin-American scholars have identified three aspects to global coloniality: coloniality of Power, coloniality of Knowledge, and coloniality of Being (Maldonado-Torres, 2007; Mignolo, 2012; Quijano, 2000).

Coloniality of Power describes the continuous asymmetrical relations of power that survive the formal withdrawal of colonial administrators from the colonies, and explains the current and ongoing racial, political, economic and social disenfranchisement of societies in the Global South (Ndlovu-Gatsheni, 2013a, 2015; Robinson, 1983; Rodney, 1973; Quijano, 2000; Wallerstein, 2004). Coloniality of Knowledge highlights "the ways in which endogenous, indigenous or traditional knowledges from non-Western worlds have often been approached with suspicion and deemed inferior, whilst at the same time appropriated to serve imperial powers and to claim white supremacy" (Ramugondo, 2018, p. 5). Coloniality of Knowledge is sustained in the Westernized University (Grosfoguel, 2013) or the post-colonial African university (Mamdani, 2017) through what is regarded powerful knowledge (Young et al., 2014) and therefore valued, who gets privileged as a thinker and what interests dominate. Coloniality of Being describes a form of existence that is devoid of ontological density such that full humanity or human dignity for the '(ex)-colonised' (Ndlovu-Gatsheni, 2015) is denied or can easily be brought into question.

To the three aspects of coloniality that Latin-American decolonial scholars describe, the University of Cape Town's Curriculum Change Working Group (UCT CCWG) has added a fourth: coloniality of Doing (UCT CCWG, 2018). Drawing from the concept, occupational consciousness, which highlights every day doing as a powerful mechanism through which dominant practices are either perpetuated or disrupted, with implications for both individual and collective health (Ramugondo, 2015), coloniality of Doing describes how the colonised subjects are assimilated into the Western-led civilizing project, even as this denies them ontological density (UCT CCWG, 2018). These four aspects to coloniality are deeply inter-twined and reinforce each other.

Per definition, biomedicine is ahistorical, and claims universality. Adopted by the World Health Organisation (WHO), health promotion was to enable people all over the world, to increase control over their own health, by focussing on social determinants of health and addressing these as root causes of ill-health, not just treatment and cure (WHO, 2016). Almost 40 years later, the hope ushered in by health promotion has not borne much fruit. The burden of disease has quadrupled in much of the world population, affecting the Global South, particularly indigenous populations, the most. This chapter delves into what may explain the failure to translate good intentions behind health promotion, into every day health practice and research. The role of biomedicine, and why it remains the dominant discourse in both health care and research is outlined from the authors' particular vantage point as indigenous scholars trained in westernised health practice. The first author is a black Tshivenda speaking South African woman, with ancestral links spanning several centuries on the African continent. The second author is Māori of Tainui and Takitimu descent. Both authors are occupational therapists and scholars in occupational science.

In order to address the global failure in delivering on the most progressive thinking about health and society in the history of humanity, this chapter argues for a decolonial shift in both discourse and praxis, illustrating how this shift often necessitates disobedience within disciplines or professions such as psychology, including community psychology and occupational therapy.

Biomedicine: A Gift that Keeps on Giving

Biomedicine is a model or framework through which disease is viewed as a deviation from normal biological functioning, fully understood and explained through measurable biological variables (Engel, 1977). In the twenty-first century, biomedicine has become a framework through which life itself is visualized at the molecular level, with diagnoses and treatments mostly approached in terms of molecular dysfunction expressed through pathologies of organs or bodily systems (Rose, 2009). In the study of disease, biological entities are often removed from their natural settings, to be analysed separately in controlled research environments. Central to biomedicine, are two principles: reductionism, a worldview and philosophy which suggests that to understand a complex matter, one needs to analyse its constituent parts separately (Mazzocchi, 2011); and mind-body dualism, a worldview and principle that treats mental phenomena as non-physical and therefore separate and independent from the body (Rosciglione, 2013).

Reductionism and mind-body dualism have over the years been advanced as the hall-mark of true science. Biomedicine, by adopting these two principles, became legitimized as *the* scientific approach to health. The benefit of science in medicine cannot be over-stated. The average life expectancy has increased in much of the world's population, thanks to massive investment in research to prevent and manage infectious diseases. Complications of child-birth and those arising from injuries have also been addressed with better efficiency and efficacy for those with optimum access to evidence-based health-care. Biomedicine has thus served parts of humanity well, and it will be a grave mistake to suggest that it ought to be done away with. The success of biomedicine has led to people's ability to live with chronic diseases, such as HIV/AIDS and cancer, as well as complex disabilities. There are however, many chronic diseases that have not been amenable to a laboratory-based approach to science. These diseases, such as diabetes, heart failure, stroke and hypertension, and mental health conditions in general, have continued to rise exponentially globally, even as molecular strategies to diagnose and treat illness have become more sophisticated.

From the vantage of indigenous perspectives, the authors assert that the failure of biomedicine in dealing with chronic illness requires honest admission in both academia and clinical practice. Engel (1977) argues that when an approach to a problem is never questioned even when it fails to account adequately for all available data, that approach is no longer science, but has become dogma. Dogma does not allow for contestation. In challenging the reductionist and dualistic nature of biomedicine, which he argued evolved in the West, Engel (1977) suggested the biopsychosocial model of disease. This would allow for health or disease to be understood as a constellation of biological, psychological, social and cultural considerations, "rather than giving primacy to biological factors alone" (Engel, 1977, p. 133). Engel (1977) further argued that the biopsychosocial model would help explain why individuals confronted with the same conditions would respond differently, with some experiencing these as 'illness' while others deem them merely as 'problems of living'. To make this point even more poignant, Engel (1977) asks the question "When is grief a disease?" (p. 133), arguing that a similar question may be asked regarding schizophrenia or diabetes. The persuasive nature of Engel's arguments can be seen in how the biopsychosocial model of disease was adopted vigorously in mental health-care discourse in both the Global North and the Global South (Babalola et al., 2017; Thornicroft et al., 2012). The terms 'Global North' and 'Global South' here are to be understood beyond narrow geographical notions, but in terms of global unequal power relations between centres/metropoles ('norths') and peripheries ('souths') (Connell, 2007; de Sousa Santos, 2014; Ndlovu-Gatsheni, 2013b).

Rhetoric and frameworks on the biopsychosocial nature of illness have however not easily translated into practice. Criticism levelled at the biopsychosocial approach has not only focussed on it being time-consuming and costly (Babalola et al., 2017; Truglio et al., 2012) but also on its inability to provide clear guidelines for clinical practice (Ghaemi et al., 2012). Building on criticism that the biopsychosocial model has failed to shed its Western Eurocentric baggage in that it remains individualistic, Babalola et al. (2017) also point to its insufficient attention to the role of culture. Noting that Engel (1977), in his original conceptualisation of the model, highlighted the importance of culture, these authors argue that this aspect was lost in how it was bound up with the notion: social. In reviving the role of culture in how illness is experienced differently across different social groupings and communities, Babalola et al. (2017) invoke Nichter, pointing to the need to attend to "local idioms of distress" (Nichter, 1981, p. 379). Babalola et al. (2017) further argue that for the biopsychosocial model to entrench its global reach beyond rhetoric, and to represent a truly holistic approach to health-care, culture needs to be understood to play a central role in how people understand problems of living or 'idioms of distress' and develop relevant interventions for different sociocultural contexts.

The current decolonial turn allows for a deeper appreciation of Engel's insistence on the need to question accepted scientific approaches when they no longer account adequately for all available data. Epistemic Injustice (Fricker, 2007) explains two instances of injustices that operate when science is universalised beyond the limits of its applicability: testimonial injustice is experienced by a legitimate knower who suffers a deficit of credibility as a speaker, owing to the operation of prejudice in the listener's judgement; hermeneutical injustice occurs when a subject who is already hermeneutically marginalised (belongs to a group which does not have equal participation in the generation of social meanings) is put at an unfair disadvantage when it comes to making sense of a significant area of their social experience. Decolonising health promotion requires that health experts are not too certain about their scientific methods such that they fail to notice and address these instances of epistemic injustice.

More critical voices have gone beyond logistical difficulties in implementing Engel's biopsychosocial model, or an attempt at fore-grounding or attending to 'local idioms of distress' but have attended to the political economy of ill health in a globalized neo-liberal capitalist system. These critical voices expose how biomedicine is a gift that keeps on giving for the wealthy, pointing out that central to the mismatch between the progressive nature of the biopsychosocial model at a conceptual level: in bringing together co-existing factors impacting health, and practice or *praxis*, is the power and interests of those who benefit economically from the dominant biomedical model of disease. Applbaum (2006) and Kirmayer and Raikhel (2009) in this regard, have identified the pharmaceutical industry as the main driver behind the persistent dominance of biomedicine, through the expansion of a global market for drugs in response to human distress. Observation and critical analysis of the protection of market interests above transformational gains in health-care are an important initial step in the work of decolonising health promotion. It is paradoxical that a 'world' health approach such as the biopsychosocial model is not at once recognised as a local idiom. Hence, decolonising health promotion is also about observing the universalism inherent in 'world' health models that explain the failure to sufficiently or effectively recognise and address 'local idioms of distress'.

The impact of colonialism on the health of indigenous peoples is well-established (Cave, 2008; Madley, 2004; Melber, 2008; Paradies, 2016; Robson, 2007; Zimmerer, 2008). In Canada, where there has been much impetus to drive health promotion at conceptual, policy and programmatic levels, failure continues to be registered at implementation levels, with evidence for this being the persistent health challenges in that country (Collins & Hayes, 2007; Hancock, 2011; Low & Thériault, 2008), affecting indigenous people the most (Raphael et al., 2008).

Conversely, for indigenous health promoters, observations and critical analysis are scaffolded by ancient indigenous epistemologies of health and health promotion praxis that may or may not incorporate biopsychosocial approaches. This affords epistemological advantage as many models, frameworks and theories can sit alongside and support what limitations are noted from a dominant yet under-developed biomedical worldview (Petticrew et al., 2004). The epistemological advantage of indigenous peoples is both a beacon of hope for imagination and a necessity for survival. Indigenous peoples are concerned with systems.

Social Determinants of Health, Discourse and Disciplinary Decadence

Health promotion efforts worldwide often neglect social determinants of health in favour of risk aversion and lifestyle-oriented approaches. Research and practice in public health has often shown a propensity to locate risk within the indigenous body instead of institutional and - social structures. Such pre-occupation is intriguing, given that the WHO has emphasized the need to redirect public health efforts from 'lifestyle choices' to issues of living conditions and social justice (WHO, 2008). This emphasis was based on substantial evidence that structural inequality in the form of social, economic, and political environments – which are amenable to political intervention – is the root cause of health inequalities globally (Commission on the Social Determinants of Health, 2008; WHO, 2008; Wilkinson & Marmot, 2003; Wilkinson & Pickett, 2009). With a world body of such stature being persuaded by available evidence about the root causes of ill-health and giving a clear mandate to health professionals "to tackle the leading causes of ill-health at their roots, even when these causes lie beyond the direct control of the health sector" (WHO, 2010, p. 1), it is a concern, and telling that the dominant discourse in health remains biomedical.

There are grave consequences for this dominant discourse on conceptualisations about health as well as practices across disciplines or professions. Globally, the triple burden of disease continues to be borne heaviest by marginalised communities in terms of social location and intersectional nature of race, class, disability status and gender. Brassolotto et al. (2014) have argued that a substantial barrier to meaningful action that addresses social determinants of health is the internalization of discourses that frame health as biomedical, individualized and depoliticized. This framing has meant that the political economy of ill health in a globalized neo-liberal capitalist system has remained largely unchallenged. Those who benefit economically from the dominant biomedical model of disease, continue to advance their own interests: mainly to make a profit. In addition, disciplines and professions associated with health-care, across the world, tend to concern themselves with their own global expansion and increased recognition by governments and world bodies, instead of partnering with communities towards self-determination. Community agency in diagnosing problems that lead to ill-health is often impeded, obstructed, obscured, ignored and denied, as disciplines and professions in health-care develop and protect their colonial power and assumed legitimacy to diagnose health problems.

Gordon (2014) refers to disciplines' susceptibility to protect and advance disciplinary interests even as they fail to articulate with locally situated societal problems, disciplinary decadence. This phenomenon translates into resistance to engage meaningfully with social determinants of health. Gordon further warns about the serious consequence that arise out of disciplinary decadence, as it ultimately threatens the survival of the discipline itself. Engaging with social reality should allow professions in health-care to acknowledge four ugly truths: the burden of disease is a global crisis, especially for indigenous populations; those with money have better access to health; ill-health is a business; and growing a discipline does not equate population health. Recognising these ugly truths and identifying internalized discourses that undergird colonial health-care practices is critical for decolonising health promotion.

Internalized Discourses as Barrier

Brassolotto et al.'s (2014) observation that the internalization of discourses that frame health as biomedical, individualized and depoliticized is central to the inability to address social determinants of health, is very helpful. Discourses are powerful, especially when sustained by interlocutory work: 'speaking between', as academics navigate between canons or prominent scholars within disciplines and the publics. Dominant discourses on health often exclude and silence the voices of marginalised populations. To explain how biomedicine came to be deeply embedded into discourses that frame health in the twenty-first century, Rose (2009) draws substantially from Fleck's notion of a 'style of thought'. An important contributor to both the sociology of knowledge (Harwood, 1986) and the philosophy of science (Babich, 2003), Fleck defined style of thought as a particular way of seeing, thinking and practicing, that is normative within a particular circle ('thought community') of professionals, or within a discipline or sub-discipline. Fleck's notion of thought communities sits uncomfortably along modern scientific thinking, which aspires for Western liberal ideals of individualism (Babich, 2003) and academic freedom. A more palatable term which has gained traction in both the sociology of knowledge and philosophy of science is 'paradigm', coined by Kuhn (Kuhn, 1970). Babich (2003) however, argues that Kuhn was heavily influenced by Fleck in his own work.

In thinking about style of thought and the related thought community, Fleck challenged the concept of 'individual scientific genius', suggesting instead that the scientist is merely a product of the prevailing thinking of their time. He argues thus:

> The ordinary scientist of the day finds that 'scientific truth' is a complex mental construction, inseparably connected with investigative techniques, statistical interpretations and manifold conventions. He knows that it may often be expressed only in a special jargon and be intelligible only after a prolonged professional training… 'scientific truth' depends on

> 'conjunctures', i.e., on the scientific opportunity, on the environment and on the personal influence of the author. (Fleck, 1986 [1947], p. 153).

Expanding on Fleck's notion of style of thought in relation to the explanatory function of scientific inquiry, Rose argues that it is not only about the form in which the inquiry takes, and what it is that needs to be explained, but "also about what *there* (original emphasis) is to explain. That is to say, it shapes and establishes the very object of explanation, the set of problems, issues, phenomena that an explanation is attempting to account for." (Rose, 2009, p.12).

Discourse, as a mechanism for legitimation, becomes a powerful tool through which to sustain a particular style of thought, and to coerce individuals into thought communities, even in practices and research otherwise dubbed as scientific inquiry. Thus, the style of thought about ill-health in the twentieth-century has become a 'molecularization of vitality' (Rose, 2009; Carvalho & Teixeira, 2017), with a variety of healing practices seeking novel ways towards molecular legitimation, and the pharmaceutical industry exploring "traditional healing practices in their quest for molecular knowledge that can be extracted, developed, patented, and commodified." (Rose, 2009, p. 13). Clinical diagnoses, as well, have been reshaped through research: on a molecular basis. It is on this basis that a biomedical framing of disease has also remained individualistic, ahistorical and apolitical.

To shift the discourse of health beyond biomedicine, at least in public health disciplines, Raphael used a Foucauldian critical discourse analysis to identify seven discourses that inform various explications of the social determinants of health concept, with implications for public health practice and research, as well as health promotion in general. These discourses are outlined in Table 1. Although Raphael focussed on practices in Canada in identifying these discourses, they are helpful in beginning to understand the extent to which internalized discourses can serve as a barrier in meaningfully addressing social determinants of health, in ways that might resonate globally.

SDH discourse 1 involves the recognition that individuals, groups and communities whose living circumstances expose them to adverse social determinants of health (such as high unemployment rates, low literacy levels, low income levels, inadequate housing etc.) are at risk of suffering a high incidence of a variety of medical illnesses and other social problems. Once such individuals, groups or communities are identified, efforts are made to make sure they receive targeted healthcare and social services. Some examples may involve the screening of children for infectious diseases and disabilities, offering early childhood development programs, visual and hearing screening and effectively managing chronic diseases for the elderly, ensuring that street-connected populations are able to access health services, or promoting primary health care services in low income communities. While it is important to respond appropriately to health needs, and to target particularly vulnerable populations for social upliftment, the danger here is that this discourse does not allow for the interrogation of why these populations are vulnerable in the first place.

Table 1 Raphael's Seven Social Determinants of Health (SDH) Discourses

SDH discourse	Key concept	Dominant research and practice paradigms	Practical implications of the discourse
1. SDH as identifying those in need of health and social services	Health and social services should be responsive to peoples' material living circumstances	Develop and evaluate services for those experiencing adverse living conditions	Focus limited to service provision with assumption that this will improve health
2. SDH as identifying those with modifiable medical and behavioral risk factors	Health behaviors (e.g. alcohol and tobacco use, physical activity, and diet) are shaped by living circumstances	Develop and evaluate lifestyle programming that targets individuals experiencing adverse living conditions	Focus limited to health behaviors with assumption that targeting for behavior change will improve health
3. SDH as indicating the material living conditions that shape health	Material living conditions operating through various pathways – Including biological – Shape health	Identify the processes by which adverse living conditions come to determine health	Identifying SDH pathways and processes reinforce concept and strengthen evidence base
4. SDH as indicating material living circumstances That differ as a function of group membership	Material living conditions systematically differ among those in various social locations, such as class, disability status, gender and race	Carry out class-, race-, and gender based analysis of differing living conditions and their health-related effects	Providing evidence of systematic differences in life experiences among citizen groups form the basis for further anti-discrimination efforts
5. SDH and their distribution as results of public policy decisions made by governments and other societal institutions	Public policy analysis and examination of the role of politics should form the basis of SDH analysis and advocacy efforts	Carry out analyses of how public policy decisions are made and how these decisions impact health (i.e. health impact analysis)	Attention is directed towards governmental policy-making as the source of social and health inequalities and the role of politics. [advocacy]
6. SDH and their distribution result from economic and political structures and justifying ideologies	Public policy that shapes the SDH reflects the operation of jurisdictional economic and political systems	Identify how the political economy of a nation fosters particular approaches to addressing the SDH	Political and economic structures that need to be modified in support of the SDH are identified. [political education]

(continued)

Table 1 (continued)

SDH discourse	Key concept	Dominant research and practice paradigms	Practical implications of the discourse
7. SDH and their distribution result from the power and influence of those who create and benefit from health and social inequalities	Specific classes and interests both create and benefit from the existence of social and health inequalities	Research and advocacy efforts should identify how imbalances in power and influence can be confronted and defeated	Identifying the classes and interests who benefit from social and health inequalities mobilizes efforts towards change. [Conscientisation and mobilisation]

Source with minor adaptation: Raphael (2011)

SDH discourse 2 focusses on the behaviour and lifestyle choices of people, drawing a link between these personal attributes and chronic conditions such as lung cancer, diabetes, hypertension and morbid obesity. There is a myriad of problems with this 'healthy lifestyle choice' discourse. Raphael (2011) sites four of these. Firstly, it does not account adequately for marked variability in health outcomes for individuals with comparable exposure to the particular risk factors. Secondly, it is founded on the assumption that everyone enjoys full agency to choose freely how to live and apportions full blame on the sufferer for their own adverse health outcome. Thirdly, there is insufficient evidence to support the effectiveness of programmes aimed at promoting healthy lifestyles. Fourthly, this discourse ignores the drivers behind unhealthy life choices, further obscuring their significance. An additional problem to be added to this discourse is that it focuses only on physical well-being, and ignores other aspects of well-being (intellectual, social, emotional and spiritual) and their role in overall health.

SDH discourse 3 is one that directly implicates adverse social determinants of health for poor health outcomes. Here there is recognition that there are various pathways through which adverse living conditions find their ways 'under the skin' (Raphael, 2011) for individuals who experience them in their everyday lives and come to shape their health. Much of the evidence supporting this discourse comes from research that identifies social factors that interact with personal attributes and track how these processes come to determine health. International programs such as the United Nations International Children's Emergency Fund's First 1000 days of life are informed by such research, emphasizing the impact living conditions surrounding a baby can have in laying the foundation for optimum health and development across the lifespan, or disrupting it. Whilst this discourse recognises the importance adverse social determinants play in shaping health, it is often diluted by health policies that continue to emphasize awareness raising programmes (such as breastfeeding awareness and growth monitoring campaigns), rather than aiming to ensure equitable distribution of resources for financial security for all.

SDH discourse 4 acknowledges that inequitable distribution of resources within society often occur as a function of identity markers such as race, gender, class and disability status. Like with the previous discourse, this acknowledgement is often watered down at policy implementation level where the focus tends to be narrow, targeted at marginalised individuals or groups themselves: such as literacy programs (Raphael, 2011) or addressing racism as a relational issue and offering antidiscrimination workshops as a solution, instead of treating it as a systemic and institutional problem.

SDH discourse 5 reflects the strong assumption underlying the conclusions of the WHO's Commission on the SDH (Commission on the Social Determinants of Health, 2008); that public policy interventions can directly impact health outcomes. Indeed, it has been shown that social determinants of health experienced during early life are a direct result of material conditions that are shaped by public policy: such as educational opportunities, adequate housing, employment security and conditions for parents and regulated child-care (Raphael, 2011).

SDH discourse 6 extends the assumption reflected in the previous discourse, that public policy is implicated in the distribution of social determinants of health across any given society, by exposing how certain governments may be more amenable than others to shift policy, given their historical traditions and economic and political structures (Raphael, 2011). Drawing from a sociological analysis of four welfare state types (social democratic, liberal, conservative and Latin), Raphael (2011) demonstrates how social democratic governments such as Sweden, Denmark, Norway and Finland, outperform liberal economies such as the United States of America, United Kingdom, Canada and Ireland, on several health and quality of life indicators. In pandering to the sensibilities of their 'central institution' – the market, liberal political economies avoid interventionist approaches towards public policy (Raphael, 2011), even in the face of rising adverse social determinants of health.

SDH discourse 7 builds on the assertions reflected in the previous two discourses, and directly implicates those who benefit from social and health inequalities. These individuals and groups, usually as part of the corporate sector and wealthy communities, lobby governments and the political elite to advance their own interests, leading to a reduced quality of social determinants of health that others are exposed to, or a skewed distribution of these (Raphael, 2011). Raphael (2011) argues that counterbalances that can oppose and defeat powerful interest groups in unduly influencing governments include mobilising the working class into effective unionization and galvanizing the marginalised into informed and active citizenship to vote political parties that would support SDH-supportive public policy into power.

Extending on Raphael (2011)'s work in identifying these seven social determinants of health discourses, Brassolotto et al. (2013), have classified these discourses in terms of health practice approaches they influence at implementation level. These authors have categorised SDH discourses 1 & 2 as functional, SDH discourses 3 & 4 as analytical, and SDH 5, 6 & 7 as structural in approach (Brassolotto et al., 2013). Although there are broad principles that define these categories as distinct, the boundaries between them are not so rigid such that there are no overlaps in-between them.

SDH discourses that are functional in approach speak about the SDH in ways that fore-ground the need to identify those in need of health and social services as well as those with modifiable medical and behavioral risk factors. As a result, service-delivery and healthy-lifestyle oriented programmes become the sort of activities that are advanced as health promotion. SDH thinking in this approach serves a functional purpose, as a 'lens,' 'framework,' or 'tool' to be used when health promotion workers go about doing their work (Brassolotto et al., 2013). In the public health units (PHUs) that Brassolotto et al. (2013) investigated using Raphael's seven SDH discourses, as a framework, they found that Functional PHUs tended to overlook qualitative data or process indicators as evidence, and also featured marked apprehension and discomfort about the political nature of SDH. This apolitical approach to health, and affinity to things that can be counted, reflects an internalized biomedical view of health; the social in SDH immediately falls away.

SDH discourses that take an analytical approach acknowledge the multiple ways living conditions impact health and see addressing these issues as the correct action to take. Brassolotto et al. (2013) found that Analytical PHUs partnered with community organizations in their work, towards addressing issues such as early childhood development, food security and housing. An important overlap between functional and analytical approaches to SDH discourses is the primacy of risk factors, often decontextualized from broader public policy implications and the role of structural inequalities. This limitation functions as an epistemological barrier to further action (Brassolotto et al., 2013).

SDH discourses that are structural in approach attend more to contextual issues: these are societal structures such as economics, and public policy that shapes discourses and customs that impact people differently according to social positioning (race, class, gender, disability or sexual orientation). Brassolotto et al. (2013) found that Structural PHUs invested in advocacy work, engaging both the public and politicians in developing community partnerships, capacity building and conducting health impact assessments related to public policies and political decisions. Social determinants of health in these discourses are often described as "indicators of structural inequalities in society" (Brassolotto et al., p. 8).

SDH discourses that take a structural approach are an important basis upon which to address health disparities in society. The Canadian example through Brassolotto et al.'s (2013) work however, suggests the structural approach to be rare in the public health sector, with many professionals trained to work mostly in clinical settings, attending to illness in as far as it affects the individual, and using primarily medical interventions. Beyond Canada, it is possible that structural approaches to addressing social determinants of health may even be more uncommon. This observation alone may already suggest one limitation in the way Raphael (2011) approached the seven SDH discourses, in that he ignores the role of geopolitics, treating nation states as isolated entities. What if the world is such that it has become impossible for governments to emulate Sweden, Denmark, Norway and Finland, in becoming social democratic states?

Another limitation is that even within the progressive analytical and structural approaches to SDH discourses (Brassolotto et al., 2013), public health professionals are treated as external experts looking in, with the power to mobilize communities, when in fact marginalized communities are acutely aware of their needs and tire of external experts who seem unable to make promised structural changes leading to self-determination. An over-emphasis on physical health in Raphael's SDH discourses and in how these are applied by Brassolotto et al. (2013), is another important observation. Linked to this, is the ongoing challenge of defining health in ways that consider the role of culture or the 'local idioms of distress' (Nichter, 1981).

Importantly, there is silence about the role of public health practitioners from marginalized communities and their allies. The potent and powerful interdisciplinary and inter-institutional networks of practitioners move seamlessly amongst their communities softening the blows of colonialism assessment by assessment, promotional event by promotional event, model by model, policy by policy. In Aotearoa New Zealand, disobedience continues with the work of allies and partnerships in decoloniality who are aware of historical privilege (Borell et al., 2018). Building and teaching critical anti-racism praxis is evidenced in power sharing (Came, 2014; Durie, 2003; Pease, 2010). As well, these kinds of partnerships and networks provide a 'correction' to the over-valued position of humans and re-place health promotion alongside concern for all fellow creature families of Earth Mother (Heke et al., 2019). Decolonised health promotion reaches beyond human beings and promotes health for all who share this planet.

Decolonial scholarship and praxis may provide a way to build on Raphael's analytical and structural approaches to SDH discourses, in ways that acknowledge the role of geopolitics as well as advance self-determination and collective self-reliance for marginalized communities that continue to bear the brunt of the political economy of ill health in a globalized neo-liberal capitalist system.

Shifting the Discourse: Building a Decolonial Epistemology on Health

To address coloniality in all its four aspects (Power, Knowledge, Being and Doing), it seems decoloniality of Being may play a central role. Decoloniality of Being is a concerted effort by the (ex)-colonised (Ndlovu-Gatsheni, 2015) to regain their full humanness or integrity of being (Lushaba, 2017). Decoloniality of Being and health are closely intertwined, if health is understood holistically beyond the narrow biomedical and individualistic approach that focuses primarily on medical diagnosis. The dominant approach to health and well-being in Western medicine centres on diagnosis: as both a process and an outcome of clinical investigation (Ramugondo et al., 2017). As a process, clinical diagnosis within Western medicine and its internalized discourses that frame health as biomedical, individualized, ahistorical and depoliticized, on one hand affords the doctor immense power: seen to "hold the necessary skills to locate the problem in the body, and to fix it" (Ramugondo et al., 2017, p. 8). Disciplines and professions in health-care as well, get to monopolise the terrain to develop and protect their own legitimacy to diagnose. On the other hand, this process strips individuals and their communities of their agency to diagnose problems that lead to ill-health. Decoloniality of Being affords individuals and collectives, as marginalized and oppressed, yet also resilient Beings, to more accurately link coloniality of Power, Knowledge, and Doing to "unequal and historically inscribed subjectivities of difference" (Ramugondo et al., 2017, p. 12), and ill-health.

Subjectivities of difference are intimately linked to lived experience, cultivating knowledge about self, others and the world. Knowledge is also about shared concerns and fears, and in many ways relates to mental and social well-being, and in so doing implicates health. In a world where stress and mental health issues are on the rise, subjectivities of difference, informed by lived experience, including historical trauma, begin to matter in more acute ways. To illustrate this point, we share the story below, as experienced by the second author.

> A regular feature of the national conference on health in Aotearoa New Zealand is an opportunity given to sponsors to showcase and market equipment or services. At a recent conference, a psychologist was given time to present his business that focuses particularly on supporting health professionals to better manage stress at work. From the podium, he announced that he was going to have the delegates simulate a stressful experience, as a means to call to mind typical reactions to stress. The exercise involved him choosing at random, one conference delegate to sing two lines of a song to the rest of the audience. The psychologist stepped down from the stage and walked quietly amongst the seated audience and a quiet thick tension seemed to fill the air. After less than a minute, he stated that no, he would not choose someone to sing in public after all. Sighs of relief and quiet chatter followed before he asked, *"Who went blank and couldn't recall any song?"* Some hands raised and more quiet laughter. *"Who could only recall nursery rhymes or childhood songs?"* More hands and more laughter. *"Who felt angry and wanted to express that?" "Who wanted to leave the room?"* A few more hands raised. The psychologist stated that these are all typical reactions to stress, that the numbers were fairly predictable, and then proceeded to briefly explain a framework that he uses on stress.
>
> What was missing was an enquiry to the audience about any other reactions or responses to his exercise. At the start of the exercise, sitting among peers in my group, I had sung a few words that quickly came to mind, *"You got to lively up yourself"* from the famous Bob Marley song. A few laughed and smiled and I stated *"He doesn't realise that as Māori we are used to singing in public. What is more, we enjoy it."* It is not to say that I didn't feel stress within that moment. I did. But my stress was triggered by something different from what the exercise intended, realising that once again, my experience was evidently neither 'predictable' nor 'typical' for the presenter. When I offered my experience to the presenter, he appeared confused.

It is often the case that when faced with perceived lack of time, resource or support, we revert to what is typical, known or predictable, as did this presenter when he 'trucked on' through his presentation, missing a teaching opportunity or at least a chance to reflect on stress as being culturally defined and shaped. What counts as stress, matters, because resources and efforts are directed to particular stress experiences and 'evidenced' ameliorators. The stress of 'feeling othered, invisible and maligned' amongst a peer group of health professionals, is common, typical and predictable as Māori, but receives little attention. Responding to this form of microaggression from peers is only one part of praxis when the 'Othered self' living within colonised skin, struggles towards decoloniality of Being. Building a decolonial epistemology of health requires this form of praxis, but also working towards marginalized and oppressed populations reclaiming their agency to diagnose problems that lead to ill-health within their communities, with disciplines and professions as humble partners.

Decoloniality as Praxis for Health

In order to decolonise health promotion, it is important that voices from marginalized and oppressed populations inform teaching, research and practice. This cannot happen through scholarship that seeks to appropriate endogenous, indigenous or traditional knowledges from non-Western worlds only "to serve imperial powers and to claim white supremacy" (Ramugondo, 2018, p. 5): what Ramon Grosfoguel calls epistemic extractivism (Grosfoguel, 2016). Health promotion research and practice should see scholars from marginalized populations, including indigenous people, increasingly setting the agenda, driven by the need for these populations to reclaim their agency to diagnose and solve problems that lead to ill-health within their communities. These scholars themselves, should resist co-option through the civilizing assimilationist Western-led project, where mimicry is the only way to gain recognition. It is this mimicry that renders the colonial subject within the Westernized University (Grosfoguel, 2013), alienated to themselves, and no longer recognisable to members of their own community.

In Aotearoa New Zealand, decolonising the academy occurs across education services from early childhood to tertiary institutions and is underpinned by Māori knowledges and movements for change (Pihama, 2011). The 1980's saw a movement by many communities (urban and rural) from mainstream education choosing instead to create safe cultural and structural spaces for the very young. Linda Tuhiwai Smith (1999) deftly described the many methods by which research maintains colonialism (and indigenous obedience) and revealed an indigenous research infrastructure that decolonises how research is viewed and treated. Graham Smith (1997) and Pihama (2010) amongst others call on indigenous communities to reclaim theorising, education and academy as norm, as everyday critical praxis that transforms the indigenous experience. For instance, in the example above about stress management, the second author could easily have fallen into the trap of mimicry, endorsing what is 'predictable' or 'typical' about responses to stress, in order not to disturb the peace. Instead, she sang in front of her peers, shared her experience of being a Māori, and how singing in public amongst her people is enjoyable, rather than a trigger for stress. Singing, in this case and many others, could easily be a resource for ameliorating stress. The lyrics to Bob Marley's song in the example, "*You got to lively up yourself*", metaphorically, could be about a refusal to die: a refusal to be denied ontological density.

It is not surprising that the arts are often invoked whenever issues of mental health surface. During the hashtag #RhodesMusFall campaign at the University of Cape Town, and as students highlighted mental health issues as a matter of deep concern, affecting black students to the point of suicide, the arts: through poetry, song, graffiti, and protest theatre, emerged as a powerful vehicle for self and collective expression, and a potential resource for healing. But art once co-opted, is no longer art, but entertainment. Similarly, scholarship as embodied performance within the academy, once co-opted, is no longer scholarship, but commodified mimicry. To the body, art as entertainment, and scholarship as mimicry, brings (dis)ease and possibly ill-health. Occupational consciousness as praxis, thus becomes a disruptive act to sustain both personal and collective health.

Consciousness in doing or the decoloniality of Doing, as a form of critical reflexivity within every day doing, including performing the academy, is a refusal to participate in the political economy of ill health in a globalized neo-liberal capitalist system. For health professionals and scholars, this refusal comes with awareness that in reducing persons to just individuals, humans into clients, and graduates into entrepreneurs capitalizing on ill-health, disciplines are in service of coloniality in all its forms. This reduction of humans into agents that drive the commodification of health is sustained through unconscious praxis: both scholarship and practice. For the first author, decoloniality of Doing, or praxis, came through disobedient scholarship and practice within her profession, occupational therapy.

In both scholarship and practice, both authors drew from their everyday commonsense or tacit knowledge, cultivated through lived experience within their communities. As a recently graduated young occupational therapist, the first author was tasked with establishing the first Occupational Therapy Department at a school for children with learning difficulties in rural Limpopo, South Africa. A key responsibility that emerged for her in this role was assessing whether applicants to the school were school-ready or not. On advice from educationalists and psychologists, the school had purchased an aptitude test, and had one of the teachers in the past administer this as part of selecting those who could be admitted to the school. The aptitude test was developed from the United Kingdom (UK). Given that this aptitude test was in English, and used puzzles, which she knew rural kids would never have seen before, the first author indicated to the school authorities that the test was not a suitable measure of aptitude for these kids. Although her reservations with the test was understood, the school authorities felt that an alternative test should still be deemed objective. After consulting with other occupational therapists nationally, she was pointed to a newly developed test, developed in South Africa, developed specifically for environmentally disadvantaged pre-school children. The first author was still sceptical, given the diversity of South Africa in terms of language, and was unconvinced about whether at the time, in the 90s, 'disadvantage' was a clearly defined construct that applies across contexts. The school authorities however, felt this test would be a suitable substitute for the previous one.

As an inexperienced occupational therapist, the first author was not about to challenge authority and place herself at risk for in-subordination. Given that she was the only occupational therapist at the school and was left to work independently most of the time, she decided to use the test only on paper, while in practice she used other means to determine whether the child was school-ready or not. Usually these kids, with various disabilities (post-polio, osteogenesis imperfecta, cerebral palsy etc.) would present at the school between ages six and nine. The first author used interviews with these children as the most reliable means to assess their school-readiness. Two questions proved particularly useful; what is your most favourite song that you dance to? Can you share a joke with me that you think is the funniest you have ever heard, or something you saw that you thought was the funniest ever? As back-up, she always had a record cassette with the latest music trending locally. She would play a song closest to the genre the child had said was their favourite song and begin to dance. Often, she did not even have to invite the child, and their caregiver to join in the dance. If the child had rhythm, and could sing or hum along, it meant they were fairly school-ready. The answer to the second question was particularly telling and gave the first author a good sense about whether the child would benefit from being in the school or not. Once the child was admitted, and the first author still had a niggly worry that perhaps she had made a mistake, she would watch the child engaged in unstructured play with others. If they were able to negotiate their own place in a game, and know when to lead or follow, she was sold on the child being school-ready.

Although the first author never disclosed to the school authorities or her colleagues in occupational therapy that she did not follow their instruction or advise, 15 years later, she looks back and realises that her respect for play as a form of intelligence and a way of being in the world, expressed through what seems frivolous at times, has informed her scholarship, leading to her most productive theorising as part of occupational science: a discipline borne out of occupational therapy. Play as often disruptive of the status quo, and a way of subverting power, and a resource for healing, resonates deeply with disobedience within disciplines. Had the first author used the UK based aptitude test, or the one developed in South Africa, but which she was still sceptical of and felt would do a disservice to the children, her sense of (dis) ease in her practice of occupational therapy may have had dire consequences to her personal health. A child who is deemed not school-ready, goes back to a community, where this diagnosis should make sense. When this is not the case, the community itself may be left with less faith in themselves to identify problems and find workable solutions.

The use of standardized tests to assess aptitude, cognitive functioning or even intelligence is pervasive in psychology, and cognate disciplines. Recently, a study published in the journal *Aging, Neuropsychology and Cognition* had to be retracted after it was exposed that it was racist in its intention. Titled, 'Age- and education-related effects on cognitive functioning in Colored South African women', the article presented data collected from a sample of 60 women (aged between 18 and 64 years), using the Montreal Cognitive Assessment, and concluded that "Young to middle-aged Colored women present with low cognitive function and (*sic*) which is significantly influenced by education" (Nieuwoudt et al., 2019, p. 1). The study also inferred that unhealthy lifestyle behaviours partly explained the low cognitive function in this population. Given that all the five authors are white women, the question remains whether a co-author with the same demographics as the population under study would have concurred with these authors or not. A colleague from the University of Cape Town, an African Feminist scholar, Associate Professor Barbara Boswell, who was the first one to raise awareness and problematize the study via twitter, described how the study exposed the "embedded worldview, epistemology, and academic structure of the university" where this study was conducted.

Universities, however, do not operate as silos, but reflect the structurally embedded unequal relations of power between nation states, between the (ex)colonised and the beneficiaries of colonial plunder. Disciplines often re-inscribe coloniality in all its forms, such that it would be naïve to think that this study is the only one of its kind that ever got published, or that it would be the last. For communities such as those who are referred to in the retracted study to ever see justice, those who benefit from skewed distribution of social determinants of health, such that these communities continue to suffer, must be identified and exposed. Scholarship and practice that seeks to promote health should reflect an understanding of coloniality in all its forms, with scholars striving for both decoloniality of Being and Doing, as a necessary praxis for health, at both personal and collective levels.

Conclusion

Decolonising health promotion requires a shift away from the internalized dominant discourse of biomedicine, and a return to indigenous epistemes. The shift must recognize that coloniality of Power, Knowledge, Being and Doing are deeply intertwined. The Westernized University, through its disciplines, is often directly complicit in advancing coloniality in all its forms. Performing the academy in a manner that is scripted by colonial tropes, bears consequences for self and others. This calls for embodied disobedient praxis within disciplines, with both decoloniality of Being and Doing as necessary disobedient praxis strengthened by partnerships, alliances and networks of allies and indigenous practitioners.

References

Applbaum, K. (2006). Educating for global mental health: The adoption of the SSRIs in Japan. In A. Petryna, A. Lakoff, & A. Kleinman (Eds.), *Global pharmaceuticals: Ethics, markets, practices* (pp. 85–110). Duke University Press.

Babalola, E., Noel, P., & White, R. (2017). The biopsychosocial approach and global mental health: Synergies and opportunities. *Indian Journal of Social Psychiatry, 33*(4), 291–296. https://doi.org/10.4103/ijsp.ijsp_13_17

Babich, B. E. (2003). From Fleck's denkstil to Kuhn's paradigm: Conceptual schemes and incommensurability. *International Studies in the Philosophy of Science, 17*(1), 75–79. https://doi.org/10.1080/02698590305236

Borell, B., Moewaka Barnes, H., & McCreanor, T. (2018). Conceptualising historical privilege: The flip side of historical trauma, a brief examination. *AlterNative: An International Journal of Indigenous People, 14*(1), 25–34. https://doi.org/10.1177/1177180117742202

Brassolotto, J., Raphael, D. & Baldeo, N. (2013). Epistemological barriers to addressing the social determinants of health among public health professionals in Ontario, Canada: a qualitative inquiry. *Critical Public Health 24*(3), 321–326. https://doi.org/10.1080/09581596.2013.820256

Brassolotto, J., Raphael, D., & Baldeo, N. (2014). Epistemological barriers to addressing the social determinants of health among public health professionals in Ontario, Canada. *Critical Public Health, 24*, 321–336. https://doi.org/10.1080/09581596.2013.820256

Came, H. A. (2014). Sites of institutional racism in public health policy making in New Zealand. *Social Science and Medicine, 106*, 214–220. https://doi.org/10.1016/j.socscimed.2014.01.055

Carvalho, S. R., & Teixeira, R. R. (2017). Politics of life itself and the future of medical practices: Dialogues with Nikolas Rose (part 3). *Interface (Botucatu), 21*(60) http://www.scielo.br/scielo.php?pid=S1414-32832017000100221&script=sci_arttext&tlng=en

Cave, A. A. (2008). Genocide in the Americas. In D. Stone (Ed.), *The historiography of genocide* (pp. 273–295). Palgrave Macmillan.

Collins, P. A., & Hayes, M. V. (2007). Twenty years since Ottawa and Epp: Researchers' reflections on challenges, gains, and future prospects for reducing health inequities in Canada. *Health Promotion International, 22*(4), 337–345. https://doi.org/10.1093/heapro/dam031

Commission on the Social Determinants of Health. (2008). *Closing the gap in a generation: Health equity through action on the social determinants of health*. World Health Organization. https://www.who.int/social_determinants/final_report/csdh_finalreport_2008.pdf

Connell, R. (2007). *Southern theory: The global dynamics of knowledge in social science*. Allen & Unwin.

de Sousa Santos, B. (2014). *Epistemologies of the south: Justice against epistemicide*. Paradigm Publishers.

Durie, J. (2003). Speaking the silence of whiteness. *Journal of Australian Studies, 27*(79), 135–144. https://doi.org/10.1080/14443050309387894

Engel, G. L. (1977). The need for a new medical model: A challenge for biomedicine. *Science, New Series, 196*(4286), 129–136.

Fleck, L. (1986 [1947]). To look, to see, to know. In R. Cohen & T. Schnelle (Eds.), *Cognition and fact: Materials on Ludwick Fleck (pp. 129–151)*. D. Reidel Publishing Company.

Fricker, M. (2007). *Epistemic injustice: Power and the ethics of knowing*. Oxford University Press.

Ghaemi, S. N., Vöhringer, P. A., & Vergne, D. E. (2012). The varieties of depressive experience: Diagnosing mood disorders. In D. L. Mintz (Ed.), *Depression: Psychiatric clinics of North America* (pp. 73–86). W. B. Saunders Company, Elsevier Inc..

Gordon, L. R. (2014). Disciplinary decadence and the decolonisation of knowledge. *Africa development, XXXIX*(1), 81–92.

Grosfoguel, R. (2013). The structure of knowledge in westernized universities: Epistemic racism/sexism and the four genocides/epistemicides of the long 16th century. *Human Architecture: Journal of the Sociology of Self-Knowledge, 11*(1), 73–90.

Grosfoguel, R. (2016). Del extravismo económico al extravismo epistémico y al extravismo ontológico: una forma destructive de conocer, ser y estar en el mundo. *Tabula Rasa, 24*. http://www.revistatabularasa.org/numero-24/06grosfoguel.pdf

Hancock, T. (2011). Health promotion in Canada: 25 years of unfulfilled promise. *Health promotion international, 26*(2), ii263–ii267. https://doi.org/10.1093/heapro/dar061

Harwood, J. (1986). Review: Ludwik Fleck and the sociology of knowledge. *Social Studies of Science, 16*(1), 173–187. https://doi.org/10.1177/030631286016001009

Heke, I., Rees, D., Swinburn, B., Waititi, R. T., & Stewart, A. (2019). Systems thinking and indigenous systems: Native contributions to obesity prevention. *AlterNative, 15*(1), 22–30. https://doi.org/10.1177/1177180118806383

Hira, S. (in press): Decolonizing the mind: Imagining a new world civilization. Amrit Publishers.

Kirmayer, L. J., & Raikhel, E. (2009). Editorial: From amrita to substance D: Psychopharmacology, political economy, and technologies of the self. *Transcultural Psychiatry, 46*(1), 5–15. https://doi.org/10.1177/1363461509102284

Kuhn, T. S. (1970). *The structure of scientific revolutions*. University of Chicago Press.

Low, J., & Thériault, L. (2008). Health promotion policy in Canada: Lessons forgotten, lessons still to learn. *Health Promotion International, 23*(2), 200–206. https://doi.org/10.1093/heapro/dan002

Lushaba, L. (2017). *Decolonise institutions of higher learning, a conversation with Dr Lwazi Lushaba*. The Forum@8 (Audio podcast). https://iono.fm/e/344635

Madley, B. (2004). Patterns of frontier genocide 1803–1910: The aboriginal Tasmanians, the Yuki of California, and the Herero of Namibia. *Journal of Genocide Research, 6*(2), 167–192.

Maldonado-Torres, N. (2007). On the coloniality of being. *Cultural Studies, 21*(2–3), 240–270. https://doi.org/10.1080/09502380601162548

Mamdani, M. (2017). *Decolonising the post-colonial university. TB Davie Memorial Lecture*. University of Cape Town.

Mazzocchi, F. (2011). The limits of reductionism in biology: What alternatives? *Electronic Journal for Philosophy, 11*, 1–19.

Melber, H. (2008). Colonialism, genocide and mass violence – Integral parts of modernity. In H. Melber & J. Y. Jones (Eds.), *Development dialogue: Revisiting the heart of darkness – Explorations into genocide and other forms of mass violence* (pp. 265–272). The Dag Hammarskjöld Centre.

Mignolo, W. D. (2012). *Local histories/global designs: Coloniality, subaltern knowledges, and border thinking*. Princeton University Press.

Ndlovu-Gatsheni, S. J. (2013a). Why Decoloniality in the 21st century? *The Thinker, 48*, 10–15. http://www.thethinker.co.za/resources/48%20Thinker%20full%20mag.pdf

Ndlovu-Gatsheni, S. J. (2013b). *Empire, global coloniality and African subjectivity*. New Berghahn Books.

Ndlovu-Gatsheni, S. J. (2015). Decoloniality as the future of Africa. *History Compass, 13*(10), 485–496. https://doi.org/10.1111/hic3.12264

Nichter, M. (1981). Idioms of distress: Alternatives in the expression of psychosocial distress: A case study from South India. *Culture, Medicine and Psychiatry, 5*(4), 379–408. https://doi.org/10.1007/BF00054782

Nieuwoudt, S., Dickie, K. E., Coetsee, C., Engelbrecht, L., & Terblanche, E. (2019). Retracted article: Age- and education related effects on cognitive functioning in Colored South African women. *Aging, Neuropsychology, and Cognition*. https://doi.org/10.1080/13825585.2019.1598538

Paradies, Y. (2016). Colonisation, racism and indigenous health. *Journal of Population Research, 33*(1), 83–96. https://doi.org/10.1007/s12546-016-9159-y

Pease, B. (2010). *Undoing privilege: Unearned advantage in a divided world*. Zed Books.

Petticrew, M., Whitehead, M., MacIntyre, S. J., Graham, H., & Egan, M. (2004). Evidence for public health policy on inequalities: 1: The reality according to policymakers. *Journal of Epidemiology & Community Health, 58*(10), 811–816. https://doi.org/10.1136/jech.2003.015289

Pihama, L. (2010). Kaupapa Māori theory: Transforming theory in Aotearoa. *He Pukenga Kōrero: A Journal of Māori Studies, 9*(2), 5–14.

Pihama, L. E. (2011). How Maori cultural knowledge can be used as a source of innovation and creativity. In Te Mahurehure marae in Point Chevalier. Retrieved from http://www.radionz.co.nz/

Quijano, A. (2000). Coloniality of power, eurocentrism, and Latin America. *Nepentla: Views from the South, 1*(3), 533–580.

Ramugondo, E. (2018). Healing work: Intersections for decoloniality. *World Federation of Occupational Therapists Bulletin, 74*(2). https://doi.org/10.1080/14473828.2018.1523981

Ramugondo, E., Lepere, R., & Nebe, W. (2017). Decolonising stigma and diagnosis as healing work. *Health Tomorrow: Interdisciplinarity & Internationality, 5*. https://ht.journals.yorku.ca/index.php/ht/article/view/40246/36428

Ramugondo, E. L. (2015). Occupational consciousness. *Journal of Occupational Science, 22*(4), 488–501. https://doi.org/10.1080/14427591.2015.1042516

Raphael, D. (2011). A discourse analysis of the social determinants of health. *Critical Public Health, 21*(2), 221–236. https://doi.org/10.1080/09581596.2010.485606

Raphael, D., Curry-Steven, A., & Bryant, T. (2008). Barriers to addressing the social determinants of health: Insights from Canada. *Health Policy, 88*(2–3), 222–235. https://doi.org/10.1016/j.healthpol.2008.03.015

Robinson, C. J. (1983). *Black Marxism: The making of the black radical tradition*. The University of North Carolina Press.

Robson, B. (2007). Economic determinants of Māori health and disparities. In M. Bargh (Ed.), *Resistance: An indigenous response to neoliberalism* (pp. 45–61). Huia Publishers.

Rodney, W. (1973). *How Europe underdeveloped Africa*. Dar-Es-Salaam, Tanzania: Bogle-L'Ouverture Publications/ London and Tanzanian Publishing House, Transcript from 6th reprint, 1983; Transcribed: by Joaquin Arriola.

Rosciglione, C. (2013). A non-reductionist physiologism: Nietzsche on body, mind and consciousness. *PRO, 12*(1), 43–60.

Rose, N. (2009). *The politics of life itself: Biomedicine, power, and subjectivity in the twenty-first century*. Princeton University Press.

Smith, G. H. (1997). *The development of Kaupapa Māori theory and praxis (unpublished doctoral dissertation)*. University of Auckland, School of Education, Auckland, New Zealand.

Smith, L. T. (1999). *Decolonising methodologies: Research and indigenous peoples*. Zed Books.

Thornicroft, G., Cooper, S., & van Bortel, T. (2012). Capacity building in global mental health research. *Harvard Review of Psychiatry, 20*(1), 13–24. https://doi.org/10.3109/10673229.2012.649117

Truglio, J., Graziano, M., Vedanthan, R., Hahn, S., Rios, C., & Hendel – Paterson, B., & Ripp, J. (2012). Global health and primary care; increasing burden of chronic diseases and need for integrated training. *Mt Sinai Journal of Medine, 79*(4), 464–474. https://doi.org/10.1002/msj.21327

UCT CCWG (2018). *Curriculum change framework*. https://www.news.uct.ac.za/images/userfiles/downloads/reports/ccwg/UCT-Curriculum-Change-Framework.pdf

Wallerstein, I. M. (2004). *World-systems analysis: An introduction*. Duke University Press.

WHO (2008). *Closing the gap in a generation: Health equity through addressing the social determinants of health*. http://www.who.int/social_determinants/final_report/csdh_finalreport_2008.pdf

WHO (2010). *Equity, social determinants, and public health programmes*. http://whqlibdoc.who.int/publications/2010/9789241563970_eng.pdf

WHO. (2016). *What is health promotion?*. https://www.who.int/features/qa/healthpromotion/en/

Wilkinson, R. & Marmot, M. (2003). *Social determinants of health: The solid facts*. http://www.euro.who.int/document/e81384.pdf

Wilkinson, R., & Pickett, K. (2009). *The spirit level: Why greater equality makes societies stronger*. Bloomsbury Press.

Young, M., Lambert, D., Roberts, C., & Roberts, M. (2014). *Knowledge and the future school: Curriculum and social justice*. Bloomsbury.

Zimmerer, J. (2008). Colonialism and the holocaust: Towards an archeology of genocide. In H. Melber & J. Y. Jones (Eds.), *Development dialogue: Revisiting the heart of darkness – Explorations into genocide and other forms of mass violence* (pp. 95–124). The Dag Hammarskjöld Centre.

Decolonising Australian Psychology: The Influences of Aboriginal Psychologists

Yvonne Clark and Tanja Hirvonen

Introduction

Context of Invasion and Colonisation in Australia

Aboriginal and Torres Strait Islander peoples (thereafter Aboriginal)[1] inhabited Australia for up to 100,000 thousand years before colonisation. Archaeological information continues to find evidence to support this (i.e. Malaspinas et al., 2016). Aboriginal culture is respectfully recognised as the oldest living culture in the world, and prior to colonisation had sophisticated and distinctive beliefs, customs, and practices that included resourceful land management practices (Hill et al., 2013). Aboriginal people were dispersed and diverse, yet they were connected by strong collective beliefs in the creation stories and the responsibilities of nurturing mother earth that in turn nurtured her people. These strong beliefs extended to the view of life where there were sacred places and actions that were passed down from generation to generation which needed to be maintained and sustained to preserve the land and life for people and animals alike. Life and existence depended on the relationship between people, animals, culture, spirituality, and the land. These beliefs and

[1] Throughout this paper, the term 'Aboriginal' will respectfully be used to refer to the diversity of both Aboriginal and Torres Strait Islander nations and peoples. The term 'Indigenous' will also be used in some contexts, particularly when referring to the Australian Indigenous Psychologists Association, in international contexts and when referenced as such.

Y. Clark (✉)
SAHMRI Women and Kids Theme, South Australian Health and Medical Research Institute, Adelaide, SA, Australia
e-mail: Yvonne.Clark@sahmri.com

T. Hirvonen
Flinders University, Adelaide, SA, Australia
e-mail: Tanja.Hirvonen@flinders.edu.au

S. Kessi et al. (eds.), *Decolonial Enactments in Community Psychology*, Community Psychology, https://doi.org/10.1007/978-3-030-75201-9_11

practices made up the holistic essence of Aboriginal people and provided purpose, belief systems, identity, and meaning.

Although Australia has been visited by various nations over time, the British invasion in 1788 along with colonial laws, processes, and behaviours brought about the most profound changes in relation to Australia's landscape, animals, its people, and ancient cultural practices. Aboriginal people's lives were altered drastically by colonial domination, racism, oppression, and ways of living that fragmented culture and the holistic way of living which had devastating impacts that are still felt today. This was done by acts such as the clearing and taking of land, massacres, the displacement of Aboriginal groups, incarceration, excluding and alienating Aboriginal people from services, society, and resources as well as the denial of wages and economic development. If Aboriginal people resisted such treatment and domination they were met with violence and further disempowerment. Yet, even in the absence of direct forms of resistance, Aboriginal people experienced exploitation, abuse, intimidation and coercion (Glover et al., 2005; Milroy et al., 2014). Hence Aboriginal people lived in a context of subjugation and oppression.

Australia's governing history in relation to Aboriginal people has been categorised in five broad and overlapping policy time frames in Australia, depending on the State or Territory. These included Protectionism (1869–1937), Assimilation (1937–1965), Integration (1965–1972), Self-determination/management (1972–2005), and Reconciliation (2005–). A brief exploration of these time frames, policies, practices and earlier political agendas below, set the tone for inter- and multi-generational trauma (Atkinson, 2002; Gee, 2016), loss and grief issues (Wanganeen, 2014), negative discourses (Fogarty et al., 2018), and violence (including family and lateral violence) (Clark, 2017). Many of the historical remnants of colonisation continue to be consequential today and are reflected in the overrepresentation of social welfare statistics including those related to mental health and wellbeing. For example, the social and emotional wellbeing (SEWB) concerns contribute to health issues, difficulty in managing chronic disease, high rates of chronic disease, difficulties in pregnancy and early childhood, and a range of family and social issues (Australian Indigenous HealthInfoNet, 2018; Milroy et al., 2014).

Protectionism had its origins in the theory of Social Darwinism which endorsed the notion of 'survival of the fittest' in society. Aboriginal people were seen as primitive and not able to compete with a 'superior' race of Europeans and therefore expected to die out as part of an evolutionary process of natural selection. There is debate about the size of the Aboriginal population pre-colonisation with estimates between 500,000 to 1,000,000 Aboriginal people (Ranzijn et al., 2009). At the turn of the nineteenth century extremely high death rates coupled with low birth rates led to an estimated 60,000 Aboriginal people remaining, a population decline of 95% (Ranzijn et al., 2009). This decline in the Aboriginal population provided justification for supporting the theory of Social Darwinism, despite the deaths being mainly due to frontier violence, displacement from land, starvation and introduced diseases rather than evolution (Broome, 1994). Such high death rates meant the loss of knowledge bearers of culture, lore and tradition (Dudgeon, Wright, et al., 2014b), which placed the coherence and stability of Aboriginal cultures under stress. In the name of protectionism, to save and protect the remaining Aboriginal people from white colonists, many were separated from the colonists and accommodated in Christian or government missions and reservations. Those not in these institutions became town fringe dwellers, living in depressed conditions on the outskirts of towns and were viewed as hopeless people by white society and merely surviving (Broome, 1994). During this time, Aboriginal people were conditioned into dependency by Governments, provided with limited survival resources such as access to education and employment and were punished if they practiced their culture or attempted to speak their respective languages (HREOC, 1997). Thus, protectionism had a broader meaning which included control and the decimation of the Aboriginal way of life. However, with strong survival skills and resiliency along with an increase in mixed relationships or sexual encounters (between Aboriginal and non-Aboriginal people) that led to a generation of children of mixed ancestry, Aboriginal people did not die out as predicted. Thus, protectionism gave way for a different policy direction of Assimilation.

The Assimilation policies focused on the notion of breeding out Aboriginal blood and merging Aboriginal people into white society (Haebich, 2000). The Government of the time sought to create one uniform white culture for Australia. There were a series of Assimilation policies that devastated Aboriginal families such as the policies which instigated the removal of Aboriginal children of mixed race to white institutions and private homes. This further disrupted and disconnected people from culture, kinship, spirituality and land. These policies were deemed as among the most profound disruptions to Aboriginal people and their way of life. Those that were removed are known as the *stolen generations* (HREOC, 1997). It is also argued that such practices of Aboriginal children being removed from families and culture continue to this day under a stringent welfare system in Australia by various government departments and state legislations within the child protection and youth detention systems (SNAICC, 2019).

Within the later policy eras, from Integration to Self-management and now Reconciliation, there have been many changes and achievements for Aboriginal people such as the 1967 Referendum[2] ground-breaking wins such as those evidenced in the Native Title Act of 1993[3] where Aboriginal people gained rights to their traditional lands and annual NAIDOC[4] celebrations. Despite these gains, there continues to be disadvantages and significant gaps in access to housing, employment, education, incarceration, and mental health status for Aboriginal people compared with non-Aboriginal Australians. The Australian government (2020) has

---

[2] This refers to the 1967 Australian referendum in which constitutional change meant that Aboriginal and Torres Strait Islander people could be included in the census. This allowed the Australian Federal Government to make laws in relation to Aboriginal peoples.

[3] The Native Title Act (1993) governs the rights of both Aboriginal and Torres Strait Islander people in Australia who have gained back rights to their traditional lands (see AHRC, 2020).

[4] NAIDOC is a widely used acronym meaning National Aboriginal and Islander Day Observance Committee. It is a week of celebrations held annually across Australia to celebrate, history, culture and achievements of Aboriginal people and encouraged to be celebrated by all Australians.

introduced seven targeted strategies to reduce disadvantage for Aboriginal people known as *Close the Gap*. These strategies include: increased life expectancy; reduction in child mortality; improved access to early childhood education, improved school attendance rates, educational attainment to year 12 or equivalent, improved literacy and numeracy, and employment outcomes. Alongside this Government initiative is also the reconciliation movement which is about strengthening relationships between Aboriginal people and non-Aboriginal people, to benefit all people (Reconciliation Australia, 2019), targeting education, employment, health and legal systems. Despite these government initiatives, after 10 years of the current political reconciliation movement within which Close the Gap initiatives sit, the disparities remain disproportionately high between Aboriginal and non-Aboriginal people (Holland, 2018). The first decade of the initiative has only managed to establish improvements in education and employment. Some positive figures from 2016 data show that a high proportion (70%) of 15–24 year old Aboriginal people are engaged in education, employment or training The trend for education over the past decade for year 12 education shows an increase from 32% (in 2006) to 47% (in 2016) of those aged 20–24 years. The number studying at universities has increased from 7000 Aboriginal people (in 2006) to 15,395 Aboriginal people in 2016 (Australian Indigenous HealthInfoNet, 2018). Similarly, there has been an increase in employment to almost half the Aboriginal population (47%). The trend shows an increase from 42.4% (in 2006) to 46.6% in 2016 (AIHW, 2020).

Invasion and colonisation along with oppressive policies and practices detrimentally impacted Aboriginal people and their unique culture in Australia. Whilst in some areas of Australia many of the traditional customs, spirituality and old ways of living are preserved, in some areas the practice of culture had been compromised, in ways that may take many generations to turn around. To improve the situation and overcome damaging colonial effects, Aboriginal people have been prioritising processes of decolonisation in various forums. This is consistent with many decolonising activities for Indigenous peoples in other countries. Some examples include the indigenisation and decolonisation of academia in colleges and institutions in Canada (Gaudry & Lorenz, 2018) and in the Caribbean world where the Taino people are taking a decolonising approach to reconstruct nationhood and identity for spiritual wellness and self-determination (Neeganagwedgin, 2015). Whilst there remain serious barriers, such as widespread racism (see Awofeso, 2011; Dudgeon, Milroy, & Walker, 2014c), in addressing the past, many Aboriginal people stand steadfast and resolute, having survived the brutalities of the past, and continue to navigate Australia today. Aboriginal people are entering and having a voice in areas such as health, law, education, politics, justice, and sovereignty, including holding leadership roles and contributing to decolonising systems and structures. Examples of this include decolonising methodologies, research methods and knowledges within academia (Nakata, 2007a; Povey & Trudgett, 2019); and decolonising the criminal justice system by increasing our understanding of institutional racism and how it affects Aboriginal people on an everyday basis (Cunneen, 2019). This chapter contributes to the dialogue regarding decolonisation for Aboriginal people, with a focus on the field of psychology. This examination of decolonisation processes within psychology highlights the actions a collective of Aboriginal psychologists and their influences within the discipline of psychology as well as the Aboriginal community, emphasising the voices and experiences of two Aboriginal clinical psychologists.

Decolonisation

Decolonisation in the Australian context is to make visible the many accounts of Aboriginal survival and resistance and tell parallel and rarely heard stories of the history of Australia where Aboriginal people resisted despite the threat of further disempowerment (Milroy et al., 2014). There are many examples of Aboriginal people's resistance to colonialism (Lowe, 1994) and accounts of how Australia became a settler-colony that need to be further privileged. Colonisation by the British had been carried out on the claim that Australia was terra nullius, meaning that the British could occupy the land denying any recognition of Aboriginal peoples' rights to the land. Colonisation delayed a significant recognition to the rights of the land and resources of Aboriginal people (Strelein, 2005). Many negative stories concerning Aboriginal people have been told by the most powerful voices in Australia such as non-Aboriginal' white' administrators, historians, media and governments, and these stories have been to the detriment of Aboriginal people throughout history (Milroy et al., 2014). Many of these stories have persevered as ongoing negative discourses (Fogarty et al., 2018; Lowitja Institute, 2018) that affect how we are viewed by ourselves and by others. A language of deficit coupled with inaccurate historical accounts can alter the trajectory of knowledge about Aboriginal people and impact on our lives, health and wellbeing (Nakata, 2007b). Telling our own stories is aligned with standpoint theory (Nakata, 2007a), a method of inquiry that knowledge stems from experience and social position. This has been important for marginalised groups whose accounts and experiences have been excluded or subjugated within intellectual knowledge production.

According to Strelein and Tran (2013), decolonisation involves engaging with and understanding former oppressive colonial laws, policies and practices and their consequences, to create a 'space' for Aboriginal people to express their distinctive identities, cultures, and ways of knowing. This engagement and understanding affirms the legitimacy of a people's ancestral culture to encourage cultural renewal and potentially, social transformation (Glover et al., 2005). In specific relation to the psychology profession in Australia, Dudgeon and Walker (2015) stress that psychology needs to recognise and integrate Aboriginal Australian cultural views and practices into mental health services, professional practice, and research. This should occur through the development and amalgamation of Aboriginal standpoints and conceptual frameworks, along with the implementation of cultural protocols and guidelines embedded in reconciliation action plans for mental health services.

Ongoing Processes of Colonising and Decolonising in Psychology

The discipline and practice of psychology has been part of the colonising processes influencing the social and emotional wellbeing (SEWB) (inclusive of mental health) of Aboriginal people in Australia. Dudgeon, Rickwood, et al. (2014a) argued that mainstream psychology's dominating discourses have been underpinned by cultural racism, ethnocentrism, and hegemony, and as such, have objectified, dehumanised, and devalued culturally different groups that don't fit the *dominant culture*. Cultural racism is generally regarded as the systematic manner in which non-Aboriginal peoples have established and imposed their primary cultural institutions (e.g., education and health) to raise and highlight mainstream and non-Aboriginal values, characteristics and achievements and to denigrate Aboriginal peoples (Oliver, 2001). This has certainly occurred in field of psychology. Examples of oppressive practices include supporting race-based evolutionary theories (i.e. Social Darwinism) and biological determinants of human behaviour which placed Aboriginal people at the bottom of an evolutionary hierarchy. Psychology and the development of mental functioning tools for assessment and testing with Aboriginal people have supported the biological determinants of 'primitive man's' behaviour and low intelligence (Dudgeon, Rickwood, et al., 2014a). Cultural bias in psychological assessments, testing, and diagnosis of mental health has been well established in the literature and in practice (Adams et al., 2014; Westerman, 2004; Westerman & Wettinger, 1998) and mainstream psychological practice and teaching continues to be ethnocentric and inaccurate, affecting mental health treatment, medication, and therapeutic support for Aboriginal people (Carey et al., 2017). Misdiagnosis can significantly alter a person's trajectory of health and wellbeing across their lifespan. For instance, a person may be prescribed medication as part of their treatment plan, and if the initial diagnosis was incorrect, the treatment including medication has the potential for harm.

Other oppressive influences and mainstream interpretations are also evident in social, developmental, and clinical psychology practice and theories. Examples include theories of racism aligned with in-group and out-group models which proposed these behaviours as inherent and normal; and the application of mainstream understandings of bonding and attachment theories to Aboriginal mother--child interactions that influenced decisions about the removal of Aboriginal children from their families and culture as part of the stolen generations (Dudgeon, Rickwood, et al., 2014a) and in the present (SNAICC, 2019). Many of these oppressive processes and practices and the perpetuation of negative representations and 'deficit discourses' (Fogarty et al., 2018; Lowitja Institute, 2018) of and towards Aboriginal people in general have been an ongoing process in health policy and practice for over a century and have become further legitimised through psychology. Their use within psychology has meant that many Aboriginal people continue to mistrust the psychology profession and psychological services (Carey et al., 2017; Dudgeon, Rickwood, et al., 2014a).

Yet given the nature of psychology as a caring and compassionate profession, it is equally well placed to greatly support and empower Aboriginal people. Psychology is the field of study that seeks to understand human behaviour, including thoughts, emotions, and actions. As part of the Aboriginal psychologists' collective, Gee et al. (2014) identified several common domains and broader determinants that Aboriginal people in Australia attribute to health and wellbeing. Social and emotional wellbeing is the widely accepted term for describing the social, emotional, spiritual and cultural health and wellbeing of Aboriginal peoples. A SEWB model was articulated based on several key domains (Social Health Reference group, 2004). This model was shaped by the strengths of one's connections to body, mind and emotions, family and kinship, community, culture, country, spirituality, and ancestry. These areas of SEWB were also recognised to be generated by social, historical, and political determinants (see also Dudgeon & Walker, 2015). The concept has further contributed to decolonising psychology in its validation from the findings of the National Empowerment Project (NEP) which used SEWB domains to explore Aboriginal individual, family, and community wellbeing (Dudgeon et al., 2017).

Dudgeon, Rickwood, et al. (2014a) provide a chronological account of psychological landmarks of individual and collective cultural responsiveness in the practice of psychology. These include the bicentennial year of colonisation in 1988, when a small group of community psychologists, congregating in Sydney for the 24th International Congress of Psychology, began a journey of understanding the detrimental effects of oppressive and racist policies within their profession and their minimal efforts to improving outcomes for Aboriginal people. Following this conference, in 1991, the Australian Psychological Society (APS) facilitated the formation of the Aboriginal and Torres Strait Islander Peoples and Psychology (ATSIPP) Interest Group (APS, 2019a). ATSIPP main focus has been the advancement of Aboriginal psychological issues and initiatives within a reconciliation framework, chaired and/or co-chaired by an Aboriginal psychologist. This group has a number of achievements that include the advancement of the *Ethical guidelines for the provision of psychological services for, and the conduct of research with, Aboriginal and Torres Strait Islander peoples* (See APS, 2019b). More recently, in 2009 ATSIPP established an Aboriginal student conference award to enable successful Aboriginal and Torres Strait Islander students to attend an annual APS conference to better understand the profession and to meet and be inspired by Aboriginal psychologists (APS, 2019a). Many of the students who have received the award over the last 10 years have gone on to become psychologists.

Aboriginal people and psychologists have led and continue to contribute to the decolonisation of psychology, and of Australian society more broadly, in many ways. Examples include tackling racism, leading research and projects in relation to Indigenising the psychology curriculum, assisting with increasing the number of Aboriginal psychologists as well as improving practice for and with Aboriginal communities in Australia. One major step was the formation of the Australian Indigenous Psychologists Association (AIPA) in March 2008 under the auspices of the APS. This group has achieved many milestones despite funding restrictions and has recently become incorporated in 2019. At AIPA's insistence, the APS committed to and developed an ongoing series of Reconciliation Action Plans (RAP) with the first launch in September 2012. The ongoing reconciliation processes identified four broad areas as priority areas for action: (1) Respectful Relationships; (2) Governance; (3) Cultural Competence; and (4) Aboriginal Education and Employment. Key achievements in 2014 highlighted improved partnerships and respectful relationships with Aboriginal people, including the AIPA, protocols for recognising traditional owners and Elders, ensuring APS staff and member groups undergo cultural awareness and safety training to name a few (APS, 2014a, b). As part of the ongoing reconciliation process, the APS delivered an apology to Aboriginal peoples in 2016 at the APS Congress in Melbourne. This was a monumental and significant event. The then APS President Professor Michael Kyrios spoke on behalf of the entire Australian psychology profession, acknowledging psychology's role in contributing to the erosion of Aboriginal culture and mistreatment of Aboriginal people. Developing and delivering the APS apology was not a small feat for the APS, and involved a lengthy, comprehensive, and culturally informed process. Future initiatives were recommended for Aboriginal and non-Aboriginal psychologists to work together in culturally appropriate ways to meet the social, emotional, and mental health needs of Aboriginal peoples (Carey et al., 2017). Going forward, for the APS Apology to have longstanding influence on psychological practice, it is critical that there is a commitment from all psychologists, as well as the APS itself, towards supporting the Apology's intent and purpose.

Another key outcome of the APS RAP was the Australian Indigenous Psychology Education Project (AIPEP), which aimed to: investigate curricular approaches to increasing cultural competence and Indigenous participation in psychology education and training; to increase recruitment and retention of Indigenous psychology students; integrate Indigenous studies in psychology courses for all students; and facilitate training pathways for Indigenous mental health workers (AIPEP, 2019; Dudgeon et al., 2016a, b, c). Outcomes for this project are still being realised and has been re-established in December 2020 (AIPEP 2) and working with a national community of practice in psychology. The project continues with improvements in cultural curriculum design in psychology by influencing and being supported by National accrediting and registration bodies authorising curriculum and qualifications.

The Australian health regulatory body is also now motivated to support culturally safe psychological work with Aboriginal and Torres Strait Islander peoples. A *Statement of Intent* (National Scheme, Aboriginal and Torres Strait Islander Health Strategy) was agreed to in June 2018 amongst 15 national health practitioner boards overseen by the Australian Health Practitioner Regulation Agency (AHPRA), accreditation authorities, and Aboriginal health sector leaders and organisations (AHPRA, 2018). A part of its own commitment to this *Statement of Intent,* AIPA agreed to various visions and strategies that prioritise cultural and clinical safety for those working in the sector and for patients. The intent is to work together to achieve equitable health outcomes to close the health gap by 2031. AIPA, through the AIPA Chair, will enact this commitment as a leader for Aboriginal psychologists on initiatives such as planning for a 'national summit' (AHPRA, 2019a). This strategy is groundbreaking, in that another aim is to eliminate racism from the health and mental health systems to protect the public.

A Collective Momentum of Aboriginal Psychologists

The Growth in Numbers of Aboriginal and Torres Strait Islander Psychologists

The number of Aboriginal psychologists has grown sporadically but steadily over the last 35 years from the first registered Aboriginal psychologist, Professor Patricia Dudgeon, in 1987 (AIPA, 2019a). It was reported that in the year 2000 there were fewer than a dozen registered Aboriginal psychologists in Australia. A decade later it was estimated that 39 Aboriginal psychologists were now registered (Rickwood et al., 2010). In 2012 this estimate grew to 50 which was based on how many Aboriginal psychologists were in contact with AIPA. At this time it was calculated that the number of Aboriginal psychologists needed to reach population parity with the total number of psychologists was 625 (Dudgeon, Rickwood, et al., 2014a). The more recent figures for Aboriginal psychologists are more accurate and based on statistics from AHPRA: in 2013 there were 137 self-identified Aboriginal psychologists, and in 2019 this increased to approximately 218 Aboriginal psychologists in Australia (AHPRA, 2019b). This was recorded as representing 0.6% of the population of 38,029 psychologists (AHPRA, 2019c), well below the population percentage of 3.3% (ABS, 2016). It is then calculated that to reach profession parity there now needs to be 1255 Aboriginal psychologists.

Even though there is a comparatively low increase, the reasons for such growth in the number of Aboriginal psychologists are unclear - the psychology workforce itself had grown in that time, as had Aboriginal participation in higher education in general. Furthermore, there was no provision for psychologists to state their cultural identity on registration prior to the establishment of AHPRA in 2010. The APS had provided this option to new members since the early 2000s, but only about two thirds of psychologists chose to join the APS. The establishment of AIPA in 2008 certainly enhanced the visibility of Aboriginal psychologists within the profession and influenced the profession with providing cultural competency to mental health professions and support for students; the AIPEP project and its influence on curriculum and recruitment of Aboriginal students within Schools of Psychology in Australia; the ongoing initiatives of ATSIPP in particular the student conference award and so on. Yet, even though there has been a steady but slow increase in the number of Aboriginal psychologists over the years, there continues to be a lack of profession parity. In fact, the parity gap appears to be increasing rather than decreasing, which is concerning, given the efforts to decolonise psychology, and this clearly shows that more work is needed by both Aboriginal and non-Aboriginal people, collectives, and systems to turn these figures around. It will be Aboriginal psychologists who, as a collective, will lead the way to improvements for the psychological services and care needed for Aboriginal people's mental health, and SEWB.

Aboriginal Psychologists and Significant Gatherings

Individually and collectively, Aboriginal psychologists in Australia have played a leading role in various and significant gatherings over the past three decades. One of the first examples was in September 1995 at the annual APS Conference in Perth, Western Australia where several Aboriginal psychologists and students gathered to support an inaugural keynote speech given by a late and renowned Aboriginal leader and activist for social justice[5]. The speech challenged the fabric of the psychology profession and its practitioners, highlighting Aboriginal self-determination as central to Aboriginal and Torres Strait Islander mental health. That challenge has underpinned progression toward improved mental and physical health outcomes for Aboriginal and Torres Strait Islander peoples (Dudgeon, Milroy, & Walker, 2014d). A significant gathering of Aboriginal psychologists at the National Aboriginal and Torres Strait Islander Suicide Prevention Conference which preceded the Second World Indigenous Suicide Prevention Conference in Perth in November 2018 is also worth highlighting. The sensitivity of the topic necessitated a healing team that was brought together to aid and debrief conference presenters and participants as well as other community attendees. The healing team consisted of highly trained and/or experienced Aboriginal health professionals with backgrounds including PhD trained researchers, Clinical Psychologists, Psychologists, Counsellors, and Traditional Healers. This culturally safe approach was supported by many organisations, counsellors, and healers including a number of AIPA members. The final conference report indicated that many attendees felt included, and culturally and emotionally safe throughout the conference. The conference report recommended that such healing and wellbeing teams should be included at all future conferences (PCIH, 2019). These two national and international Indigenous Suicide Prevention Conferences also enabled a gathering of world Indigenous psychologists. A conversation and meeting began with Indigenous psychologists from Australia, the United States of America, Canada, and New Zealand. This collective is being formalised but is still in its infancy. A number of virtual meetings took place throughout 2019 and continuing in 2020, with a proposal for a face to face gathering at the next World Indigenous Suicide Prevention Conference scheduled for 2021 in Winnipeg, Canada.

[5] With respect, the deceased activist's name has not been used, as permission was not gained from his family in time. Details of the activist can be found in Dudgeon et al. (2014a) where permission was sought.

The Australian Indigenous Psychologists Association

Historically, the collective nature and individual increase in the numbers of Aboriginal psychologists in Australia contributed to the formation of the AIPA. Preceding the decision to form AIPA, various Aboriginal psychologists gathered in a consultation forum for an APS project. It was at this forum that individuals decided to form and formalise a national body in March 2008. The AIPA has since led the way for growth and change in regard to Indigenous psychology and committed to improving the SEWB, mental health and culturally responsive care of and for Aboriginal and Torres Strait Islander peoples in Australia. Many AIPA members also formally and/or informally support Aboriginal students, and work towards achieving equity and parity within the profession of psychology, offering leadership, providing professional support, and mentoring and networking, to name a few (AIPA, 2019b).

Since its inception, the AIPA has achieved many initiatives and landmarks. Some examples include the development, delivery, and evaluation of widely acclaimed cultural competency workshops. AIPA members designed these two-day workshops to provide non-Aboriginal health participants with cultural competence to deliver appropriate mental health services within the SEWB framework. Included in the package were: contextual information and education on social, historical, and health determinants when working with Aboriginal people; guiding principles from the SEWB frameworks; and ensuring inclusion of the National Practice Standards for the Mental Health Workforce. The AIPA worked in partnership with the Department of Health and Medicare Locals to deliver 69*' Journey Towards Cultural Competence'* workshops to 931 administrators and service providers of Access to Allied Psychological Services (ATAPS) programs delivered around Australia. Snapshot evaluation data (from 909 participants) between June 2012 and June 2014 were overwhelmingly positive, with 92% of participants rating the workshop as excellent and 99% stating that they would recommend it to others (AIPA, 2019c). The AIPA as a recently incorporated organisation held its first Annual General Meeting (AGM) in November 2019 and elected its directors and office bearers (AIPA, 2019d). AIPA will move forward in strength, and there will be a plethora of future achievements and community accountability and activities.

Examples from Two Aboriginal Psychologists

As discussed thus far in this chapter there has been collective momentum evident in significant gatherings and wide-ranging work carried out by Aboriginal psychologists to decolonise psychology and make changes to improve mental health and SEWB for Aboriginal people in Australia. Often these collective processes and projects that bring people together are led or organised by individual Aboriginal psychologists, as exemplified in Aboriginal-authored psychology publications (like this chapter) that privilege Aboriginal voices (in particular the enormous contributions by Professor Patricia Dudgeon). Two Aboriginal psychologists exemplify some of that work in this chapter. Yvonne Clark and Tanja Hirvonen are both female, Aboriginal, clinical psychologists who also closely align with paradigms associated with community psychology. They are members of the AIPA and have worked in a range of psychological settings with and for the Aboriginal community since gaining their psychology qualification.

Example 1: Yvonne Clark

Yvonne Clark is a Kokatha and Wirangu Aboriginal woman from South Australia and currently lives in Adelaide. She gained a Master's in Clinical Psychology and registered as a clinical psychologist in 1997, completing her PhD in 2017. Since working as a psychologist, she has had extensive experience as a clinical psychologist, academic, and researcher within government, private, community-controlled, university, and institutional sectors. In her role as a clinician and researcher she has mainly worked with and focused on Aboriginal children, adolescents, parents, and their families within a community and empowering setting. This has been in health, mental health, perinatal and child protection sectors. She was awarded a National Health and Medical Research Council (NHMRC), early leadership investigator grant that commenced in January 2020 to conduct research on *Improving wellbeing and preventing lateral violence in the Aboriginal*[6] *community in SA: Investing in our young.* This research focuses on children and young people's wellbeing in the Aboriginal community, as an extension of her ground-breaking thesis entitled *Lateral violence in the Aboriginal community in SA: From dilemmas to strategies* Key findings from this thesis, together with the ensuing project incorporating the guiding concept of lateral violence, community awareness and education strategies, are directed towards empowering Aboriginal communities and generating positive solutions for change in Aboriginal people's SEWB in Australia.

Lateral violence has been described as the manner in which people in positions of powerlessness covertly and overtly direct their dissatisfaction toward each other, themselves, and others in powerless positions. Lateral violence can be perpetrated via physical behaviours, yet it is primarily described as being covert in nature, including behaviours such as bullying, gossiping, and stripping away others' identity. This is similar to insidious forms of modern or institutional racism which are primarily covert and therefore easier to hide, deny or silence (Clark et al., 2016). It has been well established that intergenerational, historical, and chronic forms of trauma affect Aboriginal people (see for example Atkinson, 2002; Gee, 2016), along with ongoing acts of oppression and racism toward Aboriginal people (see Awofeso,

[6] Throughout this paper, the term 'Aboriginal' will respectfully be used to refer to the diversity of both Aboriginal and Torres Strait Islander nations and peoples, unless otherwise specified.

2011; Reconciliation Australia, 2019). Lateral violence can be a cause and consequence of trauma and coincides with the interaction between intergenerational trauma and oppression which have both been occurring since invasion and colonisation. Moreover, these combined effects can assist in diminishing Aboriginal people's wellbeing and identity (Clark et al., 2016). Thus a focus on the topic of lateral violence was timely, as it had been identified as an issue with detrimental effects within Aboriginal communities, work organisations, and institutions in Australia, becoming intractable and normalised in ways which can be destructive to community relationships (Clark, 2017). Hence, it was important to name and unmask lateral violence within Aboriginal contexts and to seek solutions for change where Aboriginal people could have a voice.

Within a community context education and awareness about lateral violence was non-existent in Adelaide, South Australia (SA) and in response *Preventing Lateral Violence* (PLV) workshops were developed in 2007. The workshops were requested by and provided to Aboriginal sectors and people within various government and non-government organisations and workplaces and facilitated by two Aboriginal people. The education and awareness workshops included information about the nature of lateral violence, origins and the forms it takes. Further aims promoted prevention behaviours and strategies that potentially enhance connections with others. Feedback from ten workshops by 212 people were positive warranting its continuation. This also led to conducting exploratory and evaluative research in the form of a PhD.

A two-phased research PhD was undertaken by me (from 2013 to 2017) to understand how Aboriginal people in SA perceived, oriented, coped with and prevented lateral violence within their contexts as well as understanding the effectiveness of PLV workshops . Phase one of the research consisted of interviews with 30 adult Aboriginal people of various ages and genders. An action research component of phase two of the research evaluated a series of five PLV workshops in 2014 with 72 predominantly Aboriginal participants utilising pre, post and three-month post questionnaires . Three-month post workshop interviews were also undertaken with seven participants. The evaluation sought to understand the effectiveness of the workshops in regard to attitude and knowledge change and whether lateral violence prevention strategies were considered and actioned. The evaluation was also a means to refine and improve workshops.

Overall the findings from the quantitative (via questionnaires) and qualitative (via open questions and interviews) responses from the research with 102 participants were grouped into five broad categories. These relate to: the perception and labelling of lateral violence; the characteristics and detrimental effects of lateral violence; actions and behaviours to prevent lateral violence; an increase in knowledge and understanding of lateral violence that propelled participants in to action; and recommendations for improvements to the PLV workshops (Clark, 2017).

Thus, understanding about the perception and use of the term lateral violence was considered useful in understanding *infighting* by Aboriginal people, and provided a concrete label to what was previously unnamed, but normalised. Furthermore, it could be made tangible and visible, and thus open to challenge and prevention

(Clark & Augoustinos, 2015). The research highlighted many active coping strategies to prevent lateral violence: the act of gaining improved understanding, leading to self-reflection and change-making in their own contexts; and standing up to lateral violence which is one way to take the power out of the words (Clark et al., 2017a). Action strategies were also prominent at the workshops, which provided an avenue of awareness and education enabling an improved acceptance of Aboriginal diversity and different identities to hamper lateral violence and consequently foster a greater sense of wellbeing through more positive engagement, connections and relationships with others (Clark et al., 2017b, c). The delivery of the PLV workshops continues to be in demand.

One of the recommendations from the research was that there be exploration and resources to prevent lateral violence with the younger population in the Aboriginal community. Consultation with Aboriginal community will continue during research on the topic of lateral violence for the younger and emerging population and it is anticipated that new resources will be developed in future for *Improving wellbeing and preventing lateral violence in the Aboriginal community by investing in our young*. Such research will benefit psychology greatly, not only as an awareness and education resource but also as a means to guide the Aboriginal community on strategies to prevent lateral violence within families.

Example 2: Tanja Hirvonen

Tanja Hirvonen is a proud Jaru and Bunuba woman with family connections across most of North Australia. Tanja has family from the Kimberley region in Western Australia through to the east coast of Queensland. Tanja was born and grew up in Mount Isa and spent much time living in North West Queensland and in the Gulf of Carpentaria. Tanja's mother was born in Alice Springs, Northern Territory and whilst Queensland was home for the family's early years, many of Tanja's family have moved back to the Northern Territory and to the Kimberley. She completed her undergraduate Bachelor of Science degree with the University of Southern Queensland in 2010 and registered as a psychologist in 2011. She then continued her studies by completing her Master of Clinical Psychology with James Cook University in 2014 and gaining extensive experience as a clinical psychologist, lecturer, and researcher. She has worked for seven years in the Northern Territory in a number of areas such as research, policy development, Aboriginal Community-Controlled organisations, and in private practice. In her role as a clinician and researcher, her main focus areas are trauma, burnout/self-care, and mental health not only for the community, but for people working in front-line health service delivery.

Tanja was trained specifically in psychometric testing and assessment for people across their lifespan. The training in psychometric assessment included developing the skills to select appropriate psychometric and assessment tools and evidence-based report writing to provide a summary of recommendations to manage a person's presenting problems. It was during this time of training that Tanja found that very few psychometric tools had been culturally validated for use with Aboriginal populations, and that mainstream worldviews dominated assessment and consequently the interpretation of people's behaviour. Psychological testing and assessment with Aboriginal peoples to determine mental health, wellbeing, and general functioning is a very controversial issue among researchers and clinicians and is rarely challenged.

Aboriginal psychologists have increasingly challenged the narrow mainstream world views in psychological assessments and involved in culturally validating psychological tests and assessments or developing unique culturally appropriate tools. Tanja was involved in research which investigated and developed guidelines incorporating clinical, community, and cultural knowledge in the assessment process for Aboriginal and Torres Strait Islander peoples, titled *The BestPrAxIS study: Developing best practice guidelines for psychosocial assessment of Aboriginal and Torres Strait Islander peoples presenting to hospital with self-harm and suicidal thoughts* (Leckning et al., 2019). A number of findings emerged from this work, yet one of the most valuable findings from Tanja's perspective was that the importance of the tool is matched by the way and manner in which testing is carried out by the health professional. While there have been many gains in developing appropriate tools, there is still a long way to go in ensuring culturally safe assessment practice.

One significant way that Aboriginal health and recovery is different from mainstream health views is the holistic and whole of life view towards healing. Tanja highly recognises and values the importance of culturally responsive, strength-based approaches for people to grow and heal. Tanja was previously employed within an Aboriginal Medical Service's Social and Emotional Wellbeing team as part of an expansion of services and service delivery. As part of this new role, Tanja managed a team of Aboriginal and non-Aboriginal mental health professionals. A large component of the work entailed Tanja contributing to the clinical governance framework, supervision, and increasing the team's knowledge of social and emotional wellbeing concepts. The development of a clinical governance framework and embedding this into the work practices was critical for the SEWB team to provide culturally and clinically integrated care. Supervision and case conferences were essential for the team to engage in reflective practices in order to improve skills, work in a culturally responsive manner and facilitate professional growth. Supervision in this environment allowed for cultural knowledge to influence service delivery. The team needed to achieve a balance of clinically effective work with culturally informed understandings unique to the location and to the client and family.

Embedding Aboriginal perspectives in psychological research is an area that Tanja and many Aboriginal psychologists have embarked upon. The input of Aboriginal psychologist researchers is required to lead and embed Indigenous knowledge, beliefs and values within the field of psychology. When Aboriginal researchers are involved, the methodology and ways that research is carried out are informed by Aboriginal ways of working. With research and publications being conducted by Aboriginal psychologists, this will shift the dominant mainstream psychology discourse to ensure that Aboriginal perspectives are being reflected in the greater psychology dialogue. Aboriginal psychologists entering research are able to respond and research issues that are relevant and meaningful for ourselves and our communities. I make concerted and unashamed efforts to focus on research material and practice tools developed, written, and reviewed by Aboriginal people for Aboriginal people. In this way we able are to forge ahead with authority rather than succumb to non-Aboriginal opinions, practices and worldviews.

Shared Commitments

These two examples of Aboriginal researchers and practitioners working in the psychology space are not unique, as many Indigenous psychologists around the world are also working towards decolonising psychology in various ways. Both psychologists have invested immensely in securing a sense of personal and community identity, time, energy, skills, and knowledge toward reaching larger goals. These relate to a shared commitment to principles of community psychology by understanding and supporting the needs of Aboriginal people and communities within strength-based practices, research and models. There is also a collective dedication to decolonising psychology, and to ameliorating the consequences of intergenerational trauma and enhancing SEWB, for ourselves and others in the Aboriginal community in Australia.

Summary and Conclusion

This chapter has outlined the context of invasion and colonisation of Australia by the British along with subsequent oppressive government policies and practices toward Aboriginal people in which the profession of psychology has been implicated. Such policies and practices contributed to ongoing grief, loss and trauma which have been transmitted through the generations, and consequently continue to affect the wellbeing and mental health of many Aboriginal people today. Such effects are also reflected in the social determinants of Aboriginal peopl' health and wellbeing. These intergenerational and interacting cycles of oppression and trauma and their consequences of violence (including lateral violence), suicides and mental health, can in turn contribute to further trauma. These components interrupt significant connections to one's body, mind, and emotions, culture, family, kinship, community, country, and ancestry, which are the features described in the SEWB model.

Australian Aboriginal psychologists have grown in strength and numbers as individuals and collectively through numerous group gatherings, conferences and with the formation of the AIPA. Most, if not all, Aboriginal psychologists work toward improving the SEWB of Aboriginal people and many have been influential in the development of the SEWB framework. The influences of Aboriginal psychologists individually and collectively have been and continue to be at the forefront to decolonise psychology in Australia. As part of this decolonising process, it has also been important to bring along and work with the mainstream profession of psychology to ensure competency, solidarity, and a stronger profession particularly via reconciliation movements for better community outcomes. This process presents a range of opportunities for Aboriginal psychologists to work with people with a strong focus on Aboriginal SEWB, recovery and healing from trauma.

In decolonising psychology Aboriginal psychologists have influenced the development of appropriate practices, research, curriculum and teaching guidelines. They have led the Aboriginal Psychologist association and interest group, processes that are visible and supporting to each other as well as imminent Aboriginal psychologists and students. Aboriginal psychologists are distinctive in their experiences and leadership as well as in balancing and blending their psychology and community roles. Yet many specialise in making improvements for specific topics like racism, trauma, lateral violence, suicide, mental health, and cultural competency and safety, that aim to improve SEWB and to empower Aboriginal individuals, families and communities. The two psychologists who have authored this chapter have been part of many of these collective forums and have leadership roles within AIPA. Author one has exemplified her work with preventing lateral violence in the Adult Aboriginal community and is now developing educational and potentially healing resources for Aboriginal younger people and their families. Author two has worked in both research and practice, focuses on topics of suicide and healing. She encompasses processes that assists Aboriginal and non-Aboriginal communities to understand and curb suicide, its devastating impacts as well as traditional healing.

Aboriginal psychologists are making headway in child and adult therapy, assessment, research and teaching and as more Aboriginal psychologists enter the profession additional areas of expertise will come to fruition. Other exemplars, in this chapter, have been provided on how Aboriginal psychologists along with other Aboriginal counsellors and healers can individually and collectively work with Aboriginal people by providing a culturally and emotionally safe space for improvements to connection, healing, and wellbeing.

References

Atkinson, J. (2002). *Trauma trails: Recreating song lines*. Spinifex Press.

Adams, Y., Drew, N., & Walker, R. (2014). Principles of practice in mental health assessment with Aboriginal Australians. In P. Dudgeon, H. Milroy, & R. Walker (Eds.), *Working together: Aboriginal and Torres Strait Islander mental health and wellbeing principles and practice* (2nd ed., pp. 271–288). Commonwealth of Australia.

Australian Bureau of Statistics (ABS). (2016 December 06). *3238.0.55.001 - estimates of Aboriginal and Torres Strait Islander Australians,* June 2016. https://www.abs.gov.au/ausstats/abs@.nsf/mf/3238.0.55.001

Australian Government. (2020, April 18). *Closing the gap*. https://closingthegap.niaa.gov.au/

Australian Health Practitioner Regulation Agency (AHPRA). (2018). *Statement of intent.* https://www.ahpra.gov.au/About-AHPRA/Aboriginal-and-Torres-Strait-Islander-Health-Strategy/Statement-of-intent.aspx

Australian Health Practitioner Regulation Agency (AHPRA). (2019a, December 05). *Communique July 2019: Aboriginal and Torres Strait Islander health strategy group.* https://www.ahpra.gov.au/Search.aspx?q=communique+july+2019+Aboriginalhealth+strategy+group

Australian Health Practitioner Regulation Agency (AHPRA). (2019b, December 09). Annual Report 2018/2019. https://www.ahpra.gov.au/Publications/Annual-reports.aspx

Australian Health Practitioner Regulation Agency (AHPRA). (2019c, December 09). *Psychology Board Statistics.* https://www.psychologyboard.gov.au/About/Statistics.aspx

Australian Human Rights Commission (AHRC). (2020, April 19). *Native title reports.* https://www.humanrights.gov.au/our-work/aboriginal-and-torres-strait-islander-social-justice/publications/native-title-reports

Australian Indigenous HealthInfoNet. (2018). *Overview of Aboriginal and Torres Strait Islander health status, 2017.* Australian Indigenous HealthInfoNet.

Australian Indigenous Psychologists Association (AIPA). (2019a, December 01). *Indigenous Psychologist profiles.* http://www.indigenouspsychology.com.au

Australian Indigenous Psychologists Association (AIPA). (2019b, October 11). *Homepage.* http://www.indigenouspsychology.com.au

Australian Indigenous Psychologists Association (AIPA). (2019c, December 01). *Cultural competence.* http://www.indigenouspsychology.com.au

Australian Indigenous Psychologists Association (AIPA). (2019d, December 01). *Facebook page.* https://www.facebook.com/Australian-Indigenous-Psychologists-Association-104342444383400/?view_public_for=104342444383400

Australian Indigenous Psychology Education Project (AIPEP) (2019, October 11). *Project Overview.* http://www.indigenouspsyched.org.au/about#overview

Australian Institute of Health and Welfare (AIHW). (2020, May 12). *Indigenous Employment.* https://www.aihw.gov.au/reports/australias-welfare/indigenous-employment

Australian Psychological Society (APS). (2019a, December 1). *Aboriginal and Torres Strait Islander people and psychology (ATSIPP) interest group.* https://groups.psychology.org.au/atsipp/

Australian Psychological Society (APS). (2019b, December 5). *Ethical guidelines for Aboriginal and Torres Strait Islander peoples.* https://www.psychology.org.au/for-members/resource-finder/resources/ethics/Ethical-guidelines-Aboriginal-and-Torres-Strait

Australian Psychological Society (APS). (2014a, December 1). *APS reconciliation action plan 2011-2014.* http://www.psychology.org.au/reconciliation/APS/

Australian Psychological Society (APS). (2014b). The APS reconciliation action plan and beyond. *InPsych, 36*(6), 38–39.

Awofeso, N. (2011). Racism: A major impediment to optimal Indigenous health and health care in Australia. *Australian Indigenous Health Bulletin, 11*(3) http://healthbulletin.org.au/articles/racism-a-major-impediment-to-optimal-indigenous-health-and-health-care-in-australia

Broome, R. (1994). *Aboriginal Australians.* Allen & Unwin.

Carey, T., Dudgeon, P., Hammond, S. W., Hirvonen, T., Kyrios, M., Roufeil, L., … Smith, P. (2017). The Australian Psychological Society's apology to Aboriginal and Torres Strait Islander people. *Australian Psychologist, 52*, 261–267. https://doi.org/10.1111/ap.12300

Clark, Y. (2017). *Lateral violence in the Aboriginal community in Adelaide, SA: From Dilemmas to strategies* [Unpublished doctoral dissertation]. University of Adelaide. https://doi.org/10.4225/55/59defd9fba33b.

Clark, Y., & Augoustinos, M. (2015). What's in a name: Lateral violence within the Aboriginal community in Adelaide. *Australian Community Psychologist, 27*(2), 27–42.

Clark, Y., Augoustinos, M., & Main, M. (2016). Lateral violence in the Aboriginal community in Adelaide: It affects our identity and wellbeing. *Journal of Indigenous Wellbeing: Te Mauri Pimatisiwin, 1*(1), 43–52.

Clark, Y., Augoustinos, M., & Main, M. (2017a). Coping and prevention of lateral violence in the Aboriginal community in Adelaide. *Australian Community Psychologist, 28*(2), 105–123.

Clark, Y., Augoustinos, M., & Main, M. (2017b). Evaluation of the preventing lateral violence workshop in Adelaide, South Australia: Phase one questionnaire responses. *Journal of Indigenous Wellbeing: Te Mauri Pimatisiwin, 2*(3), 4, 41–4, 53.

Clark, Y., Augoustinos, M., & Main, M. (2017c). Evaluation of the preventing lateral violence workshop in Adelaide, South Australia: Phase two qualitative aspect. *Journal of Indigenous Wellbeing: Te Mauri Pimatisiwin, 2*(3), 5, 54–5, 66.

Cunneen, C. (2019). Institutional racism and (in)justice: Australia in the 21st century. *Decolonisation of Criminology and Justice, 1*(1), 29–51. https://doi.org/10.24135/dcj.v1i1.9

Dudgeon, P., Milroy, H., & Walker, R. (2014c). The boatshed racism roundtable declaration. In P. Dudgeon, H. Milroy, & R. Walker (Eds.), *Working together: Aboriginal and Torres Strait Islander mental health and wellbeing principles and practice* (2nd ed., pp. 575–576). Commonwealth of Australia.

Dudgeon, P., Milroy, H., & Walker, R. (2014d). A charge to mental health practitioners. In *Working together: Aboriginal and Torres Strait Islander mental health and wellbeing principles and practice* (2nd ed., pp. xvi–xviii). Commonwealth of Australia.

Dudgeon, P., Rickwood, D., Garvey, D., & Gridley, H. (2014a). A history of Indigenous psychology. In P. Dudgeon, H. Milroy, & R. Walker (Eds.), *Working together: Aboriginal and Torres Strait Islander mental health and wellbeing principles and practice* (2nd ed., pp. 39–54). Commonwealth of Australia.

Dudgeon, P., Wright, M., Paradies, Y., Garvey, D., & Walker, I. (2014b). Aboriginal social, cultural and historical contexts. In P. Dudgeon, H. Milroy, & R. Walker (Eds.), *Working together: Aboriginal and Torres Strait Islander mental health and wellbeing principles and practice (2nd ed., pp. 3–24).* Commonwealth of Australia.

Dudgeon, P., & Walker, R. (2015). Decolonising Australian psychology: Discourses, strategies and practice. *Journal of Social and Political Psychology, 3*(1), 276–297. https://doi.org/10.5964/jspp.v3i1.126

Dudgeon, P., Darlaston-Jones, D., Phillips, G., Newnham, K., Brideson, T., Cranney, J., … Page, S. (2016a). *Australian indigenous psychology education project curriculum framework.* University of Western Australia.

Dudgeon, P., Darlaston-Jones, D., Phillips, G., Newnham, K., Brideson, T., Cranney, J., … Page, S. (2016b). *Australian indigenous psychology education project guidelines for increasing the recruitment, retention and graduation of Aboriginal and Torres Strait Islander psychology students.* University of Western Australia.

Dudgeon, P., Harris, J., Newnham, K., Brideson, T., Cranney, J., Darlaston-Jones, D., … Phillips, G. (2016). *Australian Indigenous psychology education project workforce capabilities framework.* University of Western Australia.

Dudgeon, P., Bray, A., D'Costa, B., & Walker, R. (2017). Decolonising psychology: Validating social and emotional wellbeing. *Australian Psychologist, 52*, 316–325. https://doi.org/10.1111/ap.12294

Fogarty, W., Bulloch, H., McDonnell, S., & Davis, M. (2018). *Deficit discourses in Indigenous health: How narrative framings of Aboriginal and Torres Strait Islander people are reproduced in policy.* Lowitja Institute. http://ncis.anu.edu.au/research/representation_identity.php

Gaudry, A., & Lorenz, D. (2018). Indigenisation as inclusion, reconciliation, and decolonisation: Navigating the different visions for indigenising the Canadian academy. *AlterNative: An International Journal of Indigenous Peoples, 14*(3), 218–227. https://doi.org/10.1177/1177180118785382

Gee, G. (2016). *Resilience and recovery from trauma among Aboriginal help seeking clients in an urban Aboriginal community controlled health organisation* [unpublished doctoral dissertation]. University of Melbourne.

Gee, G., Dudgeon, P., Schultz, C., Hart, A., & Kelly, K. (2014). Aboriginal and Torres Strait Islander social and emotional wellbeing. In P. Dudgeon, H. Milroy, & R. Walker (Eds.),

Working together: Aboriginal and Torres Strait Islander mental health and wellbeing principles and practice (2nd ed., pp. 55–68). Commonwealth of Australia.

Glover, M., Dudgeon, P., & Huygens, I. (2005). Colonisation and racism. In G. Nelson & I. Prilleltensky (Eds.), *Community psychology: In pursuit of liberation and Well-being* (pp. 330–344). Palgrave Macmillan.

Haebich, A. (2000). *Broken circles* (1st ed.). Fremantle Press.

Hill, R., Pert, P.L., Davies, J., Robinson, C.J., Walsh, F. and Falco-Mammone, F. (2013). Indigenous Land Management in Australia. CSIRO, Ecosystems Sciences

Holland, C. (2018). *A ten-year review: Close the gap strategy and recommendations for reset.* The CTG campaign committee. https://www.humanrights.gov.au/sites/default/files/document/publication/CTG%202018_FINAL-WEB.pdf?_ga=2.27180212.84122107.1575550944-826556177.1531809881

Human Rights and Equal Opportunity Commission (HREOC). (1997). *Bringing them home: National inquiry into the separation of Aboriginal and Torres Strait Islander children from their families*. Commonwealth of Australia.

Leckning, B., Ringbauer, A., Robinson, G., Carey, T. A., Hirvonen, T., & Armstrong, G. (2019). *Guidelines for best practice psychosocial assessment of Aboriginal and Torres Strait Islander people presenting to hospital with self-harm and suicidal thoughts*. Menzies School of Health Research.

Lowe, D. (1994). *Forgotten rebels: Black Australians who fought back*. Permanent Press.

Lowitja Institute. (2018). *Summary report: Deficit discourse and Aboriginal and Torres Strait Islander health policy*. https://ncis.anu.edu.au/_lib/doc/ddih/Deficit_Discourse_summary_report_WEB.pdf

Malaspinas, A-S,, Westaway, M.C., Muller, C., Sousa, V. C., Lao, O., Alves, I.,… Willerslev, E. (2016). A genomic history of Aboriginal Australia. Nature, 538(7624), 207–214. https://doi.org/10.1038/nature18299

Milroy, H., Dudgeon, P., & Walker, R. (2014). Community life and development programs – Pathways to healing. In P. Dudgeon, H. Milroy, & R. Walker (Eds.), *Working together: Aboriginal and Torres Strait Islander mental health and wellbeing principles and practice* (2nd ed., pp. 419–436). Commonwealth of Australia.

Nakata, M. (2007a). The cultural interface. *Australian Journal of Indigenous Education, 36*(S1), 7–14. https://doi.org/10.1017/S1326011100004646

Nakata, M. (2007b). *Disciplining the savages, savaging the disciplines*. Aboriginal Studies Press.

Neeganagwedgin, E. (2015). Rooted in the land: Taino identity, oral history and stories of reclamation in contemporary contexts. *AlterNative: An International Journal of Indigenous Peoples, 11*(4), 376–388.

Oliver, W. (2001). Cultural racism and structural violence. *Journal of Human Behaviour in the Social Environment, 4*(2–3), 1–26. https://doi.org/10.1300/J137v04n02_01

Poche Centre for Indigenous Health (PCIH). (2019). *Building a stronger tomorrow: Connecting our communities through culture - Report on the 2nd National Aboriginal and Torres Strait Islander Suicide Prevention Conference and the 2nd World Indigenous Suicide Prevention Conference*. School of Indigenous Studies, University of Western Australia.

Povey, R., & Trudgett. (2019). When camp dogs run over maps: 'Proper-way' research in an Aboriginal community in the north-east of Western Australia. *Australian Aboriginal Studies, 2*, 61–72.

Reconciliation Australia. (2019). *2018 Australian reconciliation barometer*. Reconciliation Australia. https://www.reconciliation.org.au/wp-content/uploads/2019/02/ra_2019-barometer-brochure_web.single.page_pdf

Ranzijn, R., McConnoche, K., & Nolan, W. (2009). *Psychology and Indigenous Australians: Foundations of cultural competency*. Palgrave Macmillan.

Rickwood, D., Dudgeon, P., & Gridley, H. (2010). A history of psychology in Aboriginal and Torres Strait Islander mental Health. In N. Purdie, P. Dudgeon, & R. Walker (Eds.), *Working together: Aboriginal and Torres Strait islander mental health and wellbeing principles and practice* (pp. 13–24). Office of Aboriginal and Torres Strait Islander Health, Canberra: Department of Health and Ageing.

Secretariat for National Aboriginal and Islander Child Care (SNAICC). (2019). *Family matters report 2019: Measuring trends to turn the tide on the over-representation of Aboriginal and Torres Strait Islander children in out-of-home care in Australia*. Family Matters.

Social Health Reference Group. (2004). *Social and emotional Well-being framework: A national strategic framework for Aboriginal and Torres Strait Islander Peoples' mental health and social and emotional Well-being, 2004–2009*. Department of Health and Ageing.

Strelein, L. (2005). From Mabo to Yorta Yorta: Native title law in Australia. *University Washington Journal of Law and Policy, 19*, 225–272.

Strelein, L., & Tran, T. (2013). Building Indigenous governance from native title: Moving away from 'fitting in' to creating a decolonised space. *Review of Constitutional Studies, 18*(1), 19–47.

Wanganeen, R. (2014). Seven phases to integrating loss and grief. In P. Dudgeon, H. Milroy, & R. Walker (Eds.), *Working together: Aboriginal and Torres Strait Islander mental Health and wellbeing principles and practice* (2nd ed., pp. 475–492). Commonwealth of Australia.

Westerman, T. (2004). Guest editorial engagement of Indigenous clients in mental health services: What role do cultural differences play? *Australian e-Journal for the Advancement of Mental Health (AeJAMH), 3*(3), 1–7. https://doi.org/10.5172/jamh.3.3.88

Westerman, T.G., & Wettinger, M. (1998). Working with Aboriginal People. Psychologically speaking, Western Australian Psychological Society – Psychological assessment of Aboriginal People. https://indigenouspsychservices.com.au/downloads/

Towards an Expansive Conceptual/ Methodological Approach to Everyday Violence

Sarah Malotane Henkeman

> In a country such as South Africa where violence is so embedded in many processes of communication, and not just contestation, it is possible to misread violence as something else, to cease seeing some manifestations of it (Gqola, 2015, p. 152).

> The African university began as a colonial project, a top down modernist project. Its ambition being to transform society into its own image [...]. It acknowledged only a single intellectual tradition, the western tradition. At the same time it aimed to create an apartheid society, one that shut the vast majority of the colonised out of the common discourse of humanity. What would it mean to decolonise such a project? [...] We must theorise our own realities (Mahmood Mamdani, 2017).

Introduction

Protests in institutions of higher education from 2015 onwards visibly ushered in the decolonial turn in South Africa, and ripened the moment for an expanded understanding of everyday violence. The colonial academy is argued to be deeply implicated in producing graduates (particularly those in helping professions who are largely blind, deaf and silent about the ineffectiveness of interpersonal tools with which to wage peace against invisible/visible violence as a transhistorical and transnational structure described in this text. Gqola (2015) implicitly alludes to the denial we slip into, to the extent that we 'misread' and fail to 'see' the way in which violence is 'embedded' in our social fabric. Mamdani's (2017) implicit reference to epistemic violence (a form of cultural violence), as the overall act of excluding and silencing the knowledge of "the vast majority of the colonised out of the common discourse of humanity", points to the transhistorical oppressive role of the colonial academy in keeping uninterrupted aspects and forms of invisible/visible violence in society intact by omission and commission.

S. M. Henkeman (✉)
Social Law Project, University of the Western Cape, Cape Town, South Africa

S. Kessi et al. (eds.), *Decolonial Enactments in Community Psychology*, Community Psychology, https://doi.org/10.1007/978-3-030-75201-9_12

This chapter takes the form of a *bricolage* which centres transdisciplinary knowledge about recurring patterns of experiences of everyday violence by descendants of colonised, oppressed and enslaved people. As part of this category of dehumanised people, I have visceral, tacit, explicit and vicarious knowledge of the transhistorical impact of everyday violence. This embodied/conceptual knowledge of the impact side of invisible/visible violence is routinely marginalised and dismissed by the colonial academy. Once produced from below, this knowledge then becomes a counterpoint to partial knowledge about everyday violence as produced in the colonial academy. This just noticeable difference in approach decentres colonial understandings of violence by refusing to use it as a reference point against which to 'respond' and be conceptually limited by. Inclusion of banished sources of knowledge about what constitutes everyday violence shows up the limitations of disciplinarity in explaining everyday violence which is a transdisciplinary phenomenon. The exercise is not intended to present transdisciplinarity as easy; it is merely an acknowledgement that we need to build tools from below to render everyday violence visible as a structure which respects no disciplinary, nor any other artificial boundaries. As Max-Neef (2005, p. 5) suggests, "strong transdisciplinarity is still in the making, thus representing an unfinished scientific programme that offers fascinating possibilities for advanced reflection and research". According to Denzin and Lincoln (2011, p. 4) the qualitative researcher may be viewed as a "bricoleur, a maker of quilts, or in filmmaking, a person who assembles images into montages". Bricolage is thus a process that moves the researcher "beyond the blinders of particular disciplines" to a "new conceptual terrain" while highlighting "the relationship between a researcher's ways of seeing and the social location of his or her personal history" (Kincheloe et al., 2011, p. 168). Taken together as 'disrupting denial' praxis that is continuously in the process of being refined, transdisciplinarity and bricolage allow for an intersectional analysis of everyday violence as an invisible/visible structure.

Combining intimate, vicarious and academically produced knowledge of the structure of invisible/visible violence is thus first an act of self-determination to assert agency; then a counterpoint to decentre dominant discourses about everyday violence that position dehumanised people, the knowledge we bring, and the way we produce it, as largely inferior. This is by no means a one way relationship. As dehumanised people, many of us internalise the oppressor, and mimic oppressive behaviours in masked ways, thereby keeping the violent inferior/superior status quo intact. The task of the oppressed researcher is to confront these contradictions in the course of seeking to understand the complex interplay of dehumanisation as an act *and* an impact that produces both resilience/resistance *and* complicity amongst the oppressed. Where mainstream scholars are cited in the course of "theorising our own realities" in this text, it is chiefly about the judicious use of agency to "sift through" existing knowledge and knowledge production without treating it as "turnkey solutions" (Mamdani, 2017).

In the introductory remarks to this chapter, the colonial academy is presented as being implicated in shaping partial understandings of everyday violence, and of exerting its own violence by proxy, as part of the colonial project. This is juxtaposed with the need for dehumanised people to produce knowledge from the standpoint of the oppressed, first as self-determination to reclaim our agency; then as counterpoint to decentre coloniality. Following this, the immediate context for the research is provided. This refers to my experience as a practitioner within the first decade of South Africa's democracy, and my subsequent research journey to fill the gaps in my education. Thereafter, reference is made to student, staff and worker protests that ushered in the decolonial turn and how disrupting everyday violence was at the centre of these calls to decolonise the academy. A conceptual framework – the *structure of invisible/visible violence* (Fig. 1) is then displayed and discussed with the aim of deepening our understandings of violence from alternative knowledges as necessary for the decolonial project, which include excerpts of boundary crossing scholarship that provide a bridge to transdisciplinary scholarship on violence. The concluding section summarises key lessons that might contribute to decolonial praxis.

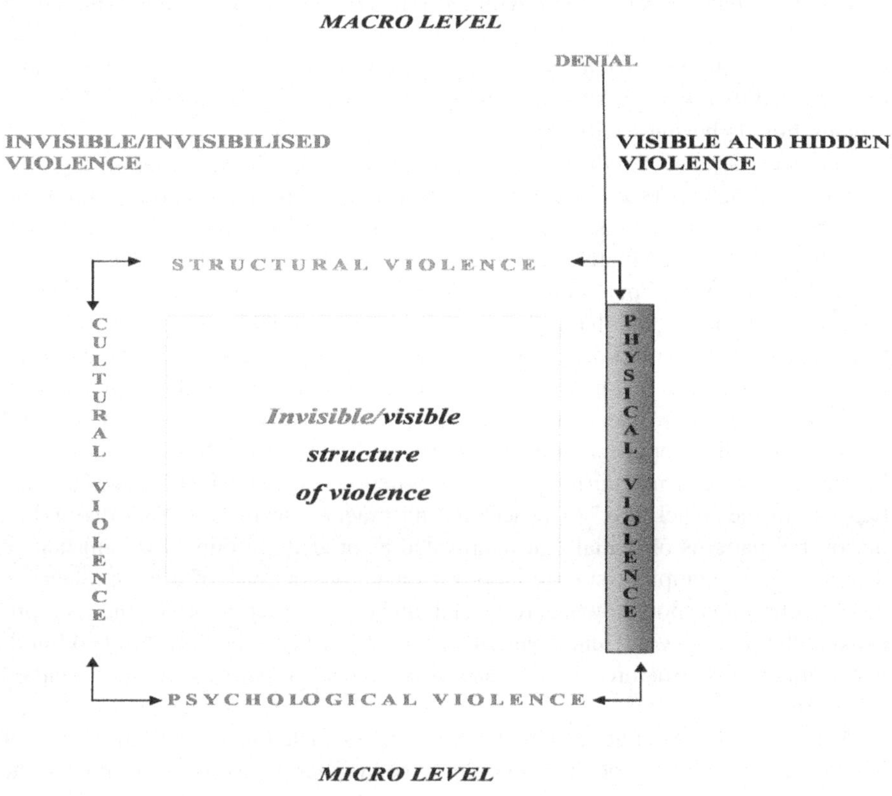

Fig. 1 The structure of invisible/visible violence

Immediate Context

The break with dominant understandings of everyday violence fed to us by the colonial academy (and complemented by our own colonised minds) originated during a conflict analysis process in the course of my work as a black conflict resolution practitioner approximately one decade after South Africa's transition to democracy in 1994. Barely one hour's drive from Cape Town, the toxic mix of uninterrupted invisible/visible violence rooted in colonialism, apartheid and market democracy, was found to be physically reproduced in how the small factory building was organised. White people sat upstairs, people of possible Khoi and San ancestry sat downstairs, and people of possible Xhosa ancestry worked outside. The factory was owned by an Afrikaner person. The South African business agents who facilitated business between this factory and a Dutch company were English and had their offices in District Six, renamed as Zonnebloem during apartheid. My first meeting with the business agents and Dutch business people, where the brief was clarified, was held in Stellenbosch, a town with close ties to apartheid ideology. This group contacted the 'left-wing' conflict resolution organisation where I was a panel member, to assist with the resolution of a key concern: that workers were unhappy, and they insisted that workers should be happy in companies that they as a 'fair trade' organisation do business with.

This was the new South Africa with all its contradictions, including my own 'progressive' naivete taken in as I was by the possibilities of conflict resolution and peacebuilding, and not understanding that the interpersonal tools I was armed with to wage peace, were no match for the structure of violence I was trained not to 'see'. Little did I know that my 'new South Africa' rainbow honeymoon was about to come crashing down, and that my 'trained blindness' would be miraculously cured, to the extent that I would dedicate the rest of my life to learn and teach about the invisible mechanisms that produce facts in the lives of dehumanised people. As Roy (2001, p. 7) suggests, "once you see it, you can't unsee it", nor can you remain blind to how internalised oppression operates. My underlying hypothesis of our trained blindness was confirmed during phase two of action research when I looked at what happens in the 'black box' of peacebuilding practice during my PhD research. I discovered patterns of denial that manifested as procedural blindness, substantive deafness and a complicit silence amongst mediators (as part of a comprehensive model of findings about how macro level denial of the structure of violence is replicated at the micro level (Henkeman, 2012, pp. 205–222)). The term 'trained blindness' about the structure of violence was coined because it mimics learned helplessness.

At the time, I did not understand the visceral response I had during the course of listening to different perspectives in order to get a fuller picture of what sparked the 'interpersonal conflict'. The workers' stories triggered painful memories of similar stories by and about other dehumanised people in my life and social circle. The dissonance I experienced caused me to eventually recuse myself from the process, stop mediating altogether, and to embark on multiple phases of empirical and secondary research to try and understand why I had such a deep grief response and was incapable of being 'neutral' and 'professional' in this instance. The journey included learning about accumulated visceral, tacit, explicit and vicarious knowledge as a descendant of colonised, oppressed and enslaved people, which forms the basis of the disrupting denial approach. It was only years later, during periods of reflection, post-hoc analyses, and writing, that I eventually came to understand how much every aspect of that assignment caused my banished knowledge to assert its presence. I was unconsciously doing what Trochim (1985, p. 575) describes as "conceptual and observational pattern-matching", and questioning why, what I and other dehumanised people 'knew' (in the embodied sense described by Moraga and Anzaldúa (2015)) was not part of existing theories of violence. They declare:

> The very act of writing then, conjuring/coming to 'see' what has yet to be recorded in history is to bring into consciousness what only the body knows to be true. The body – that site which houses the intuitive the unspoken, the viscera of our being – this is the revolutionary promise of 'theory in the flesh' for it is both the expression of evolving political consciousness and the creator of consciousness, itself. Seldom recorded and hardly honoured, our theory incarnate provides the most reliable roadmap to liberation.' (Moraga & Anzaldúa, 2015, p. xxiv).

When the decolonial turn was ushered in by protesting students, staff and workers at institutions of higher education in South Africa since 2015, as a society we could no longer routinely deny the continued violent presence of the past, and the need to adapt, develop or acknowledge our own conceptual and practical tools to negotiate this terrain. The now conscious iterative process of observational and conceptual pattern-matching (Trochim, 1985) during five phases of action research revealed that everyday violence in South Africa is at least analytically a structure, and that transhistorical patterns can be discerned in many present manifestations of everyday violence. This chapter refers to a single research project with phase one being the discovery of a gap in theory and practice in South Africa as an unequal, transitional society, phase two as PhD research during which four interlinked gaps were identified, phase three the development, ground-truthing and calibration of the conceptual framework depicted in Fig. 1; phase four co-production of knowledge about denial, and phase five co-production of knowledge which culminated in the book 'Disrupting Denial: Analysing Narratives of Invisible/Visible Violence & Trauma' with an extended network of peers from all walks of life.

Veteran peace researcher Galtung's (1996) triad of cultural-structural-direct violence was adapted to a theoretical/methodological framework, *the structure of invisible/visible violence* that includes cultural-structural-psychological-physical violence and denial. Whereas Galtung (1996) combines psychological and physical violence as 'direct' violence, the adapted framework renders psychological violence visible as an analytically distinct but interlocked aspect of a structure of invisible/visible violence depicted in Fig. 1. Both analytically and practically Galtung's (1996) conflation of two analytically distinct but related aspects of violence thus tends to obscure the role that transhistorical trauma, as unresolved historical and lifespan trauma and grief, (Brave Heart et al., 2011) play in the production of everyday violence in South Africa as a "multiply wounded, multiply traumatised" society (Cabrera, n.d., p. 1).

Used as a heuristic, analytical, and pedagogical device that is continuously being refined, the framework has enabled salient themes about the interlocked nature of everyday violence to be surfaced through empirical and secondary research from 2005 to 2012. From 2013 to date the framework is continuously groundtruthed and calibrated with a transdisciplinary spectrum of people of different educational backgrounds at different times and in different spaces. These social justice interventions have progressed to co-knowledge production activities which culminated in the development of a guide and a book on the invisible/visible structure of violence.[1] Social justice interventions in the form of workshops and writeshops in turn served to groundtruth and calibrate the conceptual framework with new information. In the process, knowledge production is democratised along the lines of Paulo Freire's praxis of liberatory pedagogy (1974, 2000). His pedagogy of the oppressed has at its root that people are 'knowers' and not empty vessels that need to be filled and othered. In exchange for entrusting their stories to strangers, participants are embraced in the knowledge production process as agents whose intimate knowledge about everyday violence has equal status to knowledge produced in the academy. The combination of lay people and academics in these workshops, allows for the exchange of different ways of knowing which includes methods to debrief visceral, tacit, explicit and vicarious knowledge as a counterpoint to the way knowledge is produced in the academy. This mitigates, in some way, the marginalisation and dismissal of knowledge from below, and collapses disciplinary, temporal, geographical, social, economic, political and other artificial boundaries which serve to obscure the impact side of everyday violence as a structure of invisible/visible violence that respects no boundaries. The ongoing groundtruthing and calibration of the adapted conceptual framework constitutes the disrupting denial process described later.

2015: Visible Disruption of Denial in South Africa

> A society beginning to move from one epoch to another requires the development of an especially flexible, critical spirit. [...] the time of epochal transition constitutes an historical-cultural 'tidal wave'. Contradictions increase between the ways of being, understanding, behaving, and valuing which belong to yesterday and other ways of perceiving and valuing which announce the future (Freire, 1974, p. 6).

There can be no doubt that student, staff and worker protests in institutions of higher learning, and particularly the intersectional approach taken and themes highlighted from 2015 onwards, popularised "other ways of perceiving and valuing which announce the future" (Freire, 1974, p. 6). Freire argued further that if people are not

---

[1] This chapter draws on my PhD research *Restorative Justice as a tool for peacebuilding* completed in 2012, *Guide to a Deeper, Wider and Longer Analysis of Violence* compiled during 2016, and the book in which knowledge was co-produced with a transdisciplinary spectrum of people: *Disrupting Denial: Analysing Narratives of Invisible/Visible Violence & Trauma* published in 2018.

able to critically perceive the "themes of their time", and if they do nothing to "intervene actively in reality", they are inevitably "carried along in the wake of change". The protests visibly ushered in the decolonial turn in South Africa. While different constituencies viewed this moment from particular social locations, the protests provided South Africa with a historical moment during which banished knowledge about the intersectional nature of invisible and visible aspects and forms of violence were excavated. The key demands were that symbolic, structural and psychological and physical vestiges of the past 'must fall', along with the statue of Cecil John Rhodes with its central place on the university campus. Uninterrupted patterns of experiences rooted in colonialism, apartheid *and* market democracy were rendered visible and denial was disrupted publicly. The aftermath of these protests can best be described by those who were present on campuses throughout this period, and will fill volumes. Despite this eruption, the tendency to individualise violence, and to view it as mainly physical and visible, remains deeply entrenched. Scholarship by people from marginalised groups, who specifically question the limitations of the 'training' they received from the colonial academy, and who describe different aspects or the intersection of these aspects, are few and far between.

Despite the urgent calls for decolonisation, scholarship on violence in key disciplines still suffer from the exclusion of knowledge and theorisation of lived experiences – about the *impact side* of transhistorical and transnational invisible/visible violence. In most instances micro level disciplinary tools are still used to deal with a structural problem. Universities, as authorised knowledge producers (Mamdani, 2017) signally fail to provide South African society with a comprehensive analysis that would explain and deal with denial, as well as the *interconnections* between transhistorical symbolic, structural, psychological and growing levels of physical violence. This situation exists even while different disciplines have in depth knowledge of their areas of competence that provide insight into different aspects of violence, albeit in isolation of each other. The result is a failure to produce local experience based theories that can 'travel' due to reliance on 'turnkey solutions' produced in the West (Mamdani, 2017). This seems to suggest that the failure to take knowledge produced from the impact side of violence into account as counterpoint, leads to normalised criminalisation of descendants of colonised, oppressed and enslaved people worldwide.

As Galtung (1996, p. 1) argues, "the peace researcher must look for causes, conditions, and contexts in various spaces – Nature, Human, Social, World, Time, Culture." He suggests that this 'transdisciplinary spectrum makes peace studies both challenging, difficult intellectually, and problematic in praxis.' While my research interest in studying the invisible mechanisms that produce visible violence was rebuffed by my alma mater in 1998, and while I pursued this question informally from 2005 and formally in PhD research from 2008 onwards, the 2015 protests and the violent institutional response to it, highlighted the intersectionality of invisible/visible violence in a way that disrupted denial more broadly.

The Disrupting Denial Approach

Lederach, Neufeldt & Cuthbertson (2007, p. 12) contend that reflective peacebuilders need to be systemic. They suggest that peace researchers should:

> Go beyond cause-and-effect to look at the wider context and history. Cause-and-effect thinking predicts that action A will produce result B; systemic thinking not only observes that, in a particular setting, A, B, and C tend to be present when a particular pattern emerges, but also asks, "What else is going on in this context?" "What visible and invisible factors are combining in the overall system to produce this result?"

My first public act of disrupting denial happened when a 'crisis of understanding' of what drives South Africa's high and growing levels of violence was argued to have developed amongst South African criminologists who focus narrowly on 'control' and not the history, biography and psyche of perpetrators (Dixon, 2012). In a response article in the same journal, the crisis was linked to denial rooted in epistemic violence and racism (both examples of cultural violence), uninterrupted inequality (structural violence) and transhistorical trauma (psychological violence), and a social justice approach was called for (Henkeman, 2013). I argued that a different reading of South Africa's prison statistics would raise the question "why are descendants of colonised, oppressed and enslaved people over- and disproportionately represented in prison" while beneficiaries of oppression were underrepresented. Implicit in this argument is that such an approach in turn would draw attention to uninterrupted symbolic, structural, psychological and physical violence. This in turn would require political, economic and social changes which would also require that attention be paid to unresolved social grief and transhistorical trauma in line with a systemic approach. Thus, denial serves a function – it is much easier to criminalise blackness, than to give the entire society an overhaul. The disrupting denialapproach,thatseekstoovercomeourtrainedblindness,resonateswith,andcombines-

1. intersectionality as an analytical framework to examine the impact of interlocking systems of power (Crenshaw, 2019);
2. transdisciplinarity as an acknowledgement of the unity-of-all-things/unity-of knowledge (Max-Neef, 2005);
3. co-production of knowledge from below is first about self-determination as advocated by Tuhiwai Smith (1999), then about counterpoint to decentre mainstream knowledge production in line with the 'decolonial turn'.
4. the idea that the peace movement is "an exercise in [making] micro-macro linkages" (Galtung, 1998); and
5. insights gained from scholarship on psychological and sociological denial (Cohen, 2001; Liem, 2007; Zerubavel, 2010).

Structure of Invisible/Visible Violence Conceptual Framework

As a visual depiction of everyday violence in South Africa, this framework suggests that by the time an individual manifests violence, she/he has already been affected by the intergenerational transmission of trauma and/or subjected to current and continuing cultural, structural, psychological and physical violence. This does not mean that individuals are deemed innocent of wrongdoing; it simply locates everyday violence by individuals within the larger context of transhistorical and transnational invisible/visible violence. The basic framework consists of four linked aspects which reflect the mutually reinforcing relationship between invisible and visible violence. An inversion of this framework provides an outline to guide social justice interventions towards everyday peace. On the left, invisible violence consists of cultural, structural and psychological aspects of violence. On top structural violence indicates the societal level at which violence occurs. At the bottom psychological violence shows the personal and relationship level at which violence is experienced. Denial appears on the dividing line between invisible and visible violence to illustrate that invisible violence is obscured when instances of everyday violence occur. On the right visible violence shows the physical aspect of violence. Horizontal and vertical arrows at each end show the links and thus the intersection and interaction of different aspects and forms of invisible/visible violence.

The words 'macro level' on top and 'micro level' at the bottom indicate that individuals are affected by the societies we live in, and that many of us in turn contribute to high levels of violence in society. This resonates with what feminists refer to as 'the personal is political' which helps us to understand what the impact of uninterrupted macro level violence is on our day-to-day existence at the micro level. While these aspects of violence are analytically distinct, the arrows indicate that they intersect, interact, combine and are compounded in ways that defy definition and that cannot be contained within single disciplines.

Cultural Violence

Cultural violence is argued to be symbolic, and is exemplified by "religion and ideology, language and art, empirical science and formal science", which serve to "legitimise or delegitimise and justify structural and direct violence" (Galtung, 1996, p. 170). Bourdieu and Wacquant (1992) suggest that symbolic violence results in unequal power relations that render some people 'inferior', and therefore not worthy of being taken into account. For example, epistemic violence, as a form of cultural violence at the transnational and micro levels plays a key role in relegating knowledge production by oppressed people to the margins. Other examples of symbolic violence are racism, inter-ethnic racism, xenophobia, patriarchy, homophobia, ableism and all forms of othering with inferiority/superiority at its root. These are all forms of violence that make structural, psychological and physical aspects and forms of violence seem reasonable and acceptable.

Structural Violence

Anthropologist Paul Farmer (2005) suggests that:

> Structural violence is one way of describing social arrangements that put individuals and populations in harm's way... The arrangements are structural because they are embedded in the political and economic organization of our social world; they are violent because they cause injury to people ... historically given (and often economically driven) processes and forces conspire to constrain individual agency. Structural violence is visited upon all those whose social status denies them access to the fruits of scientific and social progress.

According to Iadicola and Shupe (2003, p. 45), trans-national structural violence occurs "internationally in the positioning of nation-states and people within them and how these structures manifest themselves in differences in life chances". They suggest that if we focus only on interpersonal physical violence, we "fail to understand and see how the violence that is committed at the highest levels by those with the greatest power manifest itself in the violence of those with the least power in the society and the world" (p. 49). Other examples of structural violence are gender inequality that result in nested inequalities; economic globalisation that leads to poverty, unemployment, inequalities related to income and assets; unequal health outcomes that lead to lowered life expectancy, maternal mortality; infant and child mortality; and generally different forms of exclusion rooted in the examples of symbolic violence listed above.

Psychological Violence

Imbusch (2003, p. 23) argues that psychological violence is inflicted on the mind, psyche and/or soul of people. It is argued to involve some forms of "psychological cruelty and particular kinds of torture". He argues that psychological violence is based on "words, gestures, pictures, symbols, or deprivation of necessities of life, so as to force others into subjugation through intimidation and fear, or specific rewards". The key observation he makes is that "while it is difficult to detect", psychological violence is 'considerably more inhumane than physical violence'. For example, Barthes (2010, p. 51) suggests that certain photographs can generate a severe visceral response in an onlooker. According to him, this response is referred to as the "punctum of photographs [...] capable of stinging (to the point of even wounding) the viewer". This supports the argument during 2015, by some protesting students who poured faeces on the Rhodes statue and burnt artwork, that these served as painful reminders of colonialism. It also exposed the denied 'soul wound' (Duran, 2006) of colonised, oppressed and enslaved people when students publicly spoke about their transhistorical pain. In this view, despite attempts by certain sections of the media to write this off as 'the politics of pain', students placed the invisible/visible structure of violence, its uninterrupted nature, 'unresolved historical grief', and historical trauma (Brave Heart et al., 2011) firmly on the agenda.

Physical Violence

Physical violence is bodily harm that is self-inflicted (personal) or inflicted on others (interpersonal), and is intended to inflict pain and/or to end life in private and/or public settings. Physical violence can be understood to happen on a continuum of slow, attempted and successful suicide, to intentional bodily harm or homicide. This ranges from a wide variety of self-injurious behaviours at the personal level to interpersonal, intra - and intergroup, intra and international violence. I argue that physical illnesses and deaths that are directly related to symbolic, structural and psychological violence are unacknowledged forms of physical violence. A violent attack increases fear of violence. Constant awareness, fear and the threat of violence in turn feed the prison-like conditions under which many people live in gated and walled communities with electrified fences. In an unequal society like South Africa, this has led to increased surveillance by the booming private security industry. Judging by the disproportionate and over representation of black people in the criminal justice system, the 'criminalisation of blackness', in the absence of a deeper, wider and longer analysis, is already a fact.

Eduardo Duran (2006), a Native American psychologist came across the concept 'soul wound' when he conducted a needs analysis amongst Native American people. They rejected his paper-based survey, and instead described the soul wound (transhistorical trauma) in relation to instances of domestic violence, suicide, family dysfunction, community dysfunction and violence, political infighting, and epistemic violence amongst other forms of violence. The final two forms of everyday violence overlap with cultural and structural violence. In research on what he terms 'racial battle fatigue', William Smith (2008, p. 617) argues that:

> Racial battle fatigue addresses the physiological, psychological, and behavioural strain exacted on racially marginalized and stigmatized groups and the amount of energy they expend coping with and fighting against racism. The conditions that cause psychological racial battle fatigue can range from frustration to fear, and the body may respond to racial insults as if it were under a physical attack. Therefore, racial microaggressions found in the campus racial climate may be physiologically coded as violent attacks or acts of aggression.

He states that the following physiological stress responses are common among people who are subjected to racism: headaches, grinding teeth, clenched jaws, chest pain, shortness of breath, pounding heart, high blood pressure, muscle aches, indigestion, gastric distress, constipation or diarrhoea, increased perspiration, intestinal problems, hives, rashes, sleep disturbance, fatigue, insomnia, and frequent illness.

Denial

Micro and macro level denial as depicted in the conceptual framework, interfere with our ability to see the structure of invisible/visible violence that place everyday manifestations of violence in transhistorical context. The close relationship between denial and the intersection of different aspects and forms of violence, resonate with Gqola's (2015, p. 152) observation with regard to denial as 'not seeing' different aspects and forms of violence. Thus, we are trained not to "see, hear and break silence" even while we routinely experience the impact of invisible violence. This is achieved via the *null, hidden* and *official* curricula that constitute 'education'. In seeking an explanation for how denial at the macro level is replicated at the micro level, I argued that the null, hidden and official curricula in the study of violence and peace, keep the status quo intact.

It has been suggested that disciplinarity is implicated in our inability to understand the intersection and interaction of invisible/visible aspects of violence which manifest as 'trained blindness' in the colonial academy (Henkeman, 2012). Denial as alluded to by Pumla Gqola in the opening quotation, and perspectives on denial, silence and silencing by Cohen (2001), Liem (2007) and Zerubavel (2010) provide a deeper and wider understanding of the limitations placed on long-term peacebuilding in the South African context.

Cohen's (2001, pp. 278–296) argument is a general comment on free market capitalism and how it generates cultures of denial, which is worth quoting in full to drive home the fact that violence is an invisible/visible structure, and that even the most radical among us exhibit behaviours that make us complicit in unexamined ways:

> The free market of late capitalism – by definition a system that denies its immorality – generates its own cultures of denial. People are made superfluous and marginal; the deskilled, unskilled and sinking poor; the old who no longer work; the young man who cannot find work; the massive shifting populations of migrants, asylum seekers and refugees. The 'solution' to these problems now physically reproduces the conditions of denial. The strategy is exclusion and segregation: enclaves of redundant populations, living in the modern versions of ghettoes, remote enough to become 'out of sight, out of mind', separated from enclaves of winners, in their guarded shopping malls, gated communities and retirement villages.

Similarly Liem (2007, pp. 153-154) takes "political, social, and psychological mechanisms" that operate at state, community, family, and individual levels into account. This is referred to as the "politics and psychology of memory and voice" via "public remembering expressed through the arts as a means of breaking silences and fostering healing". He proposed a "multi-faceted model of the silencing of historical memories" which draws on "some of the first oral histories conducted with Korean Americans about personal and family experiences during the Korean War".

Zerubavel (2010, pp. 32–33) suggests that a "sociology of denial" would draw attention to the interpersonal dynamics of preventing information from entering public communication. This is, according to him, different from the "intrapersonal dynamics of blocking information from individual's awareness" that leads to a "conspiracy of silence" which he describes as "being both aware and (at least publicly) unaware of something". He states that, "rather than simply failing to notice something, denial too often involves an effort to actively avoid noticing it".

In this view, denial in South Africa is argued to be due to the thinking that underpins the grand narrative of a 'miracle', 'rainbow nation of God'. The narrative was traced to the knock-on effect of:

1. The discourse by transitional justice scholars and practitioners that South Africa, and other emerging democracies, are 'post-conflict' societies. This introduced built-in denial that led to, amongst others -
2. The introduction of artificial political and knowledge boundaries, that normalised a distinction between 'political' and 'criminal' violence;
3. The introduction of an artificial divide, with insidious effects, between 'ordinary' oppressed people and specific victims of what was defined as 'gross violations of human rights' within a defined period (1961–1994 then 1996) in the mandate of transitional justice instruments like South Africa's Truth and Reconciliation Commission, which arguably led to -
4. Ignoring and/or eclipsing transgenerational victimisation and trauma transmitted to the majority of South Africans from around 1652 to date (regardless of how or whether different individuals and groups processed this inflicted trauma or not);
5. Obscuring the fact that the "negotiated revolution", despite its noble goals, produced "marketisation for the minority and democratisation for the majority" as "global solutions" crafted by "Western lawyers, legal academics, economists and political scientists" (Chua, 2003, p. 145).
6. Regardless of the validity or not, of the immediate reasons why this course of action was taken with adaptations that were subject to negotiation, it served to drive transhistorical trauma and the transhistorical and transnational reasons for growing and nested inequality underground.

This expansive conception of everyday violence, lead us to invisible transhistorical and transnational mechanisms that produce, reproduce, compound and maintain everyday violence in society. A deeper, wider and longer analysis of violence inevitably leads back to colonialism, apartheid and slavery in the South African context. The next question then becomes, why do we accept this individualised definition of violence? This question leads us straight to the colonial academy, the authorised producers of knowledge, who are the centre of marginalising and dismissing knowledge produced from the standpoint of the oppressed. Yet the work of boundary-crossing scholars was found useful to scaffold the conceptual framework.

Boundary-Crossing Scholarship as a Bridge to Transdisciplinarity

If everyday violence is not theorised within the transhistorical context that produces it, then individuals who perpetrate violence are regarded as deviant and society is left off the hook. As part of secondary research, I routinely consult the work of boundary-crossing scholars from a variety of disciplines, to look for resonance, as this scholarship provides a bridge to transdisciplinary thinking. These scholars question existing ideas about everyday violence that individualises it. I summarise lessons learnt from a few of the key texts consulted in the course of constructing the transdisciplinary conceptual framework presented here.

Psychology

In a specific reference to South Africa, psychologists Stevens et al. (2003, p. 358) argue that an individualistic approach ignores the fact that "violence occurs as a relational phenomenon in a specific time, space and context", and that "the personal experiences and actions of individuals interact with temporal, spatial and contextual events and processes that are in turn shaped by broader historical, ideological and material conditions". They suggest that:

> From as early as the 1890s through until the 1960s, psycho-dynamic approaches (with strong 'racialised' overtones) dominated understandings of violence. Common to psycho-dynamic approaches violence was viewed as the conscious manifestation of unconscious wishes drives and fantasies due to poor defence mechanisms within the personality structure and an inability to repress these unconscious impulses. At other points violence was viewed as a specific deficit within the psychological constitution of individuals who were unable to control aggressive impulses due to heightened levels of frustration (p. 358).

Economics

In a reference to South Africa, Sen (2008, p. 15) effectively summarises the perspective also held by Demombynes and Özler (2006), Wilkinson and Pickett (2010), and other unorthodox economists when he cautions that there is a need to avoid 'isolationist programmes of explaining violence *only* through concerns of economic and social inequality and deprivation, or exclusively in terms of identity and cultural factors'. He argues that 'none of these *individual* influences, important as they very often are in a fuller picture, can provide an adequate understanding of the causation of widespread violence and the absence of societal peace' Instead, Sen draws attention to the *interconnections* between these influences, which, he argues 'are as important as the elements that have to be connected' (*italics mine*).

Criminology

On their part, unorthodox criminologists argue in essence that "if someone has criminogenic traits, it puts them 'at an increased risk of committing crime, but that risk may not be realised until that individual encounters conducive situational conditions". When "criminogenic situations and criminogenic traits do coincide; the risk of crime is multiplied" (Wortley, 2011, p. 113). On the other hand, Zimbardo

(2007, p. 289) casts doubt on the idea that only 'some' people have criminogenic genes. He argues that all 'good people' can be broken down over time and can commit the most atrocious deeds if prompted by social forces. In addition, it has been suggested elsewhere that, "according to scientific research", communities do not have the power to counter the "policies and market forces causing criminogenic community structures and cultures" (Sherman et al., 1996, p. 41).

Political Science

Most chillingly, Uvin (1998, p. 136), a political scientist and development worker before and during the genocide in Rwanda, has this to say about invisible mechanisms that produce visible violence in unequal societies:

> For most of us, it is hard to imagine how tense and frustration-ridden a society must be when every day the large majority of the population is shown the lifestyle of the "developed" and exhorted to achieve it but is at the same time structurally excluded from this "good life," with very little chance of achieving it.

In sum, taken together, this select sample of scholarship on violence (taken from a much longer list I consulted over the years), utilising different units and levels of analysis, confirmed the idea that an adapted transdisciplinary conceptual map of invisible/visible violence and denial is required for maximum understanding and action.

Drawing Together Key Lessons Learnt from Below

> Social justice requires systematically decolonizing, deracializing, demasculanizing, and degendering our universities. This means that we have to engage very seriously with research, scholarship, learning and teaching, curriculum, pedagogy, and a host of other issues and their meanings in a society that must overcome its apartheid past and ensure human rights and justice for all (Badat, 2015, p. 2).

In conclusion, this section draws together the context within which I became aware that my visceral, tacit, explicit and vicarious knowledge of dehumanisation was at play during a pivotal conflict analysis process; and that I did not have the conceptual tools necessary to make explicit connections between received academic knowledge, and this embodied knowledge about uninterrupted everyday violence rooted in colonialism, apartheid and market democracy. It also completes the montage by summarising how this knowledge was built from below to meet and modify an existing conceptual framework about violence, to include transhistorical trauma and denial more explicitly. The process constitutes a series of social justice interventions that include self-determination as agency to produce knowledge from the standpoint of the oppressed; then as counterpoint to decentre partial and dominant understandings about what constitutes everyday violence.

The physical reproduction of colonialism, apartheid and market democracy experienced during practice was reflected on and researched over 14 years during five phases of action research as set out above. The conceptual framework that was adapted from Galtung's (1996) triad of violence serves as a heuristic device that generates themes for investigation. This conceptual framework is continuously groundtruthed and calibrated with a transdisciplinary spectrum of people across artificial boundaries. It is able to accommodate past and present occurrences such as the impact of social, economic and political arrangements that invisibly exert violence at all levels of analysis. This is achieved through conceptual and observational pattern-matching (Trochim, 1985). It thus becomes easier to see patterns in manifestations of everyday violence, and to see cultures of violence and denial in patterns which were found to intersect and interact during phase two of action research during 2008–2012. All new information thus test and calibrate the structure of invisible/visible violence framework. As Bitsch (2005, p. 77) suggests, "the purpose of ground-truthing is calibration, testing, or validation of a model or a theory with additional data". Groundtruthing in turns provides an opportunity to co-produce knowledge with different people. Instead of being footnotes in research reports as mere 'research participants', these people are free to become part of a growing network of peers who co-publish their offerings and retain copyright thereof. This mitigates epistemic violence somewhat, and constitutes a decolonial enactment on the margins of the academy.

I lump rather than split the concepts violence, oppression and dehumanisation, and use these concepts interchangeably in the meantime. This is a deliberate pedagogical technique to accommodate different understandings of everyday violence to concretise the idea that all forms of othering and violence constitute dehumanisation as an act and an impact – regardless of how the impact is processed and articulated by victims; and defined by beneficiaries of oppression. For as long as society and individuals deny the existence of a structure of invisible/visible violence, the structure will continue to produce dehumanised people who manifest its everyday violence, and/or who resist the structure of violence by non-violent means, and/or through other resilient/resistant responses. Regardless of the chosen methods, such a society can never be described as a 'post-conflict' society.

And finally, the disrupting denial approach is rooted in transdisciplinarity which is, according to Max-Neef (2005, p. 15) "more than a new discipline or super-discipline, [it] is actually, a different manner of seeing the world, more systemic and more holistic". It is acknowledged that a transdisciplinary, social justice oriented, decolonial approach to knowledge and its production is "challenging, difficult intellectually, and problematic in praxis" as suggested by Galtung (1996, p. 1) from a peace studies perspective, and as an "enormously difficult" task as stated by criminologists Wikström and Sampson (2006, p. 1). The stark choice before people who want to contribute to a more humane society and world, is to either break free and contribute to making the decolonial road while walking it and to embrace mistakes as learning opportunities; or to remain trapped in coloniality. The biggest social justice intervention with regard to everyday violence would be the decolonisation of university disciplines that continue to define and respond mainly to manifest violence. Community psychology, as a discipline interested in bottom-up and participatory approaches to social justice, should take heed of transdisciplinary approaches that follow violence where it leads, regardless of disciplinary and other artificial boundaries. Without credible research and scholarship by the authorised knowledge producers in society to change minds, laws, policies and practices, then manifestations, patterns and cultures of violence and denial will not change. The colonial academy would have succeeded in making us all blind, deaf and silent about the routine criminalisation of blackness and otherness. We will remain colonised.

References

Badat, S. (2015). *Putting an end to causes of pain.* Graduation Speech presented at Rhodes University.

Barthes, R. (2010). *Camera lucida.* Hill & Wang.

Bitsch, V. (2005). Qualitative research: A grounded theory example and evaluation criteria. *Journal of Agribusiness, 23*, 75–91.

Bourdieu, P., & Wacquant, L. (1992). *An invitation to reflexive sociology.* University of Chicago Press.

Brave Heart, M. Y. H., Chase, J., Elkins, J., & Altschul, D. B. (2011). Historical trauma among indigenous peoples of the Americas: Concepts, research, and clinical considerations. *Journal of Psychoactive Drugs, 43*(4), 282–290.

Cabrera, M. n.d. *Living and surviving in a multiply wounded country.* https://www.medico.de/download/report26/ps_cabrera_en.pdf

Chua, A. (2003). *World on fire: How exporting free market democracy breeds ethnic hatred and global instability.* Doubleday.

Cohen, S. (2001). *States of denial: Knowing about atrocities and suffering.* Polity.

Crenshaw, K. (2019). *On intersectionality: Essential writings.* New Press.

Demombynes, G., & Özler, B. (2006). Crime and local inequality in South Africa. In H. Bhorat & R. Kanbur (Eds.), *Poverty and policy in post-apartheid South Africa* (pp. 288–320). HSRC Press.

Denzin, N. K., & Lincoln, Y. S. (Eds.). (2011). *The sage handbook of qualitative research* (4th ed.). Sage Publications.

Dixon, B. (2012). Understanding 'pointy face': What is criminology for? *SA Crime Quarterly, 41*, 3–10.

Duran, E. (2006). *Healing the soul wound: Counseling with American Indians and other native peoples.* Teachers College Press.

Farmer, P. (2005). *Pathologies of power: Health, human rights, and the new war on the poor: With a new preface by the author.* University of California Press.

Freire, P. (1974). Education for critical consciousness. Sheed Ward.

Freire, P. (2000). *Pedagogy of the oppressed.* Continuum.

Galtung, J. (1996). *Peace by peaceful means: Peace and conflict, development and civilization.* International Peace Research Institute.

Galtung, J. (1998). The peace movement: An exercise in micro-macro linkages. *International Social Science Journal, 157*, 401–405.

Gqola, D. (2015). *Rape: A south African nightmare.* MF Books.

Henkeman, S. R. (2012). *Restorative justice as a tool for peacebuilding: A South African study* [Unpublished doctoral dissertation]. University of Kwazulu-Natal, South Africa. https://ukzn-dspace.ukzn.ac.za/bitstream/handle/10413/11840/Henkeman_Sarah_Rosaline_2012.pdf?sequence=1&isAllowed=y

Henkeman, S. R. (2013). Pale face'/'pointy face': SA criminology in denial. *South African Crime Quarterly, 45*, 5–11.

Henkeman, S. R. et al. (2016). *Guide to a deeper, wider and longer analysis of violence* Unpublished guide.

Iadicola, P., & Shupe, A. (2003). *Violence, inequality, and human freedom* (2nd ed.). Rowman & Littlefield.

Imbusch, P. (2003). The concept of violence. In W. Heitmeyer & J. Hagan (Eds.), *International Handbook of Violence Research* (pp. 13–39). Kluwer Academic.

Kincheloe, J. L., McLaren, P., & Steinberg, S. R. (2011). Critical pedagogy, and qualitative research: Moving to the bricolage. In N. K. Denzin & Y. S. Lincoln (Eds.), *The sage handbook of qualitative research* (4th ed., pp. 163–177). Sage.

Lederach, J. P., Neufeldt, R., & Cuthbertson, H. (2007). *Reflective peacebuilding a planning, monitoring, and learning toolkit*. Joan B. Kroc Institute for International Peace Studies, University of Notre Dame.

Liem, R. (2007). Silencing historical trauma: The politics and psychology of memory and voice. *Peace and Conflict: Journal of Peace Psychology, 13*(2), 153–174.

Mamdani, M. (2017). *Decolonising the post-colonial university*. T. B Davie Lecture, University of Cape Town. https://www.youtube.com/watch?v=vKFAYXf05N0

Max-Neef, M. A. (2005). Foundations of transdisciplinarity. *Ecological Economics, 53*(1), 5–16.

Moraga, C., & Anzaldúa, G. (2015). *This bridge called my back: Writings by radical women of color*. SUNY Press.

Roy, A. (2001). *Power politics*. South End Press.

Sen, A. K. (2008). Violence, identity and poverty. *Journal of Peace Research, 45*(5), 5–15.

Sherman L. W., Gottfredson, D., MacKenzie, D., Eck, J., Reuter, P., & S. Bushway (1996). Communities and crime prevention in preventing crime: What works, what doesn't, what's promising. A Report to the United States Congress. https://pdfs.semanticscholar.org/7a7e/6871dfbce8940cd88f72ad87664876d8c888.pdf#page=40

Smith, L. T. (1999). *Decolonizing methodologies: Research and indigenous peoples*. Zed Books.

Smith, W. A. (2008). Higher education: Racial battle fatigue. *Encyclopedia of Race, Ethnicity, and Society, 1*, 616–618. https://doi.org/10.4135/9781412963879.n248

Stevens, G., Seedat, M., & van Niekerk, A. (2003). Understanding and preventing violence. In K. Ratele & N. Duncan (Eds.), *Social psychology: Identities and relationships* (pp. 354–371). JUTA.

Trochim, W. M. K. (1985). Pattern matching, validity, and conceptualization in program evaluation. *Evaluation Review, 9*(5), 575–604.

Uvin, P. (1998). *Aiding violence: The development enterprise in Rwanda*. Kumarian Press.

Wikström, P., & Sampson, R. J. (2006). *The explanation of crime: Context, mechanisms, and development*. Cambridge University Press.

Wilkinson, R. G., & Pickett, K. (2010). *The spirit level: Why greater equality makes societies stronger*. Bloomsbury Press.

Wortley, R. (2011). *Psychological criminology an integrative approach*. Routledge.

Zerubavel, E. (2010). The social sound of silence: Toward a sociology of denial. In E. Ben-e'ev, R. Ginio, & J. Winter (Eds.), *Shadows of war: A social history of silence in the twentieth century* (pp. 32–44). Cambridge University Press.

Zimbardo, P. G. (2007). *The Lucifer effect: Understanding how good people turn evil*. Random House.

The Past, Present, and Future Entangled: Memory-Work as Decolonial Praxis

Rosa Cordillera A. Castillo

The Past as Decolonial Battleground

"Colonialism", Frantz Fanon wrote in the *Wretched of the Earth* (Fanon, 1963, p. 209), "is not simply content to impose its rule upon the present and the future of a dominated country. Colonialism is not satisfied merely with holding a people in its grip and emptying the native's brain of all form and content. By a kind of perverted logic, it turns to the past of the oppressed people, and distorts it, disfigures and destroys it." Colonial re-presentations, repression, and erasure of the colonised's past are a form of "epistemic violence" that has been instrumental in creating and upholding the coloniality of knowledge, being, and power that divides the world's population into "degrees of being human" (Adams et al., 2015; Maldonado-Torres, 2016, p. 11; Mignolo, 2018). In this global racial/ethnic hierarchy, some bodies are in the "zone of being" who are accorded their civil and human rights, while others are relegated to the "zone of nonbeing," who are regarded as subhuman and are thus administered through war and oppression (Fanon, 1967 in Grosfoguel et al., 2014). Relatedly, European epistemologies are treated as universal while non-European epistemologies are devalued and debased (Quijano, 2007). This categorisation of peoples, which undergirds the coloniality of power, continues to structure contemporary racial, gender, and class stratifications, as well as informs the violence and oppression inflicted on certain bodies and populations (Maldonado-Torres, 2016; Quijano, 2007). Thus, the ways in which re-presentations of the past are mired in coloniality is a key battleground in decoloniality. Recovering and surfacing other pasts is part of a pluriversal project of valuing subjugated knowledges and histories that have been silenced by coloniality's logics, power, and agents (Grosfoguel, 2010). It is also a project of counter-ontology that asserts and redefines what

R. C. A. Castillo (✉)
Institute for Asian and African Studies, Humboldt Universität zu Berlin, Berlin, Germany
e-mail: rosa.castillo@hu-berlin.de

S. Kessi et al. (eds.), *Decolonial Enactments in Community Psychology*, Community Psychology, https://doi.org/10.1007/978-3-030-75201-9_13

humanness ought to be, and render present, in their own terms, those who have been erased, thereby rejecting and contestingcoloniality's categorisation of humanity.

Highlighting the power of the writing and reckoning of history in indigenous struggles, Maori scholar Linda Tuhiwai Smith (2012, p. 29) asserts that:

> A critical aspect of the struggle for self-determination has involved questions relating to our history as indigenous peoples and a critique of how we, as the Other, have been represented or excluded from various accounts. Every issue has been approached by indigenous peoples with a view to rewriting and rerighting our position in history.

The liberating potential of recovering and asserting subjugated pasts has been recognised not only in indigenous studies and movements and decoloniality but also in critical approaches to memory studies and historiography such as postcoloniality, subaltern studies, feminist theory, liberation psychology and community psychology, as well as within anthropology, the discipline in which I am trained. Writing on Homi Bhaba's work on postcoloniality, Bhambra (2014, p. 117) notes, that "bearing witness to different pasts" brings "into being new histories and from those new histories, new presents and new futures." Such a process can be liberating, according to liberation psychologist Ignacio Martín-Baró, because it can provide "models of identification that, instead of chaining and caging the people, open up the horizon for them, toward their liberation and fulfilment" (Martín-Baró, 1994, p. 30 in Adams et al., 2015, p. 217). A critical reckoning of the past and surfacing subjugated pasts is thus a part of "decolonial knowledge production and critique" (Maldonado-Torres, 2016, p. 7) and of "the critical pedagogy of decolonisation" that can lead to "alternative knowledges" and thus "alternative ways of doing things" (Smith, 2012, p. 36).

As such, it is important to look into how memory-work can be a decolonial praxis, that is "the affirmative and prospective thought-actions-reflections-actions that give shape, movement, meaning, and form to decoloniality" (Walsh, 2018, pp. 17–18). Drawing primarily on anthropology and decoloniality while conversing with community psychology, I inquire into the "decolonial attitude, posture, proposition, and force" (Walsh, 2018, p. 23) involved in memory-work particularly in struggles against violence, oppression, and dehumanisation. I consider this question by looking at Moro Islamic Liberation Front (MILF) adherents' will to remember atrocities committed against Muslims in the Philippines as well as stories of survivance, and how these shape their struggle for the right to self-determination. I focus on MILF adherents as they were my main research interlocutors, even though I recognise that remembering atrocities and survivance is also present among non-MILF Muslims (see Mawallil, 2018). I will suggest that memory-work plays a significant role in the MILF struggle, which can be seen as a movement towards what Adolfo Albán calls "re-existence" or "the mechanisms that human groups implement as a strategy of questioning and making visible the practices of racialisation, exclusion, and marginalisation, procuring the re-signifying of life in conditions of dignity and self-determination..." (Albán, 2008 in Walsh, 2018, p. 20).

Memory-Imagination Interplay

Various studies have shown that memory is significant in shaping how people make meaning of and respond to present situations and future scenarios, and in informing and shaping action and motivation (Argenti & Schramm, 2010; Seedat, 2015; Smith, 2012; Waterson & Kian-Woon, 2012). Memory is also important in identity discourse at the individual and collective level, as well as in the production of subjectivities. Indeed, one can say that memory has an "existential importance" for groups and individuals (Waterson & Kian-Woon, 2012, p. 31). But while it shapes the present and the future, memory, the meanings of the past(s), and practices of remembering are, in turn, shaped by present contexts, desires, struggles, and aspirations. These are especially significant in contexts of past and continuing violence and inequalities, in the struggle for liberation and the right to self-determination, and in building the narrative of a nation where memory serves as a moral discourse and political tool. Given these, I suggest that one way to study the decolonial potential of memory-work is to be attentive to the co-implication of the past, present, and future in people's narratives and lives, and the dynamics of collective memory formation. This attentiveness would necessitate looking into the entanglement of memory and imagination, which are both evocative of ideas, emotion, and action.

Imagination and memory are closely related concepts for Maguindanaons, an Islamized ethnic group in the Southern Philippines who also refer to themselves as Muslim, Moro, and Bangsamoro and who were my main research interlocutors. *Egkalendem* (to remember) and *egkakalendem* (to imagine) are both 'thinking-feeling.' Anxiety, euphoria, hope, sadness, anger, fear, pain, sympathy, empathy, and pity were some of the emotions people felt and invoked while remembering and imagining various aspects of their lives, such as in imagining other Muslims' experience with violence or the future of the Bangsamoro or in remembering past violence and survivance. These emotions contributed to shaping my interlocutors' moral judgments of the past, in giving them a sense of collective responsibility, and in motivating them to join the MILF's struggle for the Bangsamoro's right to self-determination. These emotions are, in other words, fellow-feeling.

A crucial aspect of the memory-imagination interplay is the role that imagination plays in engendering fellow-feeling. Privileging the knowledge of my interlocutors and putting it in critical dialogue with the theories of David Hume (2012 [1740]) and Adam Smith (2008 [1759]) on fellow-feeling, psychological and anthropological debates on empathy (Halpern, 2001; Hollan & Throop, 2011), and current kinship theory (Carsten, 2000), I take fellow-feeling to be an affective identification with the other that is imbued with moral weight. It is cogent in creating affective solidarity or feeling for the suffering of one's fellow and a sense of collective responsibility or doing something to ease their suffering.[1] It can thus have a political force and is shaped by historical, sociocultural, and political contexts, part of which

[1] I use fellow-feeling rather than empathy because the former allows for slippages between compassion, sympathy, pity, and empathy without a priori bracketing them from each other.

is how much it is valued in social and political life. For instance, among my Maguindanaon interlocutors, one of the ways in which the importance of fellow-feeling is manifested, strengthened, and reproduced is through idioms such as in the saying, "if a finger is hurt, the whole body feels the pain." That is, if a Muslim anywhere in the world is hurt, it is also felt in the self. This is a modification of the hadith:

> An-Nu'man ibn Basheer reported: The Messenger of Allah, peace and blessings be upon him, said, "The example of the believers in their affection, mercy, and compassion for each other is that of a body. When any limb aches, the whole body reacts with sleeplessness and fever." (Sahih Bukhari 5665, Sahih Muslim 2586)

This hadith pertaining to Muslim brotherhood/sisterhood and solidarity was often invoked by my interlocutors when talking about the suffering of fellow Muslims, as though it is a pain in the self, as we will see below in the story of Tahir,[2] a village leader whom I had gotten to know during my fieldwork. This message is often accompanied by reminding each other that it is a moral and religious duty to ease the suffering of their fellow. This forges a sense of solidarity and gives credence to calls for collective responsibility, since the suffering of one is the suffering of all. As such, it is a relationality that is deeply embodied and is strengthened by imagined identification.

At the same time, I suggest that imagination's productive force emanates from its *infinite temporality*. That is, one can imagine the past, present, and future (see Cocking, 1991; Levy, 2012; Sartre, 1948). An aspect of this infinite temporality is imagination's subjunctive property—the *what if, as if, once it is, maybe* (see Turner, 1990). This anticipatory character of imagination, the idea of *possibility* is provocative not only of emotions, sensations, and actions, but also of thoughts and desires. Directing our attention to imagined futures, particularly at the collective level, is to consider how the past and the present shape these imagined futures, and how these imagined futures, in turn, propel people forward, to act, to change the condition of their lives and the next generations.

The interplay of imagination and memory, and their temporality and emotionality potentially render the experience of others a part of our self and our identity. That is, imagining the other's experience with violence and suffering that has been transmitted through memory-making practices can link individuals with each other, thereby helping constitute (but also challenge) the collective, and provides a framework for understanding resistance and peace-making. This is a dynamic process that is shaped by specific contexts, the meanings attached to the past and how they relate to the present and the future, how effectively these are transmitted to members of the group, and how these are received. A decolonial approach to memory-work should, however, not merely reproduce stories of victimhood that can harden differences and hatred between people (Krondorfer, 2016; Quayle & Sonn, 2019). As examples from community psychology and indigenous studies have shown (Seedat,

---

Additionally, fellow-feeling is, unlike empathy, necessarily directed towards doing something good. For a more thorough discussion, see Castillo (2017).

[2] Names of my interlocutors are pseudonyms to protect their identities.

2015; Smith, 2012; Quayle & Sonn, 2019), and which I will also explore through the experiences and narratives of my MILF interlocutors, memories of shared suffering and survivance can instead serve as a guide or "resource" for groups in liberating themselves from oppressive conditions.

Researching Memory-Work: Convergences Between Critical Ethnography and Critical Community Psychology

Living Attentionally with Others

In understanding memory-work, which involves the memory-imagination interplay, I draw from my long-term ethnographic fieldwork between 2011 and 2018 in mainly rural parts of the Cotabato region in the Southern Philippines among politically, socially, and economically non-elite Maguindanaon adherents of the MILF. Ethnography is well-suited to understanding the meanings, importance, nuances, and dynamics of memory-work in individual and social life because it is an embodied form of knowledge production built on rapport, trust, and the researcher's empathy, that enable the development of ongoing relationships with and close involvement in people's life-worlds. It is a way of "living attentionally with others" and learning from them (Ingold, 2014, p. 389). With this approach to ethnography, I undertook participant observation of peace process-related events in addition to participant observation of everyday life and events in my field sites, one of which is Lalin, an anonymised farming village of a few hundred Maguindanaon families. Lalin is one of my main field sites because it has been in the throes of the decades-long conflict in the Cotabato region and many of its residents have been part of the liberation struggle since the late 1960s. During participant observation, I was particularly attentive to the commentaries of people on events and the conditions of their lives. I also collected narratives such as life stories and oral histories from various interlocutors. In my ethnographic interviews with ninety-nine individuals, some of whom I interviewed several times, and in my conversations and observations of people's commentaries and everyday life, which are often conducted in Maguindanaon language and, to a lesser extent, in Tagalog and Bisaya languages, I gave attention to people's narratives of suffering, violence, peace, motivations, hopes, and desires, the emotions that were engendered in remembering and imagining violence and peace, and the meanings and moral and ethical judgments they gave to events that impacted their lives.

In my broader ethnographic work (Castillo, 2017), I look at imagination's interrelatedness with memory, emotions, temporality, and action, the production of subjectivities, and the making and unmaking of an imagined community, as these are grounded in Maguindanaon concepts and everyday life, and within a context of violence, liminality, and struggle for the right to self-determination. This chapter focuses on an aspect of this broader work re-read through the lens of decoloniality, that is on how coloniality continues to shape Muslim lives in the Philippines and how my interlocutors' push back against this, part of which is their assertions of memory and history and their corresponding hopes for the present and imaginings of the future, can be seen as a decolonial act of assertion of humanness, resistance, and liberation.

Reflexivity

A key part of ethnographic fieldwork is the researcher's reflexivity. This calls for questioning the social and political conditions of knowledge production, and how our positionality or the researcher's position within hierarchies of power and our "place of enunciation" (Walsh, 2018), shape the knowledge we produce, our engagements, and responsibilities. I was both the self and the other, an insider-outsider position that was never fixed, but instead varied from situation to situation (Castillo, 2015). I am from the same country as my interlocutors and thus share similar stakes in the welfare of the country. This becomes particularly pronounced during certain critical events such as the signing of peace accords or the threat of a new war. As I citizen, I am entangled in these and have a responsibility towards these developments. At the same time, I could empathise with and relate to those who have suffered from state violence because I have also lived through some form of state terror, and continue to be anxious about the effects of political violence in the country on myself and my family. However, the stakes are different for my interlocutors and me in that the poverty, discrimination, violence, and suffering they experienced and continue to experience do not compare to mine. Nor am I a member of the MILF, and I do not belong to the same religion, class, and ethnicity. Add to this, my middle-class background, educational attainment from the University of the Philippines and Germany, and place of origin, Manila, which my interlocutors see as the centre of power and prosperity in the country and from where major political decisions that impact their lives emanate, further set me apart as an outsider. More specifically, I am a privileged outsider. Due to my educational and class background, as well as place of origin (Manila) and residence (Germany), my interlocutors saw me as a source of solidarity, hope, and aid.

Given the suffering that they live through, their expressed desire for solidarity and help, and my commitment to social justice, I adopted a stance of "anthropological witnessing," which "positions the anthropologist inside human events as a responsive, reflexive, and morally or politically committed being" (Scheper-Hughes & Bourgois, 2003, p. 26). Part of this engagement, as Skidmore (2006, p. 47) asserts, is "to advocate for the histories of terror and misery to be retained in the contemporary world" because they serve as "alternate readings of political power, legitimacy, and moral authority [that] have the power to puncture the hegemonic reading of past and present...". In other words, writing about my interlocutors' memories and experiences of suffering and resistance is an act of engagement in itself (Scheper-Hughes, 1995).

Elders in Lalin, for instance, expressed to me after I presented parts of my research results and analysis to community members for their feedback, that what I was doing was writing their *tarsila*, a record of genealogical lineage and historical events of the nobility. Not being members of the nobility means to them being absent from accounts of history. That I am writing about their lives is, for them, a recognition of their humanity and existence. I am therefore constantly aware that I am part of their memory-work, and that there is a big responsibility entailed in this. This is an engagement that was pregnant with dilemmas and challenges that I had to labour through emotionally, politically, and analytically (Castillo, 2015). As such, the anthropology and ethnography I adhere to is critical of the colonial roots and legacies of the discipline and is sensitive to the politics of knowledge production (see Uperesa, 2016). I, therefore, strived to enact fieldwork that is reflexive, ethical, participatory, and reciprocal and writing that centres but does not romanticise subjugated perspectives.

This turn to witnessing, reciprocity, and reflexivity within anthropology is not dissimilar from critical community psychology, which is "an emancipatory enterprise that is overtly political, value-laden, and intersubjective in nature...[C]ritical community psychologies subvert the notion of scientific neutrality, acknowledge the socially constructed nature of knowledge, and view power as constituted in the context of social relations" (Suffla et al., 2015, p. 10). Suffla and colleagues (2015, p. 10) further call for reflexivity where we "critically examine the assumptions, values, discourses, and practices that we deploy to portray reality and create knowledge." And while specific methods and conceptualisations might differ, critical ethnography and critical approaches to psychology such as certain strands of community psychology, liberation psychology, and cultural psychology share the importance given to everyday life and the perspectives of communities with whom we conduct our research and engagement, particularly those who have been oppressed (Adams et al., 2015; Segalo et al., 2015). Such convergences signal to the decolonial potential of an interdisciplinary approach to memory-work. That is, context-specific, intersubjective, reflexive, reciprocal, and politically-engaged research that centres subjugated perspectives, if done with care and sensitivity to the power dynamics between researcher and her interlocutors and among her interlocutors, can be a decolonial praxis.

I thus turn at this juncture to a critical event in 2015 and how, on the one hand, it demonstrates the dehumanisation of Muslims in the Philippines, and, on the other hand, how Muslims resisted this dehumanisation through assertions of memory.

In the Zone of Nonbeing

In early 2015, it seemed that Muslim parts of Mindanao were once again on the brink of war after members of the Special Armed Forces (SAF) were killed in the municipality of Mamasapano. The SAF, a government elite police unit, had entered two areas of the municipality in the early hours of January 25 in search of an alleged

Malaysian terrorist wanted by the United States. One of these areas, Tukanalipao, is a government-recognised camp of the MILF, which means that any government security operations in this area needed to be coordinated with the MILF. The SAF, however, kept their operation secret from the MILF and the military, both of which oversee the coordination of security operations in MILF areas.

Some residents of the villages were up early due to Ramadan on the day the heavily armed SAF entered their area in the cover of early morning darkness. Because the SAF did not coordinate with the local MILF command, the residents mistook the armed men as forces of an opposing clan with whom they were involved in a violent feud. A heavy gun battle ensued between the SAF and armed villagers that lasted 11 hours, leaving 44 policemen dead. It was the biggest casualty the Philippine police has suffered in a day.

Needless to say, these deaths deserve to be condemned. But the way they were condemned by politicians and the public needs to be critically examined, especially since absent in their condemnation was a recognition of the deaths of 18 MILF fighters and at least seven Muslim civilians, including a five-year old girl hit by a stray bullet and five men sleeping in a mosque whom residents said were killed by the SAF (forumZFD, 2015).

Public anger quickly escalated as news of the SAF's deaths began to be widely reported. Their deaths were labelled a "massacre," and the SAF were exalted as the "country's heroes" who laid their lives on the line for the Philippine nation against terrorist threats. Meanwhile, the MILF, and Muslims in general, were repeatedly accused by politicians, the media, opinion-makers, laypeople, and even some scholars of "overkill," "brutality," and having terrorist links. Some personalities called for war against the MILF, while others said that Muslims should be expelled from the country.

Public mourning for the SAF commenced with the hashtag #Filipinolivesmatter trending. Thereafter, a highly publicised Senate investigation was conducted. Lasting 2 months, the Senate's report also referred to the clash as a "massacre," despite the MILF and some government officials' insistence that it was a "misencounter," a euphemism for an accidental armed encounter. Meanwhile, the deaths of Muslims in the clash did not receive as much media, politician, and public attention as the death of the policemen. This is an "omission," a "refusal of discourse," that renders Muslim lives as not recognisable as life, and therefore not grievable (Butler, 2003).

As I have written elsewhere (Castillo, 2018), these discourses captured the imagination of a nation where Muslims have been stereotyped and differentiated as violent, terrorists, untrustworthy, and prone to run amok, and where there has been a hard-line position on keeping the peace in certain parts of Mindanao (PIDS, 2006). The anti-Muslim portrayals in the aftermath of the clash appealed to these preexisting constructs and scenarios about the Muslim Other in the majority Christian Filipinos' minds, scenarios which people connected to moral orders of good versus evil, terrorists versus heroes, and traitors versus nationalists. These readings were shaped by the legacies of Spanish and American colonial-era depiction of Muslims as a violent people. The proliferation of negative Muslim stereotypes by the colonisers was aimed at inciting Christian Filipinos to fight against Moro, the pejorative label used by the Spanish for people who practiced Islam in the 1500s in what is now the Philippines (Che Man, 1990). This led the former to have a sense of superiority over non-Christians, while Moro began to perceive both the Spaniards and Filipinos as foreign aggressors, with the label 'Filipino' being used until recently by Moro to refer to a different nationality (Blanchetti-Revelli, 2003; Che Man, 1990). The Moro National Liberation Front (MNLF) for instance states in its 1975 Manifesto that "the Filipino government has the birthmarks of its Spanish and American predecessors. Its most distinct mark is its colonial character" (Gowing, 1988, p. 166). More recently, Mohd. Musib Buat of the MILF asserted that "the Bangsamoro people have remained under state neocolonialism…The Bangsamoro remains a captive nation" (2007, p. 35). The old saying "the only good Moro is a dead one" (Che Man, 1990, p. 29) succinctly captures the combined prejudice, fear, and ignorance of the history of Muslims in the country and of their dehumanisation by the colonial and postcolonial governments and the broader public. This is a prevailing Islamophobia rooted in colonialism that continues to structure Muslim lives in the Philippines.

The biases against Muslims were further shaped by the communal violence of the Marcos regime in the late 1960s and 1970s when religious identity, specifically being a Muslim, became a focal point for a violent form of othering, giving rise to numerous massacres of Muslims during that era. This colonial and postcolonial context, coupled with the skewed portrayal of Muslims in Philippine history textbooks--when they are at all included for they are oftentimes excluded--and the more recent global "war on terror," which is also rooted in Islamophobia, contributed to making Muslims' reported brutality in Mamasapano easily acceptable as fact, despite the absence of credible investigations in the immediate aftermath of the clash.

The prejudices that surfaced in the aftermath of the Mamasapano clash is just one of the countless incidents that evince Muslims' social, political, and economic exclusion from the Philippine body politic and imagination (see Mawallil, 2018). This is an exclusion that is firmly rooted in coloniality and its continuing logic and matrix of power, reproduced even by those who have been colonised, such as Christian Filipinos. In this regard, the hierarchies of being human and the relegation to the zone of nonbeing is marked not along skin colour but along ethnic, religious, and cultural lines (Grosfoguel et al., 2014). The dehumanisation of Muslims and their political, economic, and social consequences are so important that, according to Buat (2007), the deep-seated biases and prejudice against Muslims have been "one of the main causes why the Bangsamoro people wanted to set up their own country and do not wish to be integrated and assimilated into the Filipino national community."

The Birth of a Liberation Struggle

Islam pre-dated the arrival of the Spanish in the archipelago and when the latter conquered the islands in the 1500s, Muslim populations were largely successful at repelling the Spanish forces. Thus, during the Spanish colonial era, the southern part of the country was populated mainly by Muslims and Lumad, the collective term for various non-Islamized indigenous peoples in Mindanao. It was only during the American colonial era, at the turn of the twentieth century, that Mindanao was effectively incorporated into the larger colonial order when the colonial strategy shifted to homestead policies for Christian Filipinos and partnering with Muslim elites who accommodated and negotiated with colonial powers. These land policies reversed the population ratio, rendering Muslims and Lumads minorities in Mindanao where they were once the majority (Anave, 2000). Along with this changed demography was the diminished economic and political position of Muslims (Abinales, 2000). The Autonomous Region of Muslim Mindanao (ARMM) was, for a long time, the most impoverished region in the country. Sixty-three percent of its population were considered poor in the first semester of 2018, despite the region being rich in natural resources (RSSO-ARMM, 2019).

By the end of the 1960s, competition over land and political power brought on by colonial and postcolonial land policies and President Ferdinand Marcos' political manoeuvring in the region erupted in communal violence (Abinales, 2000). In 1972, Marcos declared martial law, citing these instances of "chaos" in Mindanao as one of his reasons. After martial law was declared, a war ensued between government forces and Moro fighters who had, by 1972, formed the MNLF. Their avowed enemy was not Christians, but the national government. The MNLF initially fought for the secession of Bangsamoro (Moro Nation) but eventually settled for autonomy. In 1977 the MNLF split, giving rise to the MILF.

The MILF dropped its secessionist stance in the 1990s and began engaging in peace negotiations with the Philippine government, culminating in the ratification of the Bangsamoro Organic Law and the establishment of the Bangsamoro Autonomous Region in Muslim Mindanao (BARMM) in February 2019. By this time, the protracted war has killed more than 100,00 people over a span of 26 years (Busran-Lao, 2005). It further displaced 3.5 million in the last 15 years (TJRC, 2016).

The violence of the 1960s and 1970s destabilised people's livelihood, sense of security, and inter-group relationships where Muslims and Christians who used to be good neighbours became distrustful of each other. This marked a profound shift in social relations, fracturing and breaking apart communities, engendering new ways of relating, producing new subjectivities, and stamping into people's memories the vicious potential and reality of violence. This period thus figured prominently in my interlocutors' memories and imaginings of the oppression they endured and struggled against.

Re-Existence: The MILF's Will to Remember

Muslims countered the dominant portrayal of the Mamasapano clash by posting on social media about the massacres that were committed against Muslims from the colonial era up to the present, urging people to remember these as well. They mentioned, among others, the 1906 Bud Dajo massacre of 800–900 Tausug men, women, and children in Sulu by American soldiers, the 1968 Jabidah massacre of Muslim military recruits in the island of Corregidor, the 1971 Manili massacre of 70–80 men, women, and children inside a mosque, the 1974 Malisbong/Kulong-Kulong massacre of an estimated 1500 men, again inside a mosque, and the rape of women, and the massacre of six members of the Mandi family, including children, in 2009. For many Moros I interacted with, massacres and the reign of Marcos, during which most of the massacres occurred, were considered the pinnacle of violence against innocent people and became an important aspect of their collective identification as an oppressed but valiant people. They were thus deeply hurt at the description of the death of the SAF as a "massacre" because they did not consider the policemen who were fully armed to go into battle as innocent. A Bangsamoro university student and MILF member told me in our email conversation about the clash in February 2015,

> We are all Filipinos. We are not seceding from the Philippines. What is painful for them is also painful for us. But we also know that since [Marcos'] martial law our rights have never been equal because the system in the Philippines is rotten...To those who are calling for justice, I hope they also call for justice for the Muslim victims of massacres. Because they were the ones who died without any real fighting chance.

Massacres have been such a significant part of people's lived experience that the word *masaker* had entered my interlocutors' vocabulary of violence and features centrally in the MILF's will to remember. This will to remember the massacres and Marcos' dictatorship constitutes intentional and deliberate acts of remembering that have framed the memories of massacres in a political and moral way. They are part of the liberation fronts' narratives of the Bangsamoro as an oppressed but valiant people, in my interlocutors' experiences with violence, their imaginings of the threat of violence and the brutal potential of the government, and their motivation to partake of the Bangsamoro's struggle for the right to self-determination.

Transmitting and Framing the Past

The MILF, through its ustadzs and affiliated NGO workers, mediated narratives of massacres in trainings, seminars, mosque sermons, and radio shows. A key message in these narratives was for the Bangsamoro to never surrender and to never let the past happen again so they may have a peaceful life. The ustadzs tell stories of the massacres and the atrocities of the Ilaga in order to awaken and motivate Moro youth, and, according to the head of the Da'wah in the area, "to inspire the Bangsamoro to continue claiming their rights." This was echoed by another ustadz

I interviewed who said, "They should persevere so that the past will never be repeated again." Or as an MILF NGO worker who also conducts trainings for MILF members told me, "Our lesson from history is that we should learn to defend ourselves."

These messages are especially potent because massacre is not a thing of the past. It continued to be part of people's lived reality and a haunting reminder of what could potentially be. The notion that massacre is an ever-present threat, and the possibility of its recurrence, created a compulsion among liberation front leaders and adherents to remember the massacres with the aim to strengthen people's resolve to defend themselves, fight for their rights, and to not let the past happen again. For them, to remember is a way of exercising control over the present and of directing the future, a project not merely of individuals, but also of the group.

My interlocutors who were recipients of these narratives would refer to these massacres in conversations and commentaries, in stories told within families, in recalling experiences with violence, in weighing the present, and in imaginings of the future as a counterpoint to the violent past and uncertain present. They would, for instance, invoke the massacres when they remark on the violence that they as residents of the village, and that Moro or 'the Bangsamoro' as a collective, endured, and, correspondingly, how they must respond as a Bangsamoro. Such conversations are often infused with fellow-feelings provoked by remembering and imagining these suffering. These are seen in the narrative of Tahir, a Lalin village leader whose father was a local MILF commander. I interviewed him several times covering a range of topics, one of which is on stories about the Ilaga and martial law:

> Tahir: I was already born during that time [the 1970s], although I cannot recall what happened. But my father told us stories of what happened in the 1970s during the time of the Ilaga. That was a difficult period. My father narrated to us the massacre in Kulong-Kulong [Malisbong] and Manili. The Manili massacre was difficult. It was painful the way they killed Muslims there. They [the people of Lalin] also suffered during that time because soldiers should not see even their shadows. Even shadows were shot at. That was how difficult that period was. Even your shadow could not be seen.
>
> Rosa: What did you feel when you heard these stories?
>
> Tahir: It was painful. I became brave, even though I was very young. I said, that is no longer right because sometimes they [soldiers and Ilaga] would chop people's ears off. Sometimes they would chop off hands, little by little. Sometimes they would hurt them, kill them slowly. And sometimes they would shackle them, tie them. And then they would be killed slowly. They would be hit, kicked, knifed. It's painful. Difficult. Even when I was a child, I felt the pain that was inflicted on my fellow Bangsamoro. It was as though it was really inside me…You will really rebel against them. That's the reason why a Muslim, a Bangsamoro would rebel. Because of deeds done that are painful for any human being.

Tahir was told by his father to never surrender and to prepare himself so that the past would never be repeated, similar to MILF narratives of massacres that emphasised preparation, steadfastness, and resistance. Men were, however, not the sole recipients of these messages. Women were also actively part of memory-making. Not only were they told by their parents about the atrocities of the Ilaga and the massacres, in which the main message was also to "never surrender", they also told the stories to their children. One of the women I interviewed was from Manili. She was not in the village when the massacre happened, but she lost relatives to it. She said that they discussed the Manili massacre in their families so that their children "will know what they should do." Adding, "only those who are newly born do not know [about the massacre]."

This intergenerational transmission of memory connects individual stories and experiences with the collective story, and bridges generations with each other. Like Tahir, Rahib, a farmer in his twenties when I met him, was also deeply affected by stories of the massacres that he heard through MILF trainings, sermons in mosques, in community gatherings, and within his family, even though he never experienced the massacres nor Marcos' martial law. He told me, "it is as though you come to your senses, you fight," whenever he thinks and hears about the Manili massacre. For him, the stories indicate that "Our countrymen were being oppressed. We are fighting for the rights of our countrymen and our religion. That is why we are on jihad (struggle)." He was deeply affected because he also experienced violence and wars growing up as a male Moro in the Cotabato region, often targeted and harassed by soldiers. For him and many others I met during my fieldwork, it was the continuity of violence and injustice in the present, and the presence of past violence in contemporary discourse that made the suffering across generations shared. Furthermore, the *possibility* that the violence might happen again motivated many of my interlocutors to continue the memory-work about the atrocities so they may learn how to defend themselves.

Memory-work thus entails the travel via discourse of personal biographies of those who witnessed or survived a massacre to places far beyond their origins. These biographies become part of stories of violence in other Moro communities where people identified with the massacre victims just as they themselves were subjected to suffering, even if they had never directly experienced a massacre. In this way, imagination bridges the gap between those who have experienced and those who have not, such that the memory of another becomes a part of the self through the work of imagination and its embodiment through fellow-feeling. Consequently, these acts of commemoration and imagining create a sense of shared heritage, whereby the collective identity is marked by oppression and resistance.

Resistance is not revenge, because revenge is against the Qur'an, as people would tell me. The MILF ulama would teach their flock not to harbour ill feelings towards Christians and soldiers and to fight oppression, even if the perpetrators are fellow Moro, and to contribute to the struggle for self-determination through various means. This is commonly done through NGO and development work, community efforts, or supporting the struggle through financial contributions and prayers, or less commonly through armed struggle, all of which they consider as part of jihad. In this sense, there is a moral and religious duty to fight oppression of which massacre is one of the most feared forms.

Accompanying these memories of massacres is the MILF's assertion of Muslim history in the Philippines hitherto erased or distorted in mainstream historiography. They highlight Muslims' continued resistance throughout the 300 years of Spanish colonisation and assert that the Bangsamoro people are oppressed but valiant and unconquered (Jubair, 1999). On the one hand, these assertions tend to gloss over particular Muslim leaders' accommodation of colonial rule (Abinales, 2000; Hayase, 2007; McKenna, 1998). McKenna (1998, p. 84) argues, for instance, that these "'narratives entail the subjugation of a threateningly unruly history' (Spencer 1990, p. 287) in support of their [nationalist ideologues] ideological stance that a self-conscious oppositional identity as Philippine Muslims is ancient, deep, and broadly shared." On the other hand, Smith (2012) points out that for those who were colonised, the belief in an "authentic self" has political currency.

> It does appeal to an idealised past when there was no coloniser, to our strengths in surviving thus far, to our language as an uninterrupted link to our histories, to the ownership of our lands, to our abilities to create and control our own life and death, to a sense of balance among ourselves and with the environment, to our authentic selves as a people. Although this may seem overly idealised, these symbolic appeals remain strategically important in political struggles. (p. 77)

She further asserts that:

> The remembering of a people relates not so much to an idealised remembering of a golden past but more specifically to the remembering of a painful past, re-membering in terms of connecting bodies with place and experience, and, importantly, people's responses to that pain. (p. 147)

Thus, aside from "rewriting" (Smith, 2012, p. 9) colonial history, local and more recent stories of bravery also feature prominently in memory-work, such as the story of a Muslim soldier who intervened to prevent a massacre in Lalin in the 1970s. This is a memory that various Lalin residents told me several times. These and other stories of bravery and survivance during the colonial and postcolonial era, idealised as parts of these narratives are, have become foundational to my interlocutors' political identity and notion of peoplehood, as well as a source of strength, motivation, and belief in the legitimacy of the Bangsamoro struggle.

The preceding discussion suggests that MILF adherents' will to remember can be seen as a decolonial posture and force in defence of their communities, lands, and Islam. It also shapes the way they constitute themselves and their collective amidst a violent lived experience and history of oppression and struggle. As a Moro civil society training officer asserts, "It is in our past that defines and guides us in our role both in our present and future status as Bangsamoro. It also eases us as well as outsiders on the burden of defining what and why we are called Bangsamoro" (Abba, 2011, p. 11).

Unevenness of Memory-Work

However, despite the power of the collective memory related to the massacres and Marcos' martial law, some of my interlocutors also wanted to forget these aspects of the past. After the signing of a peace accord in 2012 led to improved security in

Lalin and euphoria among its residents that their imagined future of a Bangsamoro homeland was becoming a reality, one of my interlocutors told me that there were villagers who now desired to "let go" of memories of past wars so they can move forward with the tasks of everyday life in a time of diminished uncertainty. "If we don't let it go, we won't be able to plant in our gardens or plant rice," she said. Meanwhile, for those who survived and experienced the massacres, remembering and forgetting have a different contour. For them, hearing narratives about the massacres involved bringing the details of the days when the killings were happening back to life. Suharto, a resident of Palimbang where the Malisbong/Kulong-Kulong massacre happened and who was a teenager when his relatives and village-mates were massacred in 1974, told me during our interview in 2013, "I've almost memorised the event. But we don't want to hear the stories anymore. We don't want to tell the stories. It is as though it happened only yesterday. As though we see the faces of our fathers, grandfathers." A similar sentiment was aired to me by a survivor of the Manili massacre. While they did not mention the word "trauma," nor do I use it in my research because it was not commonly used by my interlocutors, one can think of this as an instance of unresolved past where "ordinary timeliness suddenly gives way to timelessness, and the painfulness of the past is felt as if it is occurring now" (Prager, 2016, p. 15).

Notwithstanding these desires to forget, there are still those who want to speak about the painful past they experienced, not only to encourage people to fight for the Bangsamoro's rights, but also so that such experiences would become part of history. Abbas, also a resident of Palimbang, was 13 years old when the massacre happened. He was the youngest person inside the mosque and survived the massacre after the month-long killings. He eventually became a development worker affiliated to the MILF and was often invited to give talks about the massacre. At one point during my interview with him in 2013, as he was recounting his experience, he broke off and cried. I cried too at seeing his pain, and gently told him that he does not have to continue telling me his story, assured him that it happened in the past, and that he is now in a safe space and time. But he composed himself and told me resolutely, "Let's continue… Even if it is painful for me to remember the massacre, as though I am seeing the people still alive, I will still tell you my story. I want the world to know what happened. And that it truly happened. It should be part of history." For Abbas, a way towards healing was to talk about his experiences so that others may know and give recognition to their suffering. In this sense, storytelling "can be a decolonial and liberation method," as community psychologists Quayle and Sonn (2019, p. 11) remind us "It is part of the process of digesting the past and its connection with the present, connecting the personal and the political, and in seeking healing and justice."

Remembering and forgetting are not a uniform process for all individuals involved, even if the collective memory is prevalent, impacted as these processes are by material, social, and political conditions and individual circumstances. The past can be contested, or have different significance, even among those who belong to the same group of oppressed people. Conversely, the imagined future is also not free from tension or contestation as visions of the future, and experiences of past suffering and survivance, can differ as well depending on one's position in the hierarchies of power. While I do not have the space to flesh out these nuances, I recognise that a decolonial approach to memory-work needs to take these tensions and contestations into consideration as well.

Despite the unevenness of memory-work, its importance in shaping the present and future renders it an integral part of decoloniality. My interlocutors' will to remember is part of their struggle to have a peaceful and prosperous life and a future of self-determination. They are also attempts at breaking the hierarchies of humanness and the colonial logics of violence and subjugation that have inflicted so much suffering on Muslim populations in the country. They are, in other words, a "disparaged community's resistance to hegemonic identity constructions… and a quest for generative collective formations" (Seedat, 2015, pp. 1–2). The recent establishment of the BARMM, which is currently headed by the MILF, is a historic and important step towards recognising the Bangsamoro's right to self-determination. It is a victory in their struggle, and is thus a step towards a potentially decolonial future, for as long as it is not co-opted by entrenched corrupt structures in Philippine politics and governance. And for as long as it addresses existing exclusions and inequalities that have been partly shaped by MILF practices and ideologies, and does not lead to the creation of a new elite and the drawing of new boundaries between peoples (see Castillo, 2014; Paredes, 2015). That is, the BARMM can be decolonial if it is to build "a transmodern, pluriversal decolonised world of multiple and diverse ethico-political projects in which there can exist a truly equal and horizontal communication and dialogue between the peoples of the world" (Grosfoguel, 2012, pp. 97–98).

Decolonial Futures

I hope to have sketched out above how memory-work, which involves the memory-imagination interplay and the co-implication of past, present, and future, is an arena of decolonial struggle for those who have been marginalised and oppressed. It is an arena where there is great collaborative potential for artists, academics, activists, and local communities to harness the productive force of the interplay of imagination and memory. How can such collaborative works help forge fellow-feeling between those who have experienced the past and those who have not? How can the gap between those who have experienced and those who have not be bridged, so that that the memory of another becomes a part of the self? How can imagined futures, as these are linked to the past and the present, be given more prominence? Can such projects also contribute to a decolonial anthropology and community psychology that "disrupt orthodox, received, and taken-for-granted knowledges" (Seedat & Suffla, 2017, p. 424)? Innovative approaches to memory projects such as that by community psychologists Segalo and colleagues (2015) who used embroidery and community mapping to surface memories and imaginings in South Africa, Israel, and Palestine, or Quayle and Sonn (2019) on the use of community arts to recover and assert memories of Aboriginal elders, or the Eldorado Park Oral History Project in South Africa, which aims to reclaim the community's past and rebuild their future (Seedat, 2015, p. 28) provide ways forward.

Successfully harnessing the decolonial potential of the imagination-memory interplay necessitates carefully and critically considering the modes, circulation, and politics of transmission of experiences, narratives, and meanings of the past and what kind of present and future this memory-work tries to build. Conversely, memory-work that challenges dominant discourses does not necessarily make for a decolonial project. Such memory-work must be accompanied by a sustained critique and pushback against the coloniality of knowledge, being, and power, and coloniality's material, political, and societal structures and consequences, towards a decolonial future of re-existence.

References

Abba, M. (2011). No title. *Bangsamoro Journal, 7*(1), 11.

Abinales, P. (2000). *Making Mindanao: Cotabato and Davao in the formation of the Philippine nation-state*. Ateneo de Manila University Press.

Adams, G., Dobles, I., Gómez, L. H., Kurtiş, T., & Molina, L. E. (2015). Decolonising psychological science: Introduction to the special thematic section. *Journal of Social and Political Psychology, 3*(1), 213–238. https://doi.org/10.23668/psycharchives.1766

Anave, J. (2000). The socio-political dimensions of the Mindanao conflict. *Third World Network*. https://twn.my/title/2057.htm. accessed 8 September 2021.

Argenti, N., & Schramm, K. (2010). Introduction. In N. Argenti & K. Schramm (Eds.), *Remembering violence: Anthropological perspectives on intergenerational transmission* (pp. 1–42). Berghahn Books.

Bhambra, G. K. (2014). Postcolonial and decolonial dialogues. *Postcolonial Studies, 17*(2), 115–121. https://doi.org/10.1080/13688790.2014.966414

Blanchetti-Revelli, L. (2003). Moro, Muslim or Filipino? Cultural citizenship as practice and process. In R. Rosaldo (Ed.), *Cultural citizenship in island Southeast Asia: Nation and belonging in the hinterlands* (pp. 44–75). University of California Press.

Buat, M. M. (2007). The right to self-determination: Impact and implications to the Mindanao peace process. *Autonomy and Peace Review, 3*(1).

Busran-Lao, Y. (2005). *Human development, economic and social costs, and spillovers of conflict: The case of the Province of Lanao del Sur*. A Background paper submitted to the Human Development Network Foundation, Inc. for the Philippine Human Development Report 2005.

Butler, J. (2003). Violence, mourning, politics. *Studies in Gender and Sexuality, 4*(1), 9–37. https://doi.org/10.1080/15240650409349213

Castillo, R. C. A. (2018). The Mamasapano clash, memories of violence, and the politics of Muslim belonging in the Philippines. *New Mandala*. http://www.newmandala.org/mamasapano-memories-muslim-belonging-philippines/

Castillo, R.C.A. (2017). *Being and becoming: Imagination, memory, and violence in the southern Philippines* [dissertation in microfilm]. Freie Universität Berlin.

Castillo, R. C. A. (2015). The emotional, political, and analytical labor of engaged anthropology amidst violent political conflict. *Medical Anthropology: Cross-Cultural Studies in Health and Illness, 34*(1), 70–83. https://doi.org/10.1080/01459740.2014.960564

Castillo, R. C. A. (2014). Perspectives on truth, justice, reparation, and reconciliation in Central Mindanao. In M. Domes & D. Jaeger (Eds.), *Dealing with the past in Mindanao: Perspectives and entry points for transitional justice* (pp. 29–39). ForumZFD.

Carsten, J. (2000). Introduction: Cultures of relatedness. In J. Carsten (Ed.), *Cultures of relatedness: New approaches to the study of kinship (pp. 1–36).* Cambridge University Press.

Che Man, W. K. (1990). *Muslim separatism: The Moros of southern Philippines and the Malays of southern Thailand.* Oxford University Press.

Cocking, J. M. (1991). *Imagination: A study in the history of ideas.* Routledge.

Fanon, F. (1963). *The wretched of the earth.* Grove Weidenfeld.

forumZFD. (2015, March 1). *Tingog Mamasapano* [video file]. https://www.youtube.com/watch?v=yeixPwSoWJE

Gowing, P. (1988). Contrasting agendas for peace in the Muslim south. In P. Gowing (Ed.), (pp. 156–173). Understanding Islam and Muslims in the Philippines, New Day Publisher.

Grosfoguel, R. (2012). Decolonising Western uni-versalisms: Pluri-versalism from Aimé Césaire to the Zapatistas. *Transmodernity: Journal of Peripheral Cultural Production of the Luso-Hispanic World, 1*(3), 88–104.

Grosfoguel, R. (2010). Epistemic islamophobia and colonial social sciences. *Human Architecture: Journal of the Sociology of Self-knowledge, 8*(2), 29–38.

Grosfoguel, R., Oso, L., & Christou, A. (2014). 'Racism', intersectionality and migration studies: Framing some theoretical reflections. *Identities: Global Studies in Culture and Power, 22*(6), 635–652. https://doi.org/10.1080/1070289X.2014.950974

Halpern, J. (2001). *From detached concern to empathy: Humanising medical practice.* Oxford University Press.

Hayase, S. (2007). *Mindanao ethnohistory beyond nations: Maguindanao, Sangir, and Bagobo societies in east maritime Southeast Asia.* Ateneo de Manila University Press.

Hollan, D., & Throop, J. (2011). The anthropology of empathy: Introduction. In D. Hollan & J. Throop (Eds.), *The anthropology of empathy: Experiencing the lives of others in Pacific societies* (pp. 1–24). Berghahn Books.

Hume, D. (2012 [1740]). *A treatise of human nature.* http://www.gutenberg.org/files/4705/4705-h/4705-h.htm

Ingold, T. (2014). That's enough about ethnography! *Hau: Journal of Ethnographic Theory, 4*(1), 383–395. https://doi.org/10.14318/hau4.1.021

Jubair, S. (1999). *Bangsamoro: A nation under endless tyranny.* IQ Marin SDN BHD.

Krondorfer, B. (2016). Unsettling empathy: Intercultural dialogue in the aftermath of historical and cultural trauma. In P. G. Madikizela (Ed.), *Breaking intergenerational cycles of repetition: A global dialogue on historical trauma and memory* (pp. 91–112). Barbara Budrich Publishers.

Levy, L. (2012). Rethinking the relationship between memory and imagination in Sarte's the imaginary. *Journal of the British Society for Phenomenology, 43*(2), 143–161. https://doi.org/10.1080/00071773.2012.11006764

Maldonado-Torres, N. (2016). *Outline of ten theses on coloniality and deconoliality.* http://caribbeanstudiesassociation.org/docs/Maldonado-Torres_Outline_Ten_Theses-10.23.16.pdf

Mawallil, A. (2018). *A constant retelling: Exploring the Bangsamoro narratives.* Balangiga Press.

McKenna, T. (1998). *Muslim rulers and rebels: Everyday politics and armed resistance in the southern Philippines.* University of California Press.

Mignolo, W. (2018). The invention of the human and the three pillars of the colonial matrix of power: Racism, sexism, and nature. In W. Mignolo & C. Walsh (Eds.), *On decoloniality* (pp. 153–176). Duke University Press.

Paredes, O. (2015). Indigenous vs. native: Negotiating the place of Lumads in the Bangsamoro homeland. *Asian Ethnicity, 16*(2), 166–185. https://doi.org/10.1080/14631369.2015.1003690

Philippine Institute for Development Studies. (2006). *Bias against Muslims: A creeping perception.* Development Research News. http://dirp4.pids.gov.ph/ris/drn/pidsdrn06-4.pdf

Prager, J. (2016). Disrupting the intergenerational transmission of trauma: Recovering humanity, repairing generations. In P. G. Madikizela (Ed.), *Breaking intergenerational cycles of repetition: A global dialogue on historical trauma and memory* (pp. 12–25). Barbara Budrich Publishers.

Quayle, A., & Sonn, C. (2019). Amplifying the voices of indigenous elders through community arts and narrative inquiry: Stories of oppression, psychosocial suffering, and survival. *American Journal of Community Psychology, 64*, 46–58. https://doi.org/10.1002/ajcp.12367

Quijano, A. (2007). Coloniality and modernity/rationality. *Cultural Studies, 21*(2), 168–178.

RSSO-ARMM. (2019, April 10). *Proportion of poor Filipinos in ARMM registered at 63.0 percent in the first semester of 2018.* http://rssoarmm.psa.gov.ph/statistics/poverty

Sartre, J. (1948). *The psychology of imagination.* Philosophical Library.

Scheper-Hughes, N. (1995). The primacy of the ethical: Propositions for a militant anthropology. *Current Anthropology, 36*(3), 409–440.

Scheper-Hughes, N., & Bourgois, P. (2003). Introduction: Making sense of violence. In N. Scheper-Hughes & P. Bourgois (Eds.), *Violence in war and peace: An anthology* (pp. 1–31). Blackwell Publishing.

Seedat, M. (2015). Oral history as an enactment of critical community psychology. *Journal of Community Psychology, 43*(1), 22–35. https://doi.org/10.1002/jcop.21690

Seedat, M., & Suffla, S. (2017). Community psychology and its (dis)contents, archival legacies, and decolonisation. *South Africa Journal of Psychology, 47*(4), 421–431.

Segalo, P., Manoff, E., & Fine, M. (2015). Working with embroideries and counter-maps: Engaging memory and imagination within decolonising frameworks. *Journal of Social and Political Psychology, 3*(1), 342–364. https://doi.org/10.5964/jspp.v3i1.145

Skidmore, M. (2006). Scholarship, advocacy, and the politics of engagement in Burma (Myanmar). In V. Sanford & A. Angel-Ajani (Eds.), *Engaged observer: Anthropology, advocacy and activism* (pp. 42–59). Rutgers University Press.

Smith, A. (2008 [1759]). *The theory of moral sentiments.* http://www.earlymoderntexts.com/assets/pdfs/smith1759.pdf

Smith, L. (2012). *Decolonising methodologies: Research and indigenous peoples* (2nd ed.). Zed Books.

Suffla, S., Seedat, M., & Bawa, U. (2015). Reflexivity as enactment of critical community psychologies: Dilemmas of voice and positionality in a multi-country photovoice study. *Journal of Community Psychology, 43*(1), 9–21. https://doi.org/10.1002/jcop.21691

Transitional Justice and Reconciliation Commission (TJRC). (2016). *Report of the transitional Justice and reconciliation commission.* Transitional Justice and Reconciliation Commission.

Turner, V. (1990). Are there universals of performance in myth, ritual, and drama? In R. Schechner & W. Appel (Eds.), *By means of performance: Intercultural studies of theatre and ritual* (pp. 8–18). Cambridge University Press.

Uperesa, L. (2016). A decolonial turn in anthropology? A view from the Pacific. https://savageminds.org/2016/06/07/a-decolonial-turn-in-anthropology-a-view-from-the-pacific/

Walsh, C. (2018). The decolonial *for*: Resurgences, shifts, and movements. In C. Walsh & W. Mignolo (Eds.), *On Decoloniality, concepts, analytics, praxis* (pp. 15–32). Duke University Press.

Waterson, R., & Kian-Woon, K. (2012). Introduction. In R. Waterson & K. Kian-Woon (Eds.), *Contestations of memory in Southeast Asia* (pp. 1–16). NUS Press.

Index

S. Kessi et al. (eds.), *Decolonial Enactments in Community Psychology*, Community Psychology, https://doi.org/10.1007/978-3-030-75201-9

E

W

Y

Z

The manufacturer's authorised representative in the EU is Springer Nature Customer Service Centre GmbH, Europaplatz 3, 69115 Heidelberg, Germany. If you have any concerns regarding our products, please contact ProductSafety@springernature.com

Printed and bound by CPI Group (UK) Ltd, Croydon, CR0 4YY
28/09/2026
02224070-0005